AF581846

Stability Problems for Stochastic Models: Theory and Applications II

Stability Problems for Stochastic Models: Theory and Applications II

Editors

Alexander Zeifman
Victor Korolev
Alexander Sipin

MDPI • Basel • Beijing • Wuhan • Barcelona • Belgrade • Manchester • Tokyo • Cluj • Tianjin

Editors

Alexander Zeifman
Vologda State University
Russia

Victor Korolev
Lomonosov Moscow State University
Russia

Alexander Sipin
Vologda State University
Russia

Editorial Office
MDPI
St. Alban-Anlage 66
4052 Basel, Switzerland

This is a reprint of articles from the Special Issue published online in the open access journal *Mathematics* (ISSN 2227-7390) (available at: https://www.mdpi.com/journal/mathematics/special_issues/Stability_Problems_Stochastic_Models_Theory_Applications_II).

For citation purposes, cite each article independently as indicated on the article page online and as indicated below:

LastName, A.A.; LastName, B.B.; LastName, C.C. Article Title. *Journal Name* **Year**, *Volume Number*, Page Range.

ISBN 978-3-0365-3815-0 (Hbk)
ISBN 978-3-0365-3816-7 (PDF)

Contents

About the Editors

Alexander Zeifman—Professor, Head of Department of Applied Mathematics, Vologda State University, Vologda, Russia; Senior Researcher, Institute of Informatics Problems, Federal Research Center "Computer Sciences and Control" of the Russian Academy of Sciences, Russia; Chief Researcher, Vologda Research Center of the Russian Academy of Sciences, Russia. Graduate of Vologda State Pedagogical Institute, 1976. Candidate of Science in Physics and Mathematics (PhD), 1981. Doctor of Science in Physics and Mathematics (1994, Institute of Control Sciences, Russian Academy of Sciences). Main research interests: stochastic models, continuous-time Markov chains, bounds on the rate of convergence, perturbation bounds, queueing models, biological models, queueing theory.

Victor Korolev—Professor, Head of Department of Mathematical Statistics, Faculty of Computational Mathematics and Cybernetics, Lomonosov Moscow State University, Moscow, Russia; Leading researcher, Institute of Informatics Problems, Federal Research Center "Computer Sciences and Control" of the Russian Academy of Sciences, Moscow, Russia. Graduate of Faculty of Computational Mathematics, Lomonosov Moscow State University, 1977. Candidate of Science in Physics and Mathematics (PhD), 1981. Doctor of Science in Physics and Mathematics (1994, Lomonosov Moscow State University). Main research interests: Limit theorems of probability theory and their applications in distribution theory, statistics, risk theory, reliability theory. Probability models of real processes in physics, meteorology, financial mathematics and other fields.

Alexander Sipin—Professor at the Department of Applied Mathematics, Vologda State University, Institute of Mathematics, Natural and Computer Sciences, Russia. Graduate of Faculty of Mathematics and Mechanics, Leningrad State University, now St.Petersburg State University, Russia in 1975. Candidate of Science in Physics and Mathematics (PhD), 1979. Doctor of Science in Physics and Mathematics (2016, St. Petersburg State University, Russia). Research interests: Monte Carlo and quasi-Monte Carlo methods, Markov chains, meshless numerical methods for solving boundary value problems.

Preface to "Stability Problems for Stochastic Models: Theory and Applications II"

Most papers published in this Special Issue of *Mathematics* are written by the participants of the XXXVI International Seminar on Stability Problems for Stochastic Models. This seminar was founded by outstanding Russian mathematician Vladimir Zolotarev (27 February 1931–7 November 2019).

The main theme of the seminar is the development of an approach to limit theorems of probability theory and related fields proposed by V.M. Zolotarev. The main point of this approach is that limit theorems of probability theory are treated as special stability theorems. Zolotarev created the theoretical foundation of the key method used within this approach, namely, the theory of probability metrics. This approach assumes that statements establishing convergence must be accompanied by statements establishing the convergence rate. Zolotarev called the conditions of convergence those that simultaneously serve as convergence rate and estimates "natural". This approach was developed in the works of Zolotarev, Kalashnikov, Kruglov, Senatov, Korolev, Khokhlov, and their colleagues. By the mid-1970s, the investigation of the continuity and stability of probability and statistical models, such as, say, characterization models for probability distributions and queuing models, grew into a kind of a separate domain of probability theory. In May and November 1974, in Leningrad and Vilnius, two compact symposia on these problems were held. These symposia were initiated and organized by Vladimir Zolotarev. In 1975–1976 the research seminar on stability problems for stochastic models and related topics headed by V. Zolotarev was held at the Steklov Mathematical Institute of the Academy of Sciences of the USSR. Later, this weekly seminar moved to the Faculty of Computational Mathematics and Cybernetics, Lomonosov Moscow State University. Together with V. Zolotarev, this seminar was coordinated by V. Kalashnikov and V. Kruglov. In parallel with the weekly seminars, annual international sessions were launched with the wide participation of mathematicians from many countries. Now, this seminar is internationally recognized for the originality and relevance of the considered problems and presented results. The seminar formed and developed a breakthrough approach to limit theorems of probability theory such as stability theorems. Within this approach, many deep results were obtained.

Now, the scope of the seminar embraces the following:

- Limit theorems and stability problems;
- Asymptotic theory of stochastic processes;
- Stable distributions and processes;
- Asymptotic statistics;
- Discrete probability models;
- Characterization of probability distributions;
- Insurance and financial mathematics;
- Applied statistics;
- Queueing theory;

and other fields.

Here is the complete list of the International sessions of the Seminar on Stability Problems for Stochastic Models:

I. May 1974, Leningrad (now Sankt-Petersburg), USSR

II. November 1974, Vilnius, Lithuanian SSR

III. 5–9 December 1977, Moscow, USSR

IV. 11–16 April 1979, Palanga, Lithuanian SSR

V. 17–22 November 1980, Panevezys, Lithuanian SSR

VI. 19–27 April 1982, Moscow, USSR

VII. 18–24 June 1984, Saratov, USSR

VIII. 23–29 September 1984, Uzhgorod, Ukrainian SSR

IX. 13–19 May 1985, Varna, Bulgaria

X. October 1986, Kuybyshev (now Samara), USSR

XI. 4–11 October 1987, Sukhumi, Abkhasian ASSR

XII. October 1988, Kharkov, Ukrainian SSR

XIII. October 1989, Kirillov, Vologda Region, USSR

XIV. 27 January–2 February 1991, Suzdal, USSR

XV. 1–6 June 1992, Perm, Russia

XVI. 29 August–3 September 1994, Eger, Hungary

XVII. 19–26 June 1995, Kazan', Russia

XVIII. 26 January–1 February 1997, Hajdúszoboszló, Hungary

XIX. 6–12 September 1998, Vologda, Russia

XX. 5–11 September 1999, Nałeczów, Poland

XXI. 28 January–3 February 2001, Eger, Hungary

XXII. 25–31 May 2002, Varna, Bulgaria

XXIII. 12–17 May 2003, Pamplona, Spain

XXIV. 10–17 September 2004, Majori (Jurmala), Latvia

XXV. 20–24 September 2005, Maiori (Salerno), Italy

XXVI. 27 August–2 September 2006, Sovata-Bai, Romania

XXVII. 22–26 October 2007, Nahariya, Israel

XXVIII. 31 May–5 June 2009, Zakopane, Poland

XXIX. 10–16 October 2011, Svetlogorsk, Russia

XXX. 24–30 September 2012, Svetlogorsk, Russia

XXXI. 23–27 April 2013, Moscow, Russia

XXXII. 15–21 June 2014, Trondheim, Norway

XXXIII. 13–18 June 2016, Svetlogorsk, Russia

XXXIV. 24–28 August, 2017, Debrecen, Hungary

XXXV. 24–28 September, 2018, Perm, Russia

XXXVI. 22–26 June, 2020, Petrozavodsk, Russia (on-line session),
21–25 June, 2021, Petrozavodsk, Russia (off-line session).

Most papers published in this Special Issue of the "Mathematics" are written by the participants of the XXXVI International Seminar on Stability Problems for Stochastic Models, 21 – 25 June, 2021, Petrozavodsk, Russia.

The scope of the seminar embraces

- Limit theorems and stability problems;
- Asymptotic theory of stochastic processes;
- Stable distributions and processes;
- Asymptotic statistics;
- Discrete probability models;
- Characterization of probability distributions;
- Insurance and financial mathematics;
- Applied statistics;
- Queueing theory;

and other fields.

This issue contains twelve papers by specialists who represent six counties: Belarus, France, Hungary, India, Italy and Russia.

Alexander Zeifman, Victor Korolev, and Alexander Sipin
Editors

Article

Comparing Distributions of Sums of Random Variables by Deficiency: Discrete Case

Vladimir E. Bening [1,2] **and Victor Y. Korolev** [1,2,3,*]

1 Faculty of Computational Mathematics and Cybernetics, Moscow State University, 119991 Moscow, Russia; bening@cs.msu.ru or bening@yandex.ru
2 Moscow Center for Fundamental and Applied Mathematics, 119991 Moscow, Russia
3 Federal Research Center "Computer Science and Control", Russian Academy of Sciences, 119333 Moscow, Russia
* Correspondence: vkorolev@cs.msu.ru

Abstract: In the paper, we consider a new approach to the comparison of the distributions of sums of random variables. Unlike preceding works, for this purpose we use the notion of deficiency that is well known in mathematical statistics. This approach is used, first, to determine the distribution of a separate random variable in the sum that provides the least possible number of summands guaranteeing the prescribed value of the $(1-\alpha)$-quantile of the normalized sum for a given $\alpha \in (0,1)$, and second, to determine the distribution of a separate random variable in the sum that provides the least possible number of summands guaranteeing the prescribed value of the probability for the normalized sum to fall into a given interval. Both problems are solved under the condition that possible distributions of random summands possess coinciding three first moments. In both settings the best distribution delivers the smallest number of summands. Along with distributions of a non-random number of summands, we consider the case of random summation and introduce an analog of deficiency which can be used to compare the distributions of sums with random and non-random number of summands. The main mathematical tools used in the paper are asymptotic expansions for the distributions of $\mathbb{R}$-valued functions of random vectors, in particular, normalized sums of independent identically distributed r.v.s and their quantiles. Along with the general case, main attention is paid to the situation where the summarized random variables are independent and identically distributed. The approach under consideration is applied to determination of the distribution of insurance payments providing the least insurance portfolio size under prescribed Value-at-Risk or non-ruin probability.

Keywords: limit theorem; sum of independent random variables; random sum; asymptotic expansion; asymptotic deficiency; kurtosis

Citation: Bening, V.E.; Korolev, V.Y. Comparing Distributions of Sums of Random Variables by Deficiency: Discrete Case. *Mathematics* **2022**, *10*, 454. https://doi.org/10.3390/math10030454

Academic Editors: Anatoliy Swishchuk and Antonella Basso

Received: 27 December 2021
Accepted: 28 January 2022
Published: 30 January 2022

1. Introduction

1.1. The Problem under Consideration and the Structure of the Paper

The problem considered in the paper is very close to the problem of stochastic ordering and even may be considered as a a version of this problem. In probability theory and statistics, a stochastic order quantifies the concept of one random variable being "bigger" or "smaller" than another. Many different orders exist, which have different applications, see, e.g., the book [1]. Here we propose an approach to establishing stochastic order for the distributions of sums of independent random variables (r.v.s) based on the notion of deficiency that is well known in asymptotic statistics, see, e.g., [2] and later publications [3–5]. Roughly speaking, in statistics the deficiency of a statistical procedure with respect to an 'optimal' procedure is the number of additional observations required to attain the same quality of inference as is guaranteed by the 'optimal' procedure.

In this paper we deal with the case where the deficiency is measured in natural-valued discrete units (number of 'additional' summands) and therefore here we deal with discrete case. The notion of deficiency can be extended to the case of the continuous parameter, say, time. This case will be considered in another work.

Along with the general case, in the paper main attention is paid to the situation where the r.v.s being summed are assumed to be independent and identically distributed.

The first problem to be considered below consists in determination of the distribution of a separate random variable in the sum that provides the least possible number of summands guaranteeing the prescribed value of the $(1-\alpha)$-quantile of the normalized sum for a given $\alpha \in (0,1)$. The second problem considered in the paper consists in determination of the distribution of a separate random variable in the sum that provides the least possible number of summands guaranteeing the prescribed value of the probability for the normalized sum to fall into a given interval. Actually, in both problems we deal with 'fine tuning' of the distribution of a separate summand since we assume that different possible distributions of random summands possess coinciding three first moments, so that they can differ only by their kurtosis. In both settings the best distribution delivers the smallest number of summands.

We also consider the problem where some additional randomization is introduced so that the number of summands in the sum can be random itself. This randomization may not be artificially induced, but also may occur when the exact number of summands is a priori unknown and only some its 'expected' value can be available as the parameter of the problem. For this case we introduce an analog of deficiency which can be used to compare the distributions of sums with random and non-random number of summands.

Both problems are closely related with the problem of quantification of the accuracy of approximations provided by limit theorems of probability theory. The main mathematical tools used in the paper are asymptotic expansions for the distributions of normalized sums of independent identically distributed r.v.s and their quantiles.

The formal settings mentioned above can be applied to solving practical problems where the models of the observed statistical regularities have the form of distributions of sums of r.v.s and the number of summands plays a substantial role. For example, consider an insurance company whose portfolio consists of a finite number of insurance contracts. Formally, the portfolio is assumed to be a finite set of r.v.s each of which characterizes the income of the company related to a separate contract. Instead of *income* we can speak of *loss* assuming that income is a negative loss or that loss is a negative income.

In these terms, the first setting concerns the problem of determination of the distribution of a possible loss within a separate insurance contract (say, the distribution of an insurance payment) providing the least possible portfolio size and guaranteeing the prescribed Value-at-Risk for the average losses. The approach considered in the paper can be used when the distributions of the summands (possible losses) are known only up to their three first moments and the exact Value-at-Risk is not known for sure. In the second setting the latter requirement is replaced by that of guaranteeing the prescribed 'non-ruin' probability. Within the framework of this example in both settings the problem consists in the description of the best strategy of the insurance company, if by a strategy we mean the choice of the terms of a contract (e.g., the amount of insurance payment related to each possible insurance event), that is, of the distribution of possible loss within a separate contract. Briefly, the problem is to choose an optimal distribution of a separate loss among the distributions that have the same first three moments so that the portfolio size is least possible.

The paper is organized as follows. Section 1.2 contains a short overview of the properties of statistical deficiency. In Section 2 we outline some results concerning the asymptotic expansions for the distributions of $\mathbb{R}$-valued measurable functions of r.v.s and, in particular, for the distributions of normalized sums of r.v.s, as well as for their quantiles. In Section 3 the problem of comparison of the distributions of two sums of independent r.v.s by their deficiency is considered. The notion of asymptotic deficiency is introduced and some formulas for the calculation of asymptotic deficiency are presented. Section 3.1 contains the solution of this problem for these distributions providing a prescribed value of the $(1-\alpha)$-quantile for a given $\alpha \in (0,1)$. In Section 3.2 this problem is considered for the distributions of sums of independent r.v.s guaranteeing a prescribed probability for an $\mathbb{R}$-valued measurable function of r.v.s, in particular, for a normalized sum of r.v.s, to fall into a given interval. Section 4 contains an example of extension of the results of Section 3 to the case of a random number of summands in the sum (random portfolio size, in terms of the example dealing with an insurance company). In Section 4.1 asymptotic expansions for the asymptotic $(1-\alpha)$-quantile (called α-reserve here) under a random portfolio size are presented and an analog of deficiency of the sum of a random number of summands (or the strategy with a random portfolio size) with respect to the distribution of the sum of a non-random number of summands (or a strategy with a non-random portfolio size) is considered. In Section 4.2 the problem of comparison of these distributions by an analog of deficiency is considered in a special case of three-point distribution of portfolio size.

Everywhere in what follows the set of real numbers is denoted by $\mathbb{R}$, the set of natural numbers is denoted by $\mathbb{N}$. The distribution function of the standard normal law will be denoted by $\Phi(x)$,

$$\Phi(x) = \frac{1}{\sqrt{2\pi}} \int_{-\infty}^{x} \varphi(y)dy, \quad \varphi(x) = \frac{1}{\sqrt{2\pi}} \exp\Big\{-\frac{x^2}{2}\Big\}, \ x \in \mathbb{R}.$$

The distribution of a random vector $(X_1, \ldots, X_n)$ will be denoted $\mathcal{L}(X_1, \ldots, X_n)$.

1.2. *Asymptotic Deficiency*

Following the classical terminology of [6], consider two decision rules (say, two statistical procedures) D_n^* and D_n whose quality is characterized by the quantities π_n^* and π_n, respectively. Here n is the number of observations $X_1, \ldots, X_n$ delivering the information underlying the decision rules. Assume that the rule D_n^* is in some sense optimal whereas the rule D_n is competing. For example, in the problem of estimation usually π_n^* and π_n are mean square deviations and $\pi_n^* \le \pi_n$. In the problem of testing hypotheses usually π_n^* and π_n are powers of tests so that $\pi_n^* \ge \pi_n$.

By $m(n)$ denote the number of observations required for the decision rule $D_{m(n)}$ based on $m(n)$ observations $X_1, \ldots, X_{m(n)}$ to attain the same quality as the 'best' rule D_n^* based on n observations $X_1, \ldots, X_n$. In what follows we will keep to the asymptotic approach assuming that $n \to \infty$. Following [7], by the asymptotic relative efficiency (a.r.e.) of the rule D_n with respect to the rule D_n^* we will mean the limit

$$e \equiv \lim_{n\to\infty} \frac{n}{m(n)}$$

(if it exists and does not depend on the sequence $m(n)$).

Instead of the ratio of the required number of observations, the difference $m(n) - n$ can be considered as well, vividly showing the additional number of observations required by the decision rule D_n. However, many authors considered the ratio $n/m(n)$, possibly, because the asymptotic analysis of its properties is simpler.

The systematic analysis of the asymptotic behavior of the difference $m(n) - n$ was first carried out by Hodges and Lehmann in 1970 [2]. They suggested to call the difference $m(n) - n$ *deficiency* of the competing decision rule D_n with respect to the rule D_n^* and introduced the notation

$$d_n = m(n) - n. \tag{1}$$

If the limit $\lim_{n\to\infty} d_n$ exists, then it is called *the asymptotic deficiency* of the competing decision rule D_n with respect to the rule D_n^* and is denoted d. The number d is often called the *deficiency* of D_n with respect to D_n^*. Note that if a.r.e. $e \neq 1$, then $d = \infty$, so that this case is not so interesting. In [2] it was also noticed that for some decision rules (statistical procedures) there typically appear cases $e = 1$ (see, e.g., the book [8]), that is, in these cases the a.r.e. cannot give an answer to the question, which rule is better, whereas the deficiency can clarify the case, because, generally speaking, in this case the asymptotic deficiency can be arbitrary.

So, the deficiency of D_n with respect to D_n^* shows, how many additional observations (that is, how much extra information) is required to attain the desired quality, if the decision rule D_n is used instead of the 'optimal' decision rule D_n^*. Therefore, the notion of deficiency provides natural grounds for the asymptotic comparison of D_n and D_n^* in the case $e = 1$. The study of the asymptotic behavior of the deficiency d_n requires more sophisticated techniques than is used to find the limit e. As a rule, this techniques employ the construction of asymptotic expansions (a.e.s) for the corresponding functions characterizing the quality of decision rules (see, e.g., the books [7–9]).

Since the rules D_n^* and D_n have the quality characteristics π_n^* and π_n, respectively, then, by the definition of the deficiency $d_n = m(n) - n$, for every n we have

$$\pi_n^* = \pi_{m(n)}. \tag{2}$$

So solve Equation (2), the integer-valued quantity $m(n)$ should be treated as a variable taking arbitrary real values. For this purpose the function $\pi_{m(n)}$ can be defined for non-integer $m(n)$ by the formula

$$\pi_{m(n)} = \big(1 - m(n) + [m(n)]\big)\pi_{[m(n)]} + \big(m(n) - [m(n)]\big)\pi_{[m(n)]+1}$$

(see [2]).

The functions π_n^* and π_n are usually unknown, so, in practice, their approximations are used. Assume that the a.e.s

$$\pi_n^* = \frac{a}{n^r} + \frac{b}{n^{r+s}} + o\big(n^{-r-s}\big), \tag{3}$$

and

$$\pi_n = \frac{a}{n^r} + \frac{c}{n^{r+s}} + o\big(n^{-r-s}\big), \tag{4}$$

hold, where a, b and c are some numbers that do not depend on n, and $r > 0$, and $s > 0$ are constants determining the rate of decrease of these quality criteria in n. The first terms in these expansions coincide which means that the a.r.e. of the corresponding rules equals one. It can be easily obtained from relations (1)–(4) that

$$d_n = \frac{c-b}{ra} n^{1-s} + o\big(n^{1-s}\big) \tag{5}$$

(see [2] or [7]). Thus, the asymptotic deficiency has the form

$$d = \begin{cases} \pm\infty, & 0 < s < 1, \\ \dfrac{c-b}{ra}, & s = 1, \\ 0, & s > 1. \end{cases} \tag{6}$$

The asymptotic deficiency possesses the following obvious property of transitivity: if there is some third decision rule $\overline{D}_n$ with the quality characteristic $\overline{\pi}_n$ admitting an a.e. of the form (4), then the deficiency $\overline{d}_n$ of the rule $\overline{D}_n$ with respect the the rule D_n^* satisfies the equality

$$\overline{d}_n = \widetilde{d}_n + d_n,$$

where $\widetilde{d}_n$ is the deficiency of the rule $\overline{D}_n$ with respect to D_n and d_n is the deficiency of D_n with respect to D_n^*.

The case where $s = 1$ is most interesting, because in this case the asymptotic deficiency is finite. In the paper [2] some simple examples are given illustrating that this case is quite natural in mathematical statistics (also see the book [8]).

2. Asymptotic Expansions for the Distributions of Normalized Sums of Random Variables

We begin with most general case. Let $n \in \mathbb{N}$. Consider a finite set of r.v.s $X_1, \ldots, X_n$. For the time being we do not assume that the r.v.s $X_1, \ldots, X_n$ are independent and identically distributed. Let $L_n = L_n(X_1, \ldots, X_n)$ be an $\mathbb{R}$-valued measurable function of $X_1, \ldots, X_n$. (In what follows when dealing with the example of the portfolio of an insurance company we will call this function *generalized loss*). In particular, L_n may be of the form $L_n = \sqrt{n}T_n$ where T_n is the arithmetic mean,

$$T_n \equiv \frac{1}{n}\sum_{i=1}^{n} X_i. \tag{7}$$

As it has been already said, the problem consists in description of the distribution of r.v.s X_i providing the least possible number of summands n and guaranteeing the prescribed value of the $(1-\alpha)$-quantile of the function L_n for a given $\alpha \in (0,1)$.

Let $\alpha \in (0,1)$ be a small number. Consider the quantity $c_\alpha(n)$ defined by the asymptotic relation

$$\mathsf{P}(L_n \geq c_\alpha(n)) = \alpha + o(n^{-1}), \quad n \to \infty. \tag{8}$$

The quantity $c_\alpha(n)$ is the asymptotic $(1-\alpha)$-quantile of L_n. If $L_n = \sqrt{n}T_n$, then $c_\alpha(n)$ can be interpreted as the threshold, the exceedance of which by L_n is undesirable and is assumed to have the prescribed small probability α. In terms of an insurance company, $c_\alpha(n)$ is the asymptotic Value-at-Risk.

By applying the Taylor formula it is not difficult to obtain the following result.

Lemma 1. *Assume that there exist distribution function $G(x)$ and functions $g_1(x)$ and $g_2(x)$ such that*

$$\sup_{x\in\mathbb{R}}\left|\mathsf{P}(L_n < x) - G(x) - \frac{1}{\sqrt{n}}g_1(x) - \frac{1}{n}g_2(x)\right| = o(n^{-1}),$$

where the functions $G(x)$, $g_1(x)$ and $g_2(x)$ are smooth enough. Then the asymptotic $(1-\alpha)$-quantile $c_\alpha(n)$ of L_n admits the a.e.

$$c_\alpha(n) = c_\alpha - \frac{g_1(c_\alpha)}{\sqrt{n}G'(c_\alpha)} - \frac{1}{n}\left[\frac{G''(c_\alpha)g_1^2(c_\alpha)}{2(G'(c_\alpha))^3} + \frac{G'(c_\alpha)g_2(c_\alpha) - g_1(c_\alpha)g_1{}'(c_\alpha)}{(G'(c_\alpha))^2}\right] + o(n^{-1}),$$

where c_α satisfies the equation $G(c_\alpha) = 1 - \alpha$.

Consider the application of this lemma to the case where $X_1, X_2, \ldots$ are independent identically distributed r.v.s such that

$$\mathsf{E}X_1 = 0,\ \mathsf{E}X_1^2 = 1,\ \mathsf{E}|X_1|^{k+\delta} < \infty,\ k \in \mathbb{N},\ k \geq 3,\ \delta > 0 \tag{9}$$

and the function L_n has the form $L_n = \sqrt{n}T_n$ with T_n defined by (7). Here the condition $\mathsf{E}X_1 = 0$ means that the separate losses are centered by their expectations. Assume that the characteristic function $f(t)$ of the r.v. X_1 satisfies the Cramér condition (C)

$$\limsup_{|t|\to\infty} |f(t)| < 1. \tag{10}$$

Under conditions (9) and (10), from Theorem 6.3.2 of [10] (also see [9]) it follows that there exist functions $Q_1(x), \ldots, Q_{k-2}(x)$ and a $C_{k,\delta} \in (0,\infty)$ such that

$$\sup_x\left|\mathsf{P}(\sqrt{n}T_n < x) - \Phi(x) - \sum_{i=1}^{k-2} n^{-i/2}Q_i(x)\right| \leq \frac{C_{k,\delta}}{n^{(k-2+\delta)/2}},\quad n \in \mathbb{N}, \tag{11}$$

For the definition of the functions $Q_1(x), \ldots, Q_{k-2}(x)$ see the book [10]. In particular,

$$Q_1(x) = -(x^2-1)\varphi(x)\frac{\mathsf{E}X_1^3}{6},$$

$$Q_2(x) = -(x^3-3x)\varphi(x)\frac{\mathsf{E}X_1^4-3}{24} - (x^5-10x^3+15x)\varphi(x)\frac{(\mathsf{E}X_1^3)^2}{72}. \qquad (12)$$

Relations (11) and (12) and Lemma 1 directly imply the a.e. for the asymptotic $(1-\alpha)$-quantile $c_n(\alpha)$ of L_n presented in the following lemma.

Lemma 2. *Let conditions* (9) *and* (10) *hold with* $k=4$, $\delta>0$. *Then the the asymptotic* $(1-\alpha)$-*quantile* $c_n(\alpha)$ *of* L_n *admits the a.e.*

$$c_\alpha(n) = u_\alpha + \frac{\mathsf{E}X_1^3}{6\sqrt{n}}(u_\alpha^2-1) + \frac{1}{12n}\Big[\frac{\mathsf{E}^2X_1^3}{3}(5u_\alpha-2u_\alpha^3) + \frac{\mathsf{E}X_1^4-3}{2}(u_\alpha^3-3u_\alpha)\Big] + o(n^{-1}),$$

where u_α *is the* $(1-\alpha)$-*quantile of the standard normal distribution:* $\Phi(u_\alpha)=1-\alpha$.

3. The Comparison of the Distributions of Two Normalized Sums of Random Variables

3.1. The Asymptotic Deficiency of the Distributions of Summands Providing a Given $(1-\alpha)$-Quantile of the Normalized Sums

In this section we will present an approach to the comparison of the distributions of two sums of r.v.s in terms of the number of summands. The distribution of the random vector $X_1, \ldots, X_n$ will be denoted $\mathcal{L}(X_1,\ldots,X_n)$. Consider an $\mathbb{R}$-valued measurable function of $X_1,\ldots,X_n$.

From Lemma 1 we can easily obtain the following result.

Lemma 3. *Consider a sequence* $\{\epsilon_n\}_{n\ge1}$ *such that* $\epsilon_n \to 0$ *as* $n\to\infty$. *Under the conditions of Lemma 1 we have*

$$\sup_{x\in\mathbb{R}}\Big|\mathsf{P}\big(L_n(X_1,\ldots,X_n)<x+\epsilon_n\big) - \mathsf{P}\big(L_n(X_1,\ldots,X_n)<x\big) -$$

$$-\epsilon_n G'(x) - \frac{\epsilon_n^2}{2}G''(x) - \frac{\epsilon_n}{\sqrt{n}}g_1{}'(x)\Big| = o\Big(\max\Big\{\epsilon_n^2, \frac{\epsilon_n}{\sqrt{n}}, n^{-1}\Big\}\Big).$$

Along with the r.v.s $X_1,\ldots,X_n$ resulting in the value $L_n(X_1,\ldots,X_n)$ of the function L_n, consider another set of r.v.s $Y_1,\ldots,Y_n$, according to which the value of the function L_n is $L_n(Y_1,\ldots,Y_n)$. For example, $L_n(X_1,\ldots,X_n)$ may have the form $L_n(X_1,\ldots,X_n)=\sqrt{n}T_n$ with T_n defined by (7) and $L_n(Y_1,\ldots,Y_n)$ may have the form $L_n(Y_1,\ldots,Y_n)=\sqrt{n}U_n$ where

$$U_n = \frac{1}{n}\sum\nolimits_{i=1}^n Y_i. \qquad (13)$$

Let to the distribution $\mathcal{L}(Y_1,\ldots,Y_n)$ there correspond the asymptotic $(1-\alpha)$-quantile $\bar{c}_\alpha(n)$ of L_n:

$$\mathsf{P}\big(L_n(Y_1,\ldots,Y_n)\ge \bar{c}_\alpha(n)\big) = \alpha + o(n^{-1}),\ \ n\to\infty. \qquad (14)$$

Assume that the a.e. for the distribution function of $L_n(Y_1,\ldots,Y_n)$ has the form

$$\mathsf{P}\big(L_n(Y_1,\ldots,Y_n)<x\big) = G(x) + \frac{1}{\sqrt{n}}g_1(x) + \frac{1}{n}\overline{g}_2(x) + o(n^{-1}), \qquad (15)$$

where the functions $G(x)$, $g_1(x)$ and $\overline{g}_2(x)$ are smooth enough. The a.e. (15) differs from the a.e. for the distribution function of $L_n(X_1,\ldots,X_n)$ established by Lemma 1 only by the term of order n^{-1}, which means that the two distributions are rather close. Define the sequence of natural numbers $\{m(n)\}_{n\geq 1}$ by the equality

$$\mathsf{P}(L_{m(n)}(Y_1,\ldots,Y_{m(n)}) \geq c_\alpha(m(n))) = \alpha + o(n^{-1}), \ \ n \to \infty. \tag{16}$$

If $m(n) - n = d + o(1)$, $d \in \mathbb{R}$, $n \to \infty$, then d is the asymptotic deficiency of the distribution $\mathcal{L}(Y_1,\ldots,Y_1)$ with respect to the distribution $\mathcal{L}(X_1,\ldots,X_n)$. In other words, d is the asymptotic number of 'additional' r.v.s be included in the set $Y_1,\ldots,Y_1$ in order that the distribution $\mathcal{L}(Y_1,\ldots,Y_{m(n)})$ provides the same quality as the distribution $\mathcal{L}(X_1,\ldots,X_n)$.

Theorem 1. *Assume that the conditions of Lemma 1 and (15) hold and $G'(c_\alpha)c_\alpha \neq 0$. Then the asymptotic deficiency d of the distribution $\mathcal{L}(Y_1,\ldots,Y_1)$ with respect to the distribution $\mathcal{L}(X_1,\ldots,X_n)$ has the form*

$$d = \frac{2[g_2(c_\alpha) - \overline{g}_2(c_\alpha)]}{G'(c_\alpha)c_\alpha} + o(1).$$

Proof. From Lemma 1 and condition (15) it directly follows that

$$\overline{c}_\alpha(n) = c_\alpha - \frac{g_1(c_\alpha)}{\sqrt{n}G'(c_\alpha)} - \frac{1}{n}\Big[\frac{G''(c_\alpha)g_1^2(c_\alpha)}{2(G'(c_\alpha))^3} + \frac{G'(c_\alpha)\overline{g}_2(c_\alpha) - g_1(c_\alpha)g_1'(c_\alpha)}{(G'(c_\alpha))^2}\Big] + o(n^{-1}) \tag{17}$$

and therefore

$$\epsilon_n \equiv \sqrt{\frac{m(n)}{n}}\overline{c}_\alpha(m(n)) - c_\alpha(m(n)) = \frac{d}{2n}c_\alpha - \frac{1}{n}\frac{(g_2(c_\alpha) - \overline{g}_2(c_\alpha))}{G'(c_\alpha)} + o(n^{-1}). \tag{18}$$

Further, with the account of the definitions of $m(n)$ (see (16)) and ϵ_n we have

$$\alpha + o(n^{-1}) = \mathsf{P}(L_{m(n)}(Y_1,\ldots,Y_{m(n)}) \geq \overline{c}_\alpha(m(n))) =$$

$$= \mathsf{P}\Big(L_{m(n)}(Y_1,\ldots,Y_{m(n)}) \geq \sqrt{\frac{n}{m(n)}}\big(c_\alpha(m(n)) + \epsilon_n\big)\Big) \tag{19}$$

Applying Lemma 3 to the right-hand side of (19) we obtain

$$\alpha + o(n^{-1}) = \mathsf{P}(L_{m(n)}(Y_1,\ldots,Y_{m(n)}) \geq c_\alpha(m(n))) - \epsilon_n G'(c_\alpha) + o(n^{-1}).$$

Now from (16) and (18) it follows that

$$d = \frac{2[g_2(c_\alpha) - \overline{g}_2(c_\alpha)]}{G'(c_\alpha)c_\alpha} + o(1).$$

The theorem is proved. □

Now consider an example of the application of Theorem 1 to the optimization of the portfolio size of an insurance company. Let the possible losses $X_1, X_2, \ldots$ related with each insurance contract in the portfolio be independent identically distributed r.v.s satisfying conditions (9) and (10). Consider another distribution, under which the possible losses $Y_1, Y_2, \ldots$ are assumed to be independent identically distributed r.v.s such that

$$\mathsf{E}Y_1 = 0, \ \mathsf{E}Y_1^2 = 1, \ \mathsf{E}|Y_1|^{4+\delta} < \infty, \ \delta > 0. \tag{20}$$

Assume that the characteristic function $p(t)$ of the r.v. Y_1 satisfies the Cramér (C) condition

$$\limsup_{|t|\to\infty} |p(t)| < 1. \tag{21}$$

For each n consider the average losses U_n defined by (13). Assume that

$$\mathsf{E}X_1^3 = \mathsf{E}Y_1^3, \tag{22}$$

(for example, the r.v.s X_i and Y_i are centered by their expectations and the distributions of these centered r.v.s are symmetric). From Lemma 2 and Theorem 1 we directly obtain the following statement.

Lemma 4. *Let conditions (9), (10) and (20)–(22) hold. Then the asymptotic (as $n \to \infty$) deficiency of the distribution $\mathcal{L}(Y_1, \ldots, Y_n)$ with respect to the distribution $\mathcal{L}(X_1, \ldots, X_n)$ (the 'additional number of contracts') d has the form*

$$d = \frac{(\mathsf{E}X_1^4 - \mathsf{E}Y_1^4)(3 - u_\alpha^2)}{12} + o(1).$$

Lemma 4 illustrates that if the distributions are close, then the deficiency is determined by the kurtosis.

3.2. The Asymptotic Deficiency of the Distributions of Summands Providing a Given Probability for the Normalized Sum to Fall into a Given Interval

To begin with, in this section we again consider the values of a measurable $\mathbb{R}$-valued function $L_n(X_1, \ldots, X_n)$ and $L_n(Y_1, \ldots, Y_n)$ on random vectors $(X_1, \ldots, X_n)$ and $(Y_1, \ldots, Y_n)$ with the the distributions $\mathcal{L}(X_1, \ldots, X_n)$ and $\mathcal{L}(Y_1, \ldots, Y_n)$, respectively. The goal is to provide that the value of L_n falls into the interval $[S_1, S_2)$ for some given numbers $S_1 < S_2$. As a quality characteristic consider the probabilities

$$\pi_n = \mathsf{P}(S_1 \leq L_n(X_1, \ldots, X_n) < S_2), \quad \overline{\pi}_n = \mathsf{P}(S_1 \leq L_n(Y_1, \ldots, Y_n) < S_2). \tag{23}$$

If $L_n(X_1, \ldots, X_n) = \sqrt{n}T_n$ (see (7)) and $L_n(Y_1, \ldots, Y_n) = \sqrt{n}U_n$ (see (22)), that is, normalized sums of r.v.s are considered, then relation (23) means that π_n and $\overline{\pi}_n$ are probabilities of that the normalized sums of r.v.s are inside the interval $[S_1, S_2)$.

From the definition of π_n we directly obtain the following result.

Lemma 5. *Assume that for some $r > 0$ and $s > 0$ there exist a distribution function $H(x)$ and functions $h_1(x)$, $h_2(x)$ and $\overline{h}_2(x)$ such that*

$$\sup_{x \in \mathbb{R}} \left| \mathsf{P}(L_n(X_1, \ldots, X_n) < x) - H(x) - \frac{1}{n^r} h_1(x) - \frac{1}{n^{r+s}} h_2(x) \right| = o(n^{-r-s}),$$

$$\sup_{x \in \mathbb{R}} \left| \mathsf{P}(L_n(Y_1, \ldots, Y_n) < x) - H(x) - \frac{1}{n^r} h_1(x) - \frac{1}{n^{r+s}} \overline{h}_2(x) \right| = o(n^{-r-s}),$$

and, moreover, the functions $h_1(x)$, $h_2(x)$ and $\overline{h}_2(x)$ are measurable. Then π_n and $\overline{\pi}_n$ admit a.e.s

$$\pi_n = H(S_2) - H(S_1) + \frac{h_1(S_2) - h_1(S_1)}{n^r} + \frac{h_2(S_2) - h_2(S_1)}{n^{r+s}} + o(n^{-r-s}),$$

$$\overline{\pi}_n = H(S_2) - H(S_1) + \frac{h_1(S_2) - h_1(S_1)}{n^r} + \frac{\overline{h}_2(S_2) - \overline{h}_2(S_1)}{n^{r+s}} + o(n^{-r-s}).$$

Corollary 1. *Let $\epsilon_n \downarrow 0$ as $n \to \infty$ and $S_2 = S_1 + \epsilon_n$. Assume that the functions $H(x)$, $h_1(x)$, $h_2(x)$ and $\overline{h}_2(x)$ are smooth enough and $h_1(S_2) \neq h_1(S_1)$. Then*

$$\epsilon_n^{-1}\pi_n = H'(S_1) + \frac{\epsilon_n}{2}H''(S_1) + \frac{\epsilon_n^2}{6}H'''(S_1) + o(\epsilon_n^2) +$$

$$+ \frac{1}{n^r}h_1'(S_1) + \frac{1}{2n^r}h_1''(S_1)\epsilon_n + o(\epsilon_n n^{-r}) + \frac{1}{n^{r+s}}h_2'(S_1) + o(n^{-r-s}\epsilon_n^{-1}),$$

$$\epsilon_n^{-1}\overline{\pi}_n = H'(S_1) + \frac{\epsilon_n}{2}H''(S_1) + \frac{\epsilon_n^2}{6}H'''(S_1) + o(\epsilon_n^2) +$$
$$+\frac{1}{n^r}h_1'(S_1) + \frac{1}{2n^r}h_1''(S_1)\epsilon_n + o(\epsilon_n n^{-r}) + \frac{1}{n^{r+s}}\overline{h}_2'(S_1) + o(n^{-r-s}\epsilon_n^{-1}).$$

Lemma 5, Corollary 1 and formula (6) directly imply the expression for the asymptotic deficiency with quality characteristics (23).

Theorem 2. *Let conditions of Lemma 5 hold with $s = 1$. Then the deficiency d_n of the distribution $\mathcal{L}(Y_1, \ldots, Y_n)$ with the quality characteristic $\overline{\pi}_n$ with respect to the distribution $\mathcal{L}(X_1, \ldots, X_n)$ with the quality characteristic π_n has the form*

$$d_n = \frac{\overline{h}_2(S_2) - h_2(S_2) + \overline{h}_2(S_1) - \overline{h}_2(S_1)}{r(h_1(S_2) - h_1(S_1))} + o(1). \tag{24}$$

If $S_2 = S_1 + \epsilon_n$ with $\epsilon_n \downarrow 0$ as $n \to \infty$ and $h_1'(S_1) \neq 0$, then the formal passage to the limit in (3.13) yields the formula

$$d_n = \frac{\overline{h}_2'(S_1) - h_2'(S_1)}{rh_1'(S_1)} + o(1).$$

Consider an example of the application of Theorem 2 to the optimization of the portfolio size of an insurance company. Let the possible losses $X_1, X_2, \ldots$ related with each insurance contract in the portfolio be independent identically distributed r.v.s satisfying conditions (9) and (10). Consider another distribution, under which the possible losses $Y_1, Y_2, \ldots$ are assumed to be independent identically distributed r.v.s satisfying conditions (20) and (21). Assume that in (9) and (20) $k = 3$. We are interested in the asymptotic behavior of the average losses T_n (see (7)) and U_n (see (13)). With the account of Lemma 5 we obtain the following statement.

Lemma 6. *Let conditions (9), (10), (19) and (20) hold with $k = 3$. Then*

$$\mathsf{P}(\sqrt{n}T_n < x) = \Phi(x) + \frac{Q_1(x)}{\sqrt{n}} + \frac{Q_2(x)}{n} + o(n^{-1}),$$

$$\mathsf{P}(\sqrt{n}U_n < x) = \Phi(x) + \frac{\overline{Q}_1(x)}{\sqrt{n}} + \frac{\overline{Q}_2(x)}{n} + o(n^{-1}),$$

uniformly in $x \in \mathbb{R}$,

$$\pi_n = \Phi(S_2) - \Phi(S_1) + \frac{Q_1(S_2) - Q_1(S_1)}{\sqrt{n}} + \frac{Q_2(S_2) - Q_2(S_1)}{n} + o(n^{-1}),$$

$$\overline{\pi}_n = \Phi(S_2) - \Phi(S_1) + \frac{\overline{Q}_1(S_2) - \overline{Q}_1(S_1)}{\sqrt{n}} + \frac{\overline{Q}_2(S_2) - \overline{Q}_2(S_1)}{n} + o(n^{-1}),$$

where the functions $Q_1(x)$ and $Q_2(x)$ are defined in (12),

$$\overline{Q}_1(x) = -(x^2 - 1)\varphi(x)\frac{\mathsf{E}Y_1^3}{6},$$

$$\overline{Q}_2(x) = -(x^3 - 3x)\varphi(x)\frac{\mathsf{E}Y_1^4 - 3}{24} - (x^5 - 10x^3 + 15x)\varphi(x)\frac{(\mathsf{E}Y_1^3)^2}{72}.$$

Corollary 2. *Let $\epsilon_n \downarrow 0$ as $n \to \infty$ and $S_2 = S_1 + \epsilon_n$. Assume that conditions of Lemma 6 hold. Then*

$$\epsilon_n^{-1}\pi_n = \varphi(S_1) + \frac{\epsilon_n}{2}\varphi'(S_1) + \frac{\epsilon_n^2}{6}\varphi''(S_1) + o(\epsilon_n^2) +$$

$$+\frac{1}{\sqrt{n}}Q_1'(S_1)+\frac{\epsilon_n}{2\sqrt{n}}Q_1''(S_1)+o(\epsilon_n n^{-1/2})\frac{1}{n}Q_2'(S_1)+o(n^{-1}\epsilon_n^{-1}),$$

$$\epsilon_n^{-1}\overline{\pi}_n=\varphi(S_1)+\frac{\epsilon_n}{2}\varphi'(S_1)+\frac{\epsilon_n^2}{6}\varphi''(S_1)+o(\epsilon_n^2)+$$

$$+\frac{1}{\sqrt{n}}\overline{Q}_1'(S_1)+\frac{\epsilon_n}{2\sqrt{n}}\overline{Q}_1''(S_1)+o(\epsilon_n n^{-1/2})+\frac{1}{n}\overline{Q}_2'(S_1)+o(n^{-1}\epsilon_n^{-1}).$$

Theorem 2, Lemma 5 and formula (5) directly imply the following statement.

Theorem 3. *Let, in addition to the conditions of Lemma 5., $\mathsf{E}X_1^3=\mathsf{E}Y_1^3$. Then the deficiency d_n of the distribution $\mathcal{L}(Y_1,\ldots,Y_n)$ with the quality characteristic $\overline{\pi}_n$ with respect to the strategy $\mathcal{L}(X_1,\ldots,X_n)$ with the quality characteristic π_n (the 'additional number of contracts') has the form*

$$d_n=2\frac{\overline{Q}_2(S_2)-Q_2(S_2)+Q_2(S_1)-\overline{Q}_2(S_1)}{Q_1(S_2)-Q_1(S_1)}n^{1/2}+o(n^{1/2}).$$

Consider an example where the asymptotic deficiency is finite.

Corollary 3. *Let $\epsilon_n=\frac{1}{n}$ and $S_2=S_1+\frac{1}{n}$, $\mathsf{E}X_1^3=\mathsf{E}Y_1^3=0$. Then under the conditions of Lemma 5 we have*

$$\pi_n=\frac{\varphi(S_1)}{n}+\frac{\varphi'(S_1)+2Q_2'(S_1)}{n^2}+o(n^{-2}),$$

$$\overline{\pi}_n=\frac{\varphi(S_1)}{n}+\frac{\varphi'(S_1)+2\overline{Q}_2'(S_1)}{n^2}+o(n^{-2}).$$

Moreover, the deficiency d_n has the form

$$d_n=\frac{2(\overline{Q}_2'(S_1)-Q_2'(S_1))}{\varphi(S_1)}+o(1)=\frac{S_1^4-6S_1^2+3}{12}(\mathsf{E}Y_1^4-\mathsf{E}X_1^4)+o(1).$$

4. Random Number of Summands

4.1. Asymptotic Expansions for the Asymptotic $(1-\alpha)$-Quantile of $\mathbb{R}$-Valued Measurable Functions of a Random Number of Random Variables

In this section we consider the case where an additional randomization can be introduced into the problem. In this case the number of summands in the sum can be considered as random. This randomization may not be artificially induced, but also may occur when the exact portfolio size can be unknown beforehand and only some 'expected' number of summands can be available as the parameter of the problem.

Let natural-valued r.v.s $N_1,N_2,\ldots$ and r.v.s $X_1,X_2,\ldots$ be defined on one and the same probability space $(\Omega,\mathcal{A},\mathsf{P})$. In what follows we will assume that n is the expected value of N_n,

$$\mathsf{E}N_n=n. \tag{25}$$

Assume that for each $n\geq 1$ the r.v. N_n is independent of the sequence $X_1,X_2,\ldots$. As above, for each $n\geq 1$, consider the value of an $\mathbb{R}$-valued measurable function $L_n=L_n(X_1,\ldots,X_n)$. For each $n\geq 1$ consider the r.v. L_{N_n} defined as

$$L_{N_n}(\omega)\equiv L_{N_n(\omega)}(X_1(\omega),\ldots,X_{N_n(\omega)}(\omega)),\ \ \omega\in\Omega.$$

Below we will assume that the following condition holds.

Condition A. *There exist $k\in\mathbb{N}\backslash\{1\}$, $\alpha_{i,n}\in\mathbb{R}$, $i=1,\ldots,k$, $\beta_n>0$, $C_k>0$, a differentiable distribution function $G(x)$ and measurable functions $g_j(x)$, $j=1,\ldots,k$ such that*

$$\beta_n\to 0,\ \ \max_{1\leq i\leq k}|\alpha_{i,n}|\to 0$$

as $n \to \infty$ and

$$\sup_x \left| \mathsf{P}(L_n < x) - G(x) - \sum_{i=1}^k \alpha_{i,n} g_i(x) \right| \le C_k \beta_n, \quad n \in \mathbb{N}.$$

Lemma 7. *Let the function $L_n = L_n(X_1, \ldots, X_n)$ satisfy Condition A. Then*

$$\sup_x \left| \mathsf{P}(L_{N_n} < x) \ - G(x) - \sum_{i=1}^k g_i(x) \mathsf{E}\alpha_{i,N_n} \right| \le C_k \mathsf{E}\beta_{N_n}.$$

The elementary proof of this lemma directly follows by the formula of total probability.

Consider an example of application of Lemma 7. Let $X_1, X_2, \ldots$ be independent identically distributed r.v.s satisfying conditions (9) and (10). Assume that the function L_n is the normalized arithmetic mean (or, which is the same, the normalized sum) $L_n = \sqrt{n}T_n$ with T_n defined in (7). Then, in accordance with what has been said in Section 2, relation (11) holds implying the validity of Condition A. From (11) playing the role of Condition A and Lemma 7 we obtain the following statement.

Lemma 8. *Assume that $L_n = \sqrt{n}T_n$ with T_n defined in (7) and conditions (9) and (10) hold. Then*

$$\sup_x \left| \mathsf{P}(\sqrt{N_n}T_{N_n} < x) - \Phi(x) - \sum_{i=1}^{k-2} Q_i(x) \mathsf{E}N_n^{-i/2} \right| \le C_{k,\delta} \mathsf{E}N_n^{-(k-2+\delta)/2},$$

where the functions $Q_i(x)$ are defined in Theorem 6.3.2 of [10].

Relation (11) and Lemma 8 imply the following statement.

Lemma 9. *Let conditions (9) and (10) hold with $k = 4$ and $\delta > 0$. Assume that condition (25) holds and*

$$\mathsf{E}N_n^{-1/2} = \frac{1}{\sqrt{n}} + \frac{a}{n} + o(n^{-1}), \quad a \in \mathbb{R},$$

$$\mathsf{E}N_n^{-1} = \frac{b}{n} + o(n^{-1}), \quad \mathsf{E}N_n^{-(2+\delta)/2} = o(n^{-1}), \quad b \in \mathbb{R}.$$

Then

$$\sup_x \left| \mathsf{P}(\sqrt{n}T_n < x) - \Phi(x) - \frac{Q_1(x)}{\sqrt{n}} - \frac{Q_2(x)}{n} \right| = o(n^{-1})$$

and

$$\sup_x \left| \mathsf{P}(\sqrt{N_n}T_{N_n} < x) - \Phi(x) - \frac{Q_1(x)}{\sqrt{n}} - \frac{bQ_2(x) + aQ_1(x)}{n} \right| = o(n^{-1}).$$

We will use Lemma 9 in order to determine the asymptotic $(1-\alpha)$-quantile of L_n and calculate the asymptotic deficiency.

Recall that, for $\alpha \in (0,1)$, the asymptotic $(1-\alpha)$-quantile of L_n is the quantity $c_\alpha(n)$ satisfying the asymptotic equality

$$\mathsf{P}(L_n \ge c_\alpha(n)) = \alpha + o(n^{-1}), n \to \infty. \tag{26}$$

Correspondingly, we define the the asymptotic $(1-\alpha)$-quantile $\tilde{c}_\alpha(n)$ of L_{N_n} by the equation

$$\mathsf{P}(L_{N_n} \ge \tilde{c}_\alpha(n)) = \alpha + o(n^{-1}), n \to \infty. \tag{27}$$

From Lemmas 1 and 9 we directly obtain the a.e.s for these asymptotic $(1-\alpha)$-quantiles.

Lemma 10. *Under the conditions of Lemma 8, we have*

$$c_\alpha(n) = u_\alpha + \frac{\mathsf{E}X_1^3}{6\sqrt{n}}(u_\alpha^2 - 1) + \frac{1}{12n}\Big[\frac{\mathsf{E}^2X_1^3}{3}(5u_\alpha - 2u_\alpha^3) + \frac{\mathsf{E}X_1^4 - 3}{2}(u_\alpha^3 - 3u_\alpha)\Big] + o(n^{-1}),$$

$$\tilde{c}_\alpha(n) = u_\alpha + \frac{\mathsf{E}X_1^3}{6\sqrt{n}}(u_\alpha^2 - 1) +$$

$$+\frac{1}{12n}\Big[\frac{\mathsf{E}^2X_1^3}{3}(5u_\alpha - 2u_\alpha^3) + \frac{b(\mathsf{E}X_1^4 - 3)}{2}(u_\alpha^3 - 3u_\alpha) + 2a\mathsf{E}X_1^3(u_\alpha^2 - 1)\Big] + o(n^{-1}),$$

where u_α satisfies the equation $\Phi(u_\alpha) = 1 - \alpha$.

Now define the sequence $m(n)$ of natural numbers by the relation

$$\mathsf{P}\big(\sqrt{n}L_{N_{m(n)}} \geq \sqrt{m(n)}c_\alpha(m(n))\big) = \alpha + o(n^{-1}), n \to \infty. \quad (28)$$

If

$$m(n) = n + d + o(1), \quad (29)$$

$n = 1, 2, \ldots$, then d can have the meaning of the expected additional number of summands to be included in the sum in order that the function L_{N_n} exceeds $c_\alpha(n)$ for the loss under a non-random number n of summands. The quantity d will be called the *asymptotic deficiency*.

In the same way that Theorem 1 was proved, we can establish the following statement.

Theorem 4. *Assume that*

$$\mathsf{E}N_n = n,\ \mathsf{E}N_n^{-1/2} = \frac{1}{\sqrt{n}} + \frac{a}{n} + o(n^{-1}),\ a \in \mathbb{R},$$

$$\mathsf{E}N_n^{-1} = \frac{b}{n} + o(n^{-1}),\ \mathsf{E}N_n^{-(2+\delta)/2} = o(n^{-1}),\ b \in \mathbb{R},$$

and there exist $\delta > 0$, a differentiable distribution function $G(x)$ and measurable functions $g_1(x)$ and $g_2(x)$ such that

$$\sup_x\Big|\mathsf{P}(L_n < x) - G(x) - \frac{g_1(x)}{\sqrt{n}} - \frac{g_2(x)}{n}\Big| \leq \frac{C}{n^{(2+\delta)/2}}$$

and $G'(c_\alpha)c_\alpha \neq 0$. Then the expected number d of additional summands (see (28) and (29)) in the normalized random sum L_{N_n} with respect to the normalized sum L_n has the form

$$d = \frac{2\big[g_2(c_\alpha)(1-b) - a g_1(c_\alpha)\big]}{G'(c_\alpha)c_\alpha} + o(1),$$

where c_α satisfies the equation $G(c_\alpha) = 1 - \alpha$.

Theorem 4 implies the following statement.

Corollary 4. *Under the conditions of Lemma 8 the expected additional number of summands d (see (28) and (29)) corresponding to the normalized sum $\sqrt{N_n}T_{N_n}$ with a random number of summands with respect to the normalized sum $\sqrt{n}T_n$ has the form*

$$d = \frac{2((1-b)Q_2(u_\alpha) - aQ_1(u_\alpha))}{\varphi(u_\alpha)u_\alpha} + o(1).$$

If additionally $\mathsf{E}X_1^3 = 0$*, then*

$$d = \frac{(1-b)(3-u_\alpha^2)(\mathsf{E}X_1^4 - 3)}{12} + o(1).$$

4.2. An Example of Three-Point Distribution of the Number of Summands

In this section, keeping to the terminology of the example related to optimization of the portfolio size of an insurance company, we will use Corollary 4 to obtain a.e.s for the asymptotic Value-at-Risk (asymptotic $(1-\alpha)$-quantile of the normalized average loss, or asymptotic normalized α-reserve) in the case where the portfolio size N_n has a special distribution concentrated in three points so that is symmetric around the central point.

Assume that the portfolio size N_n has the distribution of the form

$$\mathsf{P}(N_n = n - h_n) = \mathsf{P}(N_n = n) = \mathsf{P}(N_n = n + h_n) = \tfrac{1}{3}, \tag{30}$$

where $h_n \in \mathbb{N}$, $h_n < n$, $n = 1, 2, \ldots$, and

$$\lim_{n\to\infty} \frac{h_n}{n} = 0. \tag{31}$$

Lemma 11. *Let the random portfolio size N_n have distribution (30) and let condition (31) hold. Then $\mathsf{E}N_n = n$ and, as $n \to \infty$,*

$$\mathsf{E}N_n^{-1/2} = \frac{1}{\sqrt{n}} - \frac{1}{4\sqrt{n}}\Big(\frac{h_n}{n}\Big)^2 + O\Big(\frac{1}{\sqrt{n}}\Big(\frac{h_n}{n}\Big)^3\Big),$$

$$\mathsf{E}N_n^{-1} = \frac{1}{n} + \frac{2}{3n}\Big(\frac{h_n}{n}\Big)^2 + O\Big(\frac{1}{n}\Big(\frac{h_n}{n}\Big)^4\Big), \quad \mathsf{E}N_n^{-3/2} = \frac{1}{n^{3/2}} + O\Big(\frac{1}{n^{3/2}}\Big(\frac{h_n}{n}\Big)^2\Big).$$

Proof. The desired statements follow from the relations

$$\mathsf{E}N_n^{-1} = \frac{3n^2 - h_n^2}{3n(n^2 - h_n^2)} = \frac{1}{n}\Big(1 - \frac{h_n^2}{3n}\Big)\Big(1 + \frac{h_n^2}{n^2} + O\Big(\frac{h_n^4}{n^4}\Big)\Big) = \frac{1}{n} + \frac{2}{3n}\Big(\frac{h_n}{n}\Big)^2 + O\Big(\frac{1}{n}\Big(\frac{h_n}{n}\Big)^4\Big),$$

$$\mathsf{E}N_n^{-3/2} = \frac{1}{3n^{3/2}}\Big(\frac{1}{(1-h_n/n)^{3/2}} + 1 + \frac{1}{(1+h_n/n)^{3/2}}\Big) = \frac{1}{n^{3/2}} + O\Big(\frac{1}{n^{3/2}}\Big(\frac{h_n}{n}\Big)^2\Big).$$

The formula for $\mathsf{E}N_n^{-1/2}$ is established in a similar way. □

Lemmas 10 and 11 imply the following statement.

Theorem 5. *Assume that the normalized average loss has the form $L_n = \sqrt{n}T_n$ with T_n defined in (7). Let the r.v. N_n be distributed according to (30) and condition (31) hold. Under the conditions of Lemma 9, for the asymptotic α-reserve $\tilde{c}_\alpha(n)$ corresponding to the normalized average loss $\sqrt{N_n}T_{N_n}$ there holds the relation*

$$\tilde{c}_\alpha(n) = c_\alpha(n) - \frac{\mathsf{E}X_1^3(u_\alpha^2 - 1)}{24\sqrt{n}}\Big(\frac{h_n}{n}\Big)^2 + o(n^{-1}), n \to \infty.$$

Remark 1. *In addition to the conditions of Theorem 5, let*

$$h_n = \gamma n^\beta + o(n^\beta), \ \ \gamma \geq 0, \ \ 0 \leq \beta < 1.$$

Then, as $n \to \infty$,

$$n^{5/2-2\beta}\big(c_\alpha(n) - \tilde{c}_\alpha(n)\big) \to \frac{\gamma^2}{24}\mathsf{E}X_1^3(u_\alpha^2 - 1).$$

Applying Lemma 9, by simple calculations we obtain the following statement.

Lemma 12. *Assume that conditions (9) and (10) hold with $k = 4$ and $0 < \delta \leq 1$. Let conditions (30) and (31) hold. Then*

$$\sup_x \left| \mathsf{P}(\sqrt{N_n} T_{N_n} < x) - \Phi(x) - \left(1 - \frac{h_n^2}{4n^2}\right) \frac{Q_1(x)}{\sqrt{n}} - \left(1 + \frac{2h_n^2}{3n^2}\right) \frac{Q_2(x)}{n} \right| = O\left(\frac{h_n^{(4+2\delta)/3}}{n^{7(2+\delta)/6}} \right).$$

Corollary 5. *Let conditions of Lemma 12 hold and $h_n = n^{3/4}$. Then*

$$\sup_{x \in \mathbb{R}} \left| \mathsf{P}(\sqrt{N_n} T_{N_n} < x) - \Phi(x) - \frac{1}{\sqrt{n}} Q_1(x) - \frac{1}{n}\left(Q_2(x) - \frac{1}{4} Q_1(x) \right) \right| = o(n^{-1}).$$

Relations (12), Lemmas 10 and 11 yield the following theorem.

Theorem 6. *Let the conditions of Corollary 5 hold. Then the asymptotic α-reserves $c_\alpha(n)$ and $\tilde{c}_\alpha(n)$ related to the normalized average losses $\sqrt{n} T_n$ and $\sqrt{N_n} T_{N_n}$ have the form*

$$c_\alpha(n) = u_\alpha + \frac{\mathsf{E}X_1^3}{6\sqrt{n}}(u_\alpha^2 - 1) + \frac{1}{12n}\left[\frac{\mathsf{E}^2 X_1^3}{3}(5u_\alpha - 2u_\alpha^3) + \frac{\mathsf{E}X_1^4 - 3}{2}(u_\alpha^3 - 3u_\alpha) \right] + o(n^{-1}),$$

$$\tilde{c}_\alpha(n) = u_\alpha + \frac{\mathsf{E}X_1^3}{6\sqrt{n}}(u_\alpha^2 - 1) +$$

$$+ \frac{1}{12n}\left[\frac{\mathsf{E}^2 X_1^3}{3}(5u_\alpha - 2u_\alpha^3) + \frac{\mathsf{E}X_1^4 - 3}{2}(u_\alpha^3 - 3u_\alpha) - \frac{1}{2}\mathsf{E}X_1^3(u_\alpha^2 - 1) \right] + o(n^{-1}),$$

where u_α satisfies the equation $\Phi(u_\alpha) = 1 - \alpha$. The corresponding expected additional number d of contracts has the form

$$d = \frac{Q_1(u_\alpha)}{2\varphi(u_\alpha) u_\alpha} + o(1) = \frac{(1 - u_\alpha^2)\mathsf{E}X_1^3}{12 u_\alpha} + o(1).$$

5. Conclusions

The paper deals with an approach to the comparison of distributions of sums of a finite number of independent random variables by deficiency. The notion of asymptotic deficiency of the distribution of a measurable $\mathbb{R}$-valued function of a random vector with respect to the distribution of the same function of another random vector was introduced. Some formulas for the calculation of asymptotic deficiency were presented in the cases where the function has the form of a normalized sum of independent identically distributed r.v.s. The formulas for the asymptotic deficiency were obtained as the solution of two problems, one of which deals with the description of the distribution of a separate summand minimizing the number of summands and providing a prescribed value of the $(1 - \alpha)$-quantile of the normalized sum for a given $\alpha \in (0, 1)$. The second problem deals with minimization of the number of summands and guaranteeing a prescribed probability for a normalized sum of r.v.s to fall into a given interval. These results were extended to the case of a random number of summands in the sum (or random portfolio size, in terms of the example dealing with an insurance company). For this case, an analog of deficiency of the sum of a random number of summands with respect to the distribution of the sum of a non-random number of summands was introduced. The problem of comparison of these distributions by an analog of deficiency was considered in a special case of three-point distribution of portfolio size. The main mathematical tools used in the paper were asymptotic expansions for the distributions of average losses and their quantiles.

Author Contributions: Conceptualization, V.E.B. and V.Y.K.; Formal analysis, V.Y.K.; Funding acquisition, V.Y.K.; Investigation, V.E.B. and V.Y.K.; Writing – original draft, V.E.B. and V.Y.K. All authors have read and agreed to the published version of the manuscript.

Funding: The research was supported by the Ministry of Science and Higher Education of the Russian Federation, project No. 075-15-2020-799.

Institutional Review Board Statement: Not applicable.

Informed Consent Statement: Not applicable.

Data Availability Statement: Not applicable.

Acknowledgments: The authors thank the anonymous referees for their comments and suggestions that improved the paper. We also thank A. K. Gorshenin for his help in formatting the paper.

Conflicts of Interest: The authors declare no conflict of interest.

References

1. Müller, A.; Stoyan, D. *Comparison Methods for Stochastic Models and Risks*; J. Wiley & Sons: Chichester, UK, 2002; 352p, ISBN 978-0-471-49446-1.
2. Hodges, J.L.; Lehmann, E.L. Deficiency. *Ann. Math. Stat.* **1970**, *41*, 783–801._50. [CrossRef]
3. Torgersen, E. *Comparison of Statistical Experiments*; Printed online: May 2013; Cambridge University Press: Cambridge, UK, 1991; doi:10.1017/CBO9780511666353.007. [CrossRef]
4. Xiang, X. Deficiency of the sample quantile estimator with respect to kernel quantile estimators for censored data. *Ann. Stat.* **1995**, *23*, 836–854. [CrossRef]
5. Bening, V.E.; Korolev, V.Y.; Zeifman, A.I. Calculation of the deficiency of some statistical estimators constructed from samples with random sizes. *Colloq. Math.* **2019**, *157*, 157–171. [CrossRef]
6. Blackwell, D.; Girshick, M.A. *Theory of Games and Statistical Decisions. Wiley Publications in Statistics*; J. Wiley & Sons: New York, NY, USA; Chapman & Hall: London, UK, 1954; pp. XI, 355.
7. Lehmann, E.L.; Casella, G. *Theory of Point Estimation*; Springer: Berlin, Germany, 1998; 589p.
8. Bening, V.E. *Asymptotic Theory of Testing Statistical Hypotheses: Efficient Statistics, Optimality, Power Loss, and Deficiency*; Walter de Gruyter: Berlin, Germany, 2011; 277p, ISBN 978-3-11-093599-8.
9. Cramér, H. *Mathematical Methods of Statistics*; Princeton University Press: Princeton, NJ, USA, 1946; 647p.
10. Petrov, V.V. *Limit Theorems of Probability Theory: Sequences of Independent Random Variables*; Clarendon Press: Oxford, UK, 1985; 437p.

 *mathematics*

Article

Equilibrium in a Queueing System with Retrials

Julia Chirkova [1,*,†,‡], **Vladimir Mazalov** [1,2,†,‡] and **Evsey Morozov** [1,2,3,†,‡]

[1] Institute of Applied Mathematical Research, Karelian Research Centre, Russian Academy of Sciences, 185910 Petrozavodsk, Russia; vmazalov@krc.karelia.ru (V.M.); emorozov@karelia.ru (E.M.)
[2] Institute of Mathematics and Information Technologies, Petrozavodsk State University, 185035 Petrozavodsk, Russia
[3] Moscow Center for Fundamental and Applied Mathematics, Moscow State University, 119991 Moscow, Russia
* Correspondence: julia@krc.karelia.ru
† Current address: IAMR KarRC RAS, 11, Pushkinskaya Str., 185910 Petrozavodsk, Russia.
‡ These authors contributed equally to this work.

Abstract: We find an equilibrium in a single-server queueing system with retrials and strategic timing of the customers. We consider a set of customers, each of which must decide when to arrive to a queueing system during a fixed period of time. In this system, after completion of service, the server seeks a customer blocked in a virtual orbit (orbital customer) to be served next, unless a new customer captures the server. We develop, in detail, a setting with two and three customers in the set, and formulate and discuss the problem for the general case with an arbitrary number of customers. The numerical examples for the system with two and three customers included as well.

Keywords: equilibrium arrivals; one-server queueing system; orbit; retrials

Citation: Chirkova, J.; Mazalov, V.; Morozov, E. Equilibrium in the Queueing System with Retrials. *Mathematics* **2022**, *10*, 428. https://doi.org/10.3390/math10030428

Academic Editor: János Sztrik

Received: 31 December 2021
Accepted: 26 January 2022
Published: 28 January 2022

1. Introduction

The retrial queues have been attracting increasing interest because of their importance in modeling modern wireless telecommunication systems. Many papers have been devoted to steady-state performance analysis of such queues with the most important sources mentioned here [1–3]. Firstly, we outline the main settings that describe the dynamics of a wide class of the retrial queueing systems.

There are many practical situation that can be modeled as a queueing system in which customers are allowed to have a few attempts to be served. For instance, in call centers with a callback option, as well as customers who cannot connect immediately with the operator, thus register their numbers and go back at a later time. These customers can be called *orbital* because it seems natural that registered customers are waiting for service in a so-called *orbit-queue (orbit)*. As these customers cannot be picked up immediately when the operator becomes available, some seeking time (called *retrieval time*) is needed to access a registered customer. We note that sometimes the operator may make an *outgoing call*, not being aware of the presence of the registered customers in the orbit. A similar situation (with retrial attempts) arises in many service systems where a ticket is issued upon the arrival of a customer who will be served in a later time when the server is available.

Now we touch upon service disciplines that are considered in the retrial queueing systems. The most traditional discipline is the so-called *classical retrials*, when the customers blocked in the orbits make retrial attempts *independently*, in which case the retrial rate increases (linearly) as the orbit size increases. Stability analysis of such systems are considered in the mentioned books (mainly in the Markovian setting), and in a more general setting, this analysis has been performed in the papers [4–6].

The latter approach, in a generalized form, has been applied to the stability analysis of a wide class of queueing systems (including many retrial systems) in the recent book [7]. The main ingredient of the analysis is to establish the negative drift of the remaining work (workload) when the orbit size becomes large. Moreover, a key observation is that, from the point-of-view of stability, such a retrial system approaches the classic buffered system when the orbit size increases unlimitedly. (We call it *an asymptotically work-conserving* discipline in [5].) The following step to simplify the regenerative stability analysis has been realized in the paper [6], in which the positive drift of the *idle time* of the servers is used (instead of the positive drift of the workload) under the assumption that the orbit size increases to infinite in probability. (For more on regenerative stability analysis, see [7].)

Another wide and important class of the retrial models is the queueing systems with the *constant retrial rate*, see Chapter 7 in [7]. These models play an important role in the analysis of modern wireless telecommunication systems. In this regard, we mention the paper [8] in which, to the best of our knowledge, such a model has been used for the first time to model a telephone exchange system. A retrial queueing system with a constant retrial rate is suitable to describe the behavior of the multiple access protocols [9]. The retrial queues with a constant retrial rate have been applied to model TCP (Transmission Control Protocol) traffic related to short HTTP (HyperText Transfer Protocol) connections and to describe an optical-electrical hybrid contention resolution scheme [10]. There is also a modification of the retrial system in which, after each departure, the server seeks the customer in orbit (this is the above-mentioned retrieval time) to be served next. For instance, such a system has been considered in the paper [11], where a logarithmic asymptotic of a large deviation probability of the orbit size during the regeneration period is obtained. Moreover, in this paper we also consider the system with the retrieval time.

The most interesting and newest setting related to a retrial system is the so-called *retrial systems with coupled orbits* [12], which have potential applications to model the wireless multiple access systems. In particular, they can model the relay-assisted cognitive cooperative wireless systems in which the users transmit packets to a common destination node, and there are a finite number of relay nodes (i.e., orbits) that assist to retransmit the blocked packets; when a direct source user transmission is blocked, it forwards the blocked packet at a relay node [13]. Recent progress in the analysis of the queueing systems with coupled orbits is presented in book [7]. (See Chapters 7–9, where motivation and further references can be also found.)

Among the most important previous results in the analysis of the retrial systems, we mention the explicit expression for the stationary remaining service time of the server, which has been obtained in the recent paper [14]. It is easy to show that this result is also applicable for the stationary retrial queueing system. The mentioned result has a potential application in the setting in which the customers (both in the input process and in orbit) can select the instant to capture the server. An analysis of such a setting is very close to the main purpose of present paper. In the context of the present paper, it is important to emphasize that in the conventional queueing theory setting, customers following an input process are unable to make their own decision, while now we allow this possibility. To best of our knowledge, this setting is completely new, and this is the main contribution of our paper.

In conventional queueing theory, the structure of the input process and service process are usually assumed to be predefined and specified by the input rate and service times of the customers. However, there exists a different approach to the queueing which is based on the assumption that the customers (or users) logging into the system are *strategic* ([15–25]). Namely, it is assumed that the user strategy is to select the arrival instant to the system on a time interval $[0, T]$. In this setting, the queue in the system is determined after each player selects (at the initial instant $t = 0$) their random arrival instant in the system. Thus, each user spends some time in the system, and this time is their personal utility function. As a result, a *non-zero-sum game* is obtained, in which we need to find the *Nash equilibrium*. To the best of our knowledge, the paper [15] (by Glazer and Hassin) is the first work that considers the queue as a result of the user's behavior, and the authors denote this system as $?/M/1$. They further formulate a *non-cooperative game* in which a Poisson-distributed number of the customers determines their arrival instances in the queue of a single-server system, over a (limited) admission interval $[0, T]$. The purpose of the customers is to minimize their waiting time in the system. It is shown in [15] that the *symmetric Nash equilibrium strategy* is mixed. In particular, it was revealed that this strategy is the (absolutely continuous) uniform distribution over time interval $[0, T]$, except a singularity at zero, and the density function decreases between zero and T. A similar model $?/M/m/c$ with $m \geq 1$ identical (exponential) servers and the buffer size $c \geq 0$ for the waiting customers is considered in the paper [22]. Note that the arrival times game with the *batch service* has been investigated in [16]. A single-server bufferless system in which the customers have a *time-sensitivity function* that they want to minimize, instead of their own waiting costs, has been studied in [18]. The paper [20] establishes conditions under which the customers cannot queue until an opening time, and shows that, in the equilibrium, there is a singularity at instant $t = 0$, and that the density is positive only since an instant $t_e > 0$. A model where the customers may incur tardiness costs in addition to the waiting costs is considered in [21]. The paper [25] considers a model combining the tardiness costs, waiting costs, and restrictions on the opening and closing times. (An overview of the existing literature on this problem can also be found in [25].)

In this paper, we apply a game-theoretic approach to a callback queueing system with one server. The queue is formed by the strategic players. (In what follows, we use the terms 'customer', 'user', and 'players' as synonyms.) The player's strategy is to choose a moment to enter the system. If the server is busy, then the user is blocked and joins a (virtual) orbit queue. Otherwise, if the server is free, it seeks a blocked user from the orbit during an (exponential) retrieval time. In this setting, we will find the optimal strategies of the players. First, we consider the case of two users, then consider in detail the case of three players and finally formulate this problem for an arbitrary number of players.

Thus, the found optimal strategies of the players in the described retrial system is the main contribution of this paper.

The paper is organized as follows. In Section 2, we describe the model in details. Then in Section 3, we study the setting with two players. The simplest setting allows one to highlight the main ideas and steps of our analysis. In Section 4, we focus on the system with three players, and the analysis in this case turns out to be much more involved. The analysis of the game with three players is continued in Section 5 where an algorithm on how to find an approximation of the equilibrium is given. The developed approach is then extended to a general setting with $N + 1$ players in Section 6. Moreover, we consider a few numerical examples in Sections 3 and 6.

2. Description of the Model

Now we describe our model in more details in a general setting. We assume that there exists a single server that serves N ('exogenous') customers presented in the system at the initial instant $t = 0$. Unlike the conventional queueing theory setting, these customers use some *strategy* to choose an instant to enter the server. By a symmetry, this strategy is the same for each user. This strategy is determined by a distribution function, which is the main purpose of the analysis, and it determines the instant of the attempt to enter the server, see (1) below. This is an important difference with the standard queueing theory setting where customers are not allowed to have their own decisions but follow the predefined rules describing the dynamics of the system. After the departure of a served customer, the server starts to seek the customer blocked in orbit (if any) to be served next. As mentioned above, this seeking time is exponential with parameter γ and mean τ, and it is called retrieve time in the retrial queueing terminology [11]. If an exogenous customer finds a server busy, then they join a (virtual) orbit, and such orbital customers constitute a virtual *orbit queue* (see [12]), which is served in FIFO (First-In-First-Out) order. If, during a retrieval time (when the server is idle), some customer arrives then they capture server for the exponentially distributed time with parameter μ. Recall that the arrival time is selected in according to distribution (1). Thus, the present setting combines some features of both the classic retrial system and a *gated* queue, in which the input gate remains closed until all N

customers leave the system. In addition, a similar situation arises in a *polling system* (see, for instance, [26]), in which different queues are served in an order, and server, returning to a fixed queue, as a rule finds there a few new customers to be served.

Remark 1. *In our setting, the system has initially a finite number of players, and in this case the system is stable in a traditional sense: The number of customers is bounded. However, in our analysis the requirement $\mu > \gamma$ on the parameters μ and γ appears (see (8)), which indeed can be treated as a stability condition in the framework of the considered game setting. Moreover, when the number of players is N (see Section 6) then the classic stability analysis could be critically important in the asymptotic setting as $N \to \infty$.*

3. Two Players Scenario

Consider the case of two players. To find an equilibrium in this two-person game, we will use the following approach. Suppose that one of the players (for the sake of definiteness, the second player) uses, as a strategy, a random arrival time with the distribution function $F(t)$ (having density $f(t)$) of the following form:

$$F(t) = \begin{cases} p, & 0 \leq t < t_e, \\ p + \int_{t_e}^{t} f(x)dx, & t_e \leq t \leq T. \end{cases} \tag{1}$$

That is, the second player enters the system at the initial moment $t = 0$ with probability $p \in (0, 1)$, otherwise, they arrive at instant $t \in [t_e, T]$ following distribution $F(t)$, where $t_e > 0$ and $T < \infty$ are predefined constants.

Now, we find the best response of the first player to the described strategy used by the second player. As a cost function of the first player, we will consider their average *sojourn time* (the average time the player spends in the system). Thus, the objective of the first player is to choose a strategy that minimizes the average *sojourn time,* that is the total time the player spends in the system. Due to the symmetry of the problem, in the equilibrium, the optimal strategy of the first player must coincide with the chosen strategy of their opponent. To do this, it is sufficient that the strategy of the second player is chosen in such a way that the cost function of the first player takes a constant value over the interval $[t_e, T]$ and at the initial instant $t = 0$ (see [24]). Then the payoff of the first player will not depend on their own strategy.

Next, we find the best response of the first player to the strategy of the second player defined by the relation (1). First, we find its cost function. The average sojourn time, provided the first player enters the system at the instant $t = 0$, is:

$$C(0) = (1-p)\frac{1}{\mu} + p(\frac{1}{2}\frac{1}{\mu} + \frac{1}{2}(\tau + \frac{2}{\mu})) = (1-p)\frac{1}{\mu} + p(\frac{3}{2\mu} + \frac{\tau}{2}).$$

In this expression, we take into account that, with the probability $1 - p$, the second player does not arrive in the system at the instant $t = 0$. Then the first player will be served first, and the average sojourn time equals the average service time $1/\mu$. If the second player arrives at the instant $t = 0$ (with the probability p), then, with probability $1/2$, the first player can be selected for service, and their average service time equals $1/\mu$. However, with probability $1/2$, the server chooses the second player, and then the first player joins the orbit and waits until the second player ends service.

If $0 < t < t_e$, then the average sojourn time of a customer that arrives at instant t satisfies:

$$\begin{aligned} C(t) &= (1-p)\frac{1}{\mu} + p\Big((1 - e^{-\mu t})\frac{1}{\mu} + e^{-\mu t}(\frac{1}{\mu} + \tau + \frac{1}{\mu})\Big) \\ &= (1-p)\frac{1}{\mu} + p\Big(\frac{1}{\mu} + e^{-\mu t}(\tau + \frac{1}{\mu})\Big). \end{aligned}$$

To obtain this expression, we take into account that, with the probability $1-p$, the second player does not arrive at instant $t=0$. In this case, the first player is served first, and the average (sojourn) service time equals $1/\mu$. If the second player arrives at $t=0$ (with the probability p), then with probability $1-\exp\{-\mu t\}$, it will be served in the time interval $[0, t]$, and then the first player will be served immediately. However, with the probability $\exp\{-\mu t\}$, the second player is still in the server at instant t, and then the first player joins the orbit. This player will wait until the second player leaves the server, and then they occupy server after time τ. We note that function $C(t)$ decreases in t and then, in the limit as $t \to 0+$, we obtain:

$$C(0+) = (1-p)\frac{1}{\mu} + p\left(\tau + \frac{2}{\mu}\right) > C(0).$$

We require the fulfillment of the following condition on t_e: $C(0) = C(t_e)$, that is:

$$\frac{1}{\mu} + e^{-\mu t_e}(\tau + \frac{1}{\mu}) = \frac{3}{2\mu} + \frac{\tau}{2},$$

which yields:

$$t_e = \frac{\log 2}{\mu}. \tag{2}$$

Similarly, for $t \geq t_e$, we obtain the average sojourn time in the form:

$$C(t) = p\left((1-e^{-\mu t})\frac{1}{\mu} + e^{-\mu t}(\frac{1}{\mu} + \tau + \frac{1}{\mu})\right) + \left(\int_{t_e}^{t} dF(\theta)\left((1-e^{-\mu(t-\theta)})\frac{1}{\mu} + e^{-\mu(t-\theta)}(\frac{1}{\mu} + \tau + \frac{1}{\mu})\right) + \int_{t}^{T} \frac{1}{\mu} dF(\theta)\right),$$

implying,

$$C(t) = p\left(\frac{1}{\mu} + e^{-\mu t}(\tau + \frac{1}{\mu})\right) + \left(\frac{1-p}{\mu} + \int_{t_e}^{t} e^{-\mu(t-\theta)}(\tau + \frac{1}{\mu})dF(\theta)\right). \tag{3}$$

Now, we find the exact shape of the target distribution function $F(t)$ using condition $C'(t) = 0$. Differentiating, we obtain:

$$-\mu p(\tau + \frac{1}{\mu})e^{-\mu t} - \mu(\tau + \frac{1}{\mu})\int_{t_e}^{t} e^{-\mu(t-\theta)}dF(\theta) + (\tau + \frac{1}{\mu})f(t) = 0,$$

implying,

$$\int_{t_e}^{t} e^{\mu\theta} f(\theta)d\theta = \frac{1}{\mu}f(t)e^{\mu t} - p.$$

Denoting:

$$g(t) = f(t)e^{\mu t},$$

we can write the equation for function $g(t)$ in the following form:

$$\int_{t_e}^{t} g(\theta)d\theta = \frac{1}{\mu}g(t) - p. \tag{4}$$

In addition, we have the following relation:

$$g'(t) = \mu g(t),$$

implying,

$$g(t) = const \cdot e^{\mu t} \text{ and } f(t) = K = const.$$

Substituting $g(t)$ to (4) we obtain:

$$K = \mu p e^{-\mu t_e} = \frac{\mu p}{2}. \tag{5}$$

On the other hand, condition:

$$1 - p = \int_{t_e}^{T} K dt,$$

yields,

$$p = 1 - K(T - t_e),$$

and then, using expression (5), we finally obtain the probability p:

$$p = \frac{1}{1 + \mu(T - t_e)e^{-\mu t_e}} = \frac{2}{2\mu + \mu T - \log 2}. \tag{6}$$

Now it follows from (3) that:

$$C(t) = C(t_e) = p e^{-\mu t_e}(\tau + \frac{1}{\mu}) + \frac{1}{\mu} = \frac{1}{2\mu + \mu T - \log 2}(\tau + \frac{1}{\mu}) + \frac{1}{\mu}.$$

The previous analysis can be summarized as the following statement.

Proposition 1. *An equilibrium in the two-person queueing game with retrials satisfies relation (1), where $f(t) = K = const$ for $t \in [t_e, T]$, and parameters p, t_e, K satisfy conditions (2), (5), and (6).*

Example 1. *Consider a system which accepts the customers within interval $[0, 2]$, and assume that the service rate $\mu = 2$ and the retrieval rate $\gamma = \frac{1}{\tau} = 1$. Then it is easy to calculate that:*

$$t_e \approx 0.347, \ f(t) \approx 0.377 \text{ for } t \geq t_e,$$

the probability (to arrival at instant 0) $p \approx 0.377$ and the average sojourn time is:

$$C(0) = C(t) \approx 0.783 \text{ for } t \geq t_e.$$

4. Three-Player Solution

Consider the case of three players. Suppose that two of the players (for definiteness, the second and third players) use, as a strategy, a random arrival time with the distribution $F(t)$ satisfying (1). We assume that customers are taken from the orbit by the server in the order they entered the orbit, i.e., using FIFO discipline. If two customers entered the orbit simultaneously, then the order is assigned at random (that is, each one is selected with probability 1/2).

Now we find the best response of the first player provided they arrive at instant $t = 0$. In this case, the payoff is:

$$C(0) = (1-p)^2 \frac{1}{\mu} + 2p(1-p)(\frac{1}{\mu} + \frac{1}{2}W_1^1(0)) + p^2(\frac{1}{\mu} + \frac{2}{3}W_2^0(0)),$$

where,

$$W_1^0(0) = \tfrac{1}{\mu} + \tfrac{1}{\gamma},$$
$$W_2^0(0) = \tfrac{1}{\mu} + \tfrac{1}{\gamma} + \tfrac{1}{2}(\tfrac{1}{\mu} + \tfrac{1}{\gamma}) = \tfrac{3}{2\mu} + \tfrac{3}{2\gamma},$$
$$W_1^1(0) = \tfrac{1}{1-p} \int_{t_e}^{T} \int_{0}^{\theta} \mu e^{-\mu s} ds \int_{0}^{\theta - s} \gamma e^{-\gamma u}(s + u) du + \int_{0}^{\theta} \mu e^{-\mu s} ds \int_{\theta - s}^{\infty} \gamma e^{-\gamma u}(s + u + \tfrac{1}{\mu}) du +$$
$$\int_{\theta}^{\infty} \mu e^{-\mu s} s ds dF(\theta) = (\tfrac{1}{\mu} + \tfrac{1}{\gamma}) + \tfrac{1}{\gamma(\mu - \gamma)(1-p)} \int_{t_e}^{T} (\gamma e^{-\gamma\theta} - \mu e^{-\mu\theta}) dF(\theta),$$

and by $W_i^j(t)$ we denote the average time spent in the orbit by a customer arriving at instant t, provided that there are i customers in the orbit, and j (exogenous) customers remains outside the system.

Now we find the value $C(0+)$ for a customer who arrives in the system at instant $t = 0+$. Then they occupy the server if both remaining players arrive after t_e. If one player arrives at instant $t = 0$, then the customer arriving at $t = 0+$, evidently, joins the orbit. Finally, if both other customers arrive at $t = 0$, then the customer arriving at $t = 0+$ joins the end of the orbit queue. This results in:

$$C(0+) = (1-p)^2\frac{1}{\mu} + p(1-p)(\frac{1}{\mu} + \frac{1}{1-p}W_1^1(0)) + p^2(\frac{1}{\mu} + W_2^0(0+)) > C(0), \quad (7)$$

where,

$$W_2^0(0+) = \frac{1}{\mu} + \frac{1}{\gamma} + \frac{1}{\mu} + \frac{1}{\gamma} = \frac{2}{\mu} + \frac{2}{\gamma} > W_2^0(0).$$

The playoff of a customer arriving at instant t_e, provided that no customers arrive in the time interval $(0, t_e)$, satisfies:

$$C(t_e) = (1-p)^2\frac{1}{\mu} + 2p(1-p)\left(\frac{1}{\mu} + \int_{t_e}^{T} \frac{dF(\theta)}{1-F(t_e)} \left(\int_{t_e}^{\theta} \mu e^{-\mu s} ds \int_{0}^{\theta - s} \gamma e^{-\gamma\tau}(s - t_e + \tau) d\tau + \right.\right.$$
$$\left.\left.\int_{t_e}^{\theta} \mu e^{-\mu s} ds \int_{\theta - s}^{\infty} \gamma e^{-\gamma\tau}(\theta - t_e + \frac{1}{\mu} + \frac{1}{\gamma}) d\tau + \int_{\theta}^{\infty} \mu e^{-\mu s}(s - t_e + \frac{1}{\gamma}) ds\right)\right) +$$
$$p^2\left(\frac{1}{\mu} + \int_{0}^{t_e} \mu e^{-\mu s} ds \int_{0}^{t_e - s} \gamma e^{-\gamma\tau} d\tau \int_{t_e - s - \tau}^{\infty} \mu e^{-\mu v}(s + \tau + v - t_e + \frac{1}{\gamma}) dv + \right.$$
$$\left.\int_{t_e}^{\infty} \mu e^{-\mu s} ds (s + \frac{1}{\mu} + \frac{2}{\gamma} - t_e)\right) = (1-p)^2\frac{1}{\mu} +$$
$$2p(1-p)\left(\frac{1}{\mu} + (\frac{1}{\mu} + \frac{1}{\gamma})e^{-\mu t_e} + \frac{e^{-\mu t_e}}{(1-p)(\mu-\gamma)} \int_{t_e}^{T} (e^{-\gamma(\theta - t_e)} - e^{-\mu(\theta - t_e)}) dF(\theta)\right) +$$
$$p^2\left(\frac{1}{\mu} + 2(\frac{1}{\mu} + \frac{1}{\gamma})e^{-\mu t_e} + (\frac{1}{\mu} + \frac{1}{\gamma})\frac{\gamma}{(\mu-\gamma)^2}(e^{-\gamma t_e} - e^{-\mu t_e}) - (\frac{1}{\mu} + \frac{1}{\gamma})\frac{\gamma}{\mu - \gamma} t_e e^{-\mu t_e}\right).$$

We assume that:

$$\mu > \gamma. \quad (8)$$

Then the payoff value $C(t_e)$ decreases by t_e, provided that there are no customers arriving into the system at the interval $(0, t_e)$.

The obtained latter equality and the inequality $C(0+) > C(0)$ (see (7)) confirm that it is more profitable for a customer to arrive with a delay following the instant $t = 0$ (but not at the instant $0+$). Since, in the equilibrium, the payoff should be the same on the strategy support, then the instant when a new customer decides to arrive in the system satisfies the equation:

$$C(0) = C(t_e).$$

Now we show that an arrival at the instant $t \in (0, t_e)$ is unprofitable even for one player, when other players make an attempt to occupy the server since instant t_e. For such t we obtain:

$$C(t) = (1-p)^2\frac{1}{\mu} + 2p(1-p)\left(\frac{1}{\mu} + \int_{t_e}^{T} dF(\theta)\left(\frac{1}{\mu} + \int_{t}^{\theta} \mu e^{-\mu s}ds \int_{0}^{\theta-s} \gamma e^{-\gamma\tau}(s-t+\tau)d\tau + \right.\right.$$

$$\left.\left.\int_{t}^{\theta} \mu e^{-\mu s}ds \int_{\theta-s}^{\infty} \gamma e^{-\gamma\tau}(\theta - t + \frac{1}{\mu} + \frac{1}{\gamma})d\tau + \int_{\theta}^{\infty} \mu e^{-\mu s}(s - t + \frac{1}{\gamma})ds\right)\right) +$$

$$p^2\left(\frac{1}{\mu} + \int_{0}^{t} \mu e^{-\mu s}ds \int_{0}^{t-s} \gamma e^{-\gamma\tau}d\tau \int_{t-s-\tau}^{\infty} \mu e^{-\mu v}(s+\tau+v-t+\frac{1}{\gamma})dv + \right.$$

$$\left.\int_{t}^{\infty} \mu e^{-\mu s}ds(s + \frac{1}{\mu} + \frac{2}{\gamma} - t)\right) =$$

$$(1-p)^2\frac{1}{\mu} + p^2\left(\frac{1}{\mu} + 2(\frac{1}{\mu} + \frac{1}{\gamma})e^{-\mu t} + \frac{1}{\mu-\gamma}(e^{-\gamma t} - e^{-\mu t}) - te^{-\mu t}\right) +$$

$$2p(1-p)\left(\frac{1}{\mu} + (\frac{1}{\mu} + \frac{1}{\gamma})e^{-\mu t} + \frac{e^{-\mu t}}{(1-p)(\mu-\gamma)}\int_{t_e}^{T}(e^{-\gamma(\theta-t)} - e^{-\mu(\theta-t)})dF(\theta)\right).$$

Then it is easy to check that function $C(t)$ decreases in $t < t_e$, confirming that, even for one player, it is better to avoid an attempt to enter the server earlier the instant t_e.

Now we consider the situation when the first player enters the system in the interval $t \in [t_e, T]$. To study it, we define, for instant t, the (time-dependent) state probabilities, denoted by $p_{ijk}(t)$, that $i \in \{0,1,2\}$ customers arrived in the system, in the interval $[t_e, T]$, $j \in \{0,1\}$ customers are in the server, and $k \in \{0,1\}$ customers are in the orbit, at instant t. The arrival rate at instant t (that is the rate of exogenous customers) depends on the chosen strategy and the number k of customers who have already entered the system up to instant t. In an evident notation, these rates are equal to:

$$\lambda_0(t) = 2\frac{f(t)}{1-F(t)}, \quad \lambda_1(t) = \frac{f(t)}{1-F(t)},$$

respectively. Now we can write down the corresponding Kolmogorov backward equations for the state probabilities:

$$\begin{aligned}
p'_{000}(t) &= -\lambda_0(t)p_{000}(t),\\
p'_{100}(t) &= -\lambda_1(t)p_{100}(t) + \mu p_{110}(t),\\
p'_{110}(t) &= -(\mu + \lambda_1(t))p_{110}(t) + \lambda_0(t)p_{000}(t),\\
p'_{201}(t) &= -\gamma p_{201}(t) + \mu p_{211}(t),\\
p'_{210}(t) &= -\mu p_{210}(t) + \gamma p_{201}(t) + \lambda_1(t)p_{100}(t),\\
p'_{211}(t) &= -\mu p_{211}(t) + \lambda_1(t)p_{110}(t),\\
p'_{200}(t) &= \mu p_{210}(t).
\end{aligned} \tag{9}$$

Now we find the state probabilities for $t = t_e$ as follows:

$$
\begin{aligned}
&p_{000}(t_e) = (1-p)^2 \\
&p_{100}(t_e) = 2p(1-p)(1-e^{-\mu t_e}) \\
&p_{110}(t_e) = 2p(1-p)e^{-\mu t_e} \\
&p_{201}(t_e) = p^2 \int_0^{t_e} \mu e^{-\mu\theta} e^{-\gamma(t_e-\theta)} d\theta = p^2 \mu \tfrac{e^{-\gamma t_e} - e^{-\mu t_e}}{\mu - \gamma} \\
&p_{210}(t_e) = p^2 \int_0^{t_e} \mu e^{-\mu\theta} \int_0^{t_e-\theta} \gamma e^{-\gamma\tau} e^{-\mu(t_e-(\theta+\tau))} d\tau d\theta = p^2 \gamma \tfrac{e^{-\gamma t_e} - e^{-\mu t_e}(\mu t_e - \gamma t_e + 1)}{(\mu-\gamma)^2} \\
&p_{211}(t_e) = p^2 e^{-\mu t_e} \\
&p_{200}(t_e) = p^2 \int_0^{t_e} \mu e^{-\mu\theta} \int_0^{t_e-\theta} \gamma e^{-\gamma\tau}(1 - e^{-\mu(t_e-(\theta+\tau))}) d\tau d\theta = \\
&\quad = p^2(1 - e^{-\mu t_e} - \gamma \tfrac{e^{-\gamma t_e} - e^{-\mu t_e}(\mu t_e - \gamma t_e + 1)}{(\mu-\gamma)^2} - \mu \tfrac{e^{-\gamma t_e} - e^{-\mu t_e}}{\mu - \gamma}).
\end{aligned}
\tag{10}
$$

Expressions (10) indeed are the initial boundary conditions for the Cauchy problem for differential Equations (9). The probability p of an arrival at instant $t = 0$ can be found from the normalization condition:

$$
p + \int_{t_e}^{T} f(t)dt = 1.
$$

Then the average sojourn time of a player entering the system at instant t is:

$$
\begin{aligned}
C(t) &= \frac{1}{\mu}\Big(p_{000}(t) + p_{100}(t) + p_{201}(t) + p_{200}(t)\Big) \\
&+ p_{110}(t)(\frac{1}{\mu} + W_1^1(t)) + p_{210}(t)(\frac{1}{\mu} + W_1^0(t)) + p_{211}(t)(\frac{1}{\mu} + W_2^0(t)) \\
&= \frac{1}{\mu} + p_{110}(t)W_1^1(t) + p_{210}(t)W_1^0(t) + p_{211}(t)W_2^0(t),
\end{aligned}
$$

where,

$$
\begin{aligned}
&W_1^0(t) = \tfrac{1}{\mu} + \tfrac{1}{\gamma}, \\
&W_2^0(t) = \tfrac{2}{\mu} + \tfrac{2}{\gamma}, \\
&W_1^1(t) = \tfrac{1}{1-F(t)} \int_t^T dF(\theta) \Bigg(\int_0^{\theta-t} \mu e^{-\mu s} ds \int_0^{\theta-t-s} \gamma e^{-\gamma\tau}(s+\tau)d\tau + \\
&\int_0^{\theta-t} \mu e^{-\mu s} \int_{\theta-t-s}^{\infty} \gamma e^{-\gamma\tau}(\theta - t + \tfrac{1}{\mu} + \tfrac{1}{\gamma})d\tau + \int_{\theta-t}^{\infty} \mu e^{-\mu s}(\tfrac{1}{\gamma} + s)ds \Bigg) \\
&= \tfrac{1}{\mu} + \tfrac{1}{\gamma} + \tfrac{1}{(1-F(t))(\mu-\gamma)} \int_t^T e^{-\gamma(\theta-t)} - e^{-\mu(\theta-t)} dF(\theta).
\end{aligned}
$$

In the equilibrium, the equality $C(t) = C(t_e) = const,\ t \in [t_e, T]$ is satisfied, implying:

$$
\begin{aligned}
&(\frac{1}{\mu} + \frac{1}{\gamma})\Big(p_{110}(t) - p_{110}(t_e) + p_{210}(t) - p_{210}(t_e) + 2(p_{211}(t) - p_{211}(t_e)\Big) + \\
&\frac{1}{\mu-\gamma}\left(\frac{p_{110}(t)}{1-F(t)} \int_t^T (e^{-\gamma(\theta-t)} - e^{-\mu(\theta-t)})dF(\theta) - \frac{p_{110}(t_e)}{1-p} \int_{t_e}^T (e^{-\gamma(\theta-t_e)} - e^{-\mu(\theta-t_e)})dF(\theta)\right) \\
&= 0.
\end{aligned}
\tag{11}
$$

If condition (11) is met, then the function $C(t)$ must be a constant on the time interval $[t_e, T]$. It remains now to require that the condition $C(t_e) = C(0)$ is satisfied. In other words,

$$(1-p)^2\frac{1}{\mu} + p^2(\frac{2}{\mu} + \frac{1}{\gamma})$$
$$2p(1-p)(\frac{1}{\mu} + \frac{1}{2}(\frac{1}{\mu} + \frac{1}{\gamma}) + \frac{1}{2\gamma(\mu-\gamma)(1-p)}\int\limits_{t_e}^{T}(\gamma e^{-\gamma\theta} - \mu e^{-\mu\theta})dF(\theta)) =$$
$$(1-p)^2\frac{1}{\mu} +$$
$$2p(1-p)\left(\frac{1}{\mu} + (\frac{1}{\mu} + \frac{1}{\gamma})e^{-\mu t_e} + \frac{e^{-\mu t_e}}{(1-p)(\mu-\gamma)}\int\limits_{t_e}^{T}(e^{-\gamma(\theta-t_e)} - e^{-\mu(\theta-t_e)})dF(\theta)\right) +$$
$$p^2\left(\frac{1}{\mu} + 2(\frac{1}{\mu} + \frac{1}{\gamma})e^{-\mu t_e} + (\frac{1}{\mu} + \frac{1}{\gamma})\frac{\gamma}{(\mu-\gamma)^2}(e^{-\gamma t_e} - e^{-\mu t_e}) - (\frac{1}{\mu} + \frac{1}{\gamma})\frac{\gamma}{\mu-\gamma}t_e e^{-\mu t_e}\right)$$

or:

$$\begin{aligned}&(\frac{1}{\mu} + \frac{1}{\gamma})(1 - 2e^{-\mu t_e}) + \\ &\frac{1}{(\mu-\gamma)}(\frac{1}{\gamma}\int\limits_{t_e}^{T}(\gamma e^{-\gamma\theta} - \mu e^{-\mu\theta})dF(\theta) - 2e^{-\mu t_e}\int\limits_{t_e}^{T}(e^{-\gamma(\theta-t_e)} - e^{-\mu(\theta-t_e)})dF(\theta)) + \\ &p(\frac{1}{\mu} + \frac{1}{\gamma})(\frac{\gamma}{\mu-\gamma}t_e e^{-\mu t_e} - \frac{\gamma}{(\mu-\gamma)^2}(e^{-\gamma t_e} - e^{-\mu t_e})) = 0.\end{aligned} \tag{12}$$

The analysis performed above is summarized in the following statement.

Proposition 2. *An equilibrium in the three-person queueing game with retrievals has the form (1), where $f(t)$ for $t \in [t_e, T]$ is determined as a solution of Equations (9), (11), and (12).*

5. Computing the Equilibrium

In this section, we describe in detail the numerical solution of the above obtained equations. Let us fix t_e and T. We divide the interval $[t_e, T]$ into $k-1$ equal segments. Then we find an approximate solution in the nodes of the grid $K = \{t_1 = t_e, t_2 = t_1 + \Delta, \ldots, t_k = t_{k-1} + \Delta\}$, where $\Delta = (T - t_e)/(k-1)$. We represent the values of the unknown function $f(t)$ at the initial moment and at the nodes of the grid as the arguments of the problem:

$$\{x_0 = f(0) = p, x_1 = f(t_1), \ldots, x_k = f(t_k) = f(T)\}.$$

Then the derived conditions for the equilibrium can be represented as the difference equations. More precisely, condition (11) becomes:

$$\begin{aligned}&(\frac{1}{\mu} + \frac{1}{\gamma})(p_{110}(t_i) - p_{110}(t_e) + p_{210}(t_i) - p_{210}(t_e) + \\ &+ 2(p_{211}(t_i) - p_{211}(t_e))) + \frac{p_{110}(t_i)}{(\mu-\gamma)(1-p-\sum\limits_{j=1}^{i-1}x_j\Delta)}\sum_{j=i}^{k-1}\left(e^{-\gamma(t_j-t_i)} - e^{-\mu(t_j-t_i)}\right)x_j\Delta - \\ &- \frac{p_{110}(t_e)}{(\mu-\gamma)(1-p)}\sum_{j=1}^{k-1}\left(e^{-\gamma(t_j-t_e)} - e^{-\mu(t_j-t_e)}\right)x_j\Delta = 0, \quad i = 2, \ldots, k.\end{aligned} \tag{13}$$

Condition (12) takes the form:

$$
\begin{aligned}
&(\frac{1}{\mu}+\frac{1}{\gamma})(1-2e^{-\mu t_e})+\\
&\frac{1}{(\mu-\gamma)}(\frac{1}{\gamma}\sum_{j=1}^{k-1}(\gamma e^{-\gamma t_j}-\mu e^{-\mu t_j})x_j\Delta-2e^{-\mu t_e}\sum_{j=1}^{k-1}(e^{-\gamma(t_j-t_e)}-e^{-\mu(t_j-t_e)})x_j\Delta)+\\
&p(\frac{1}{\mu}+\frac{1}{\gamma})(\frac{\gamma}{\mu-\gamma}t_e e^{-\mu t_e}-\frac{\gamma}{(\mu-\gamma)^2}(e^{-\gamma t_e}-e^{-\mu t_e}))=0.
\end{aligned}
\tag{14}
$$

The Kolmogorov backward (difference) equations become:

$$
\begin{aligned}
&p_{000}(t_{i+1})=p_{000}(t_i)\left(1-\Delta\frac{2x_i}{1-p-\Delta\sum\limits_{j=1}^{i-1}x_j}\right),\\
&p_{100}(t_{i+1})=p_{100}(t_i)\left(1-\Delta\frac{x_i}{1-p-\Delta\sum\limits_{j=1}^{i-1}x_j}\right)+\Delta\mu p_{110}(t_i),\\
&p_{110}(t_{i+1})=p_{110}(t_i)\left(1-\Delta(\mu+\frac{x_i}{1-p-\Delta\sum\limits_{j=1}^{i-1}x_j})\right)+\Delta\frac{2x_i}{1-p-\Delta\sum\limits_{j=1}^{i-1}x_j}p_{000}(t_i),\\
&p_{201}(t_{i+1})=(1-\Delta\gamma)p_{201}(t_i)+\Delta\mu p_{211}(t_i),\\
&p_{210}(t_{i+1})=(1-\Delta\mu)p_{210}(t_i)+\Delta\gamma p_{201}(t_i)+\Delta\frac{x_i}{1-p-\Delta\sum\limits_{j=1}^{i-1}x_j}p_{100}(t_i),\\
&p_{211}(t_{i+1})=(1-\Delta\mu)p_{211}(t_i)+\Delta\frac{x_i}{1-p-\Delta\sum\limits_{j=1}^{i-1}x_j}p_{110}(t_i),\\
&p_{200}(t_{i+1})=p_{200}(t_i)+\Delta\mu p_{210}(t_i),\\
&i=1,\dots,k-1.
\end{aligned}
\tag{15}
$$

Finally, the normalization condition takes the form:

$$
p+\Delta\sum_{i=1}^{k-1}x_i-1=0. \tag{16}
$$

Now we find a solution as follows. We iterate over the values of p on the interval $[0,1]$. For each p, we iterate over the values of t_e belonging to interval $[0,T]$. For each given p and t_e, we solve the system of the difference Equations (13), where the state probabilities satisfy system (15). We are looking for a pair p, t_e, in such a way that conditions (14) and (16) are satisfied.

To find p and t_e, we first use a partition of $[0,1]\times[0,T]$ into a grid with a small rank and look for a node, passing through which the left sides of Equations (14) and (16) change sign. We seek for a solution in a neighborhood of this node. More precisely, for each given p and t_e, the solution x is found with Algorithm 1. First we specify the uniform distribution over the interval $[t_e,T]$ as the initial approximation of the solution x, taking into account that there is an atom at the point $t=0$ with a given probability p. Then we search for a solution x that satisfies (13). Step 1 of Algorithm 1 is repeated until the solution is stabilized with a given accuracy ϵ. In practice, the algorithm converges in 2–3 runs.

At each iteration $i=2,\dots,k$ at the beginning of the execution, we have x_j, where $j=1,\dots,i-2$, and state probabilities for all steps from 1 to $i-1$, found on previous iterations. On iteration i, we solve Equation (13) for x_{i-1}, each time calculating the probabilities of states at step i using an approximation of x_{i-1} according to (15). In this case, the sum in the two last terms on the left-hand side of Equation (13) is partially calculated from the old values x^{prev} from the solution at the previous iteration, and all other components of the equation are calculated by new ones, according to the solution x at the current iteration. However, as calculation experiments confirm, with each pass of the algorithm, the difference between the solutions decreases.

Algorithm 1 Finding the solution x for p and t_e.

Step 0. Initialization.
for $i = 1$ to k **do**
  $x_i \leftarrow \frac{1-p}{T-t_e}$
end for
Calculate $p000(t_1), \ldots, p200(t_1)$ from (10).
Step 1. Next approximations.
do
  $x^{prev} \leftarrow x$
  for $i = 2$ to k **do**
    Find x_{i-1} from (13) with new $p000(t_i), \ldots, p200(t_i)$ re-calculated with (15).
  end for
while $(|x - x^{prev}| > \epsilon)$

After completing the algorithm, we get a set of values x_i, where $i = 1, \ldots, k-1$, which is enough to find $F(T) \approx p + \sum_{i=1}^{k-1} x_i \Delta$. If conditions (14) and (16) are satisfied with a given accuracy ϵ, the current p, t_e and x give a solution, otherwise we change t_e and p.

Example 2. *Let $T = 2, \mu = 2, \gamma = 1$. The computations give the optimal values in the equilibrium: $p \approx 0.412$, $t_e \approx 0.380$. The density of the optimal arrival time at the interval $[t_e, T]$ is presented at Figure 1. In the figure, the function $f(t)$ first decreases in the interval $[t_e, 0.833]$, then increases in the interval $[0.833, 1.676]$, then decreases again in the interval $[1.676, 2]$. The value at the equilibrium is equal to $C(t) \approx 1.133$.*

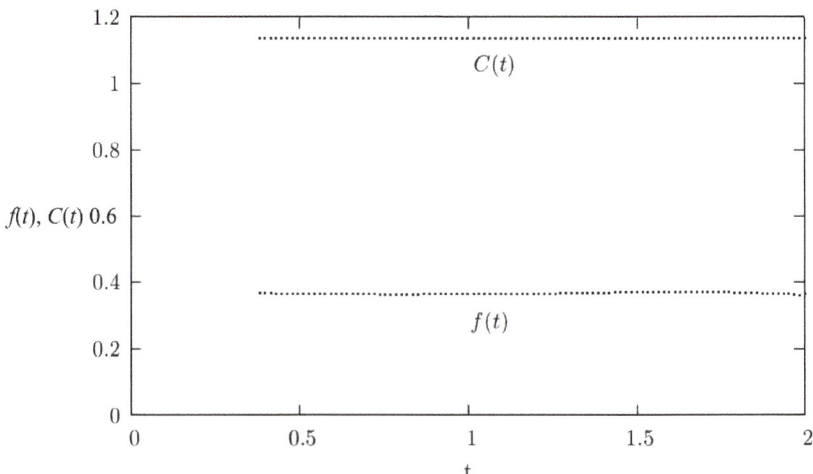

Figure 1. The equilibrium density $f(t)$ and cost $C(t)$ for $T = 2, \mu = 2, \gamma = 1$.

Example 3. *If $T = 4, \mu = 2, \gamma = 1$ then $p \approx 0.232$, $t_e \approx 0.369$, and $C(t) \approx 0.857$. The shape of the equilibrium density is similar to that presented in Figure 1, and $C(t) \approx 0.857$.*

Example 4. *If $T = 4, \mu = 4, \gamma = 1$ then the shape of the equilibrium density is similar to that as given on Figure 1, and $p \approx 0.121$, $t_e \approx 0.179$, and $C(t) \approx 0.404$.*

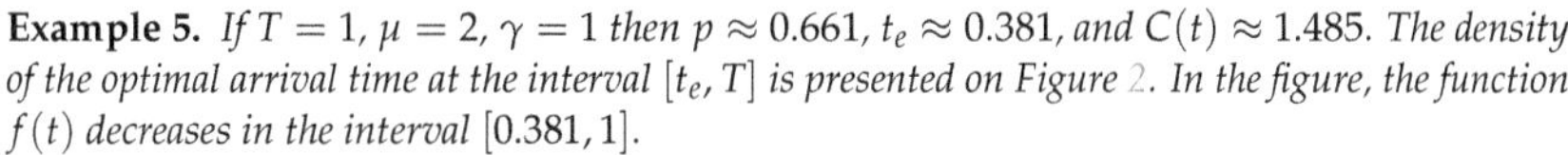

Example 5. *If $T = 1$, $\mu = 2$, $\gamma = 1$ then $p \approx 0.661$, $t_e \approx 0.381$, and $C(t) \approx 1.485$. The density of the optimal arrival time at the interval $[t_e, T]$ is presented on Figure 2. In the figure, the function $f(t)$ decreases in the interval $[0.381, 1]$.*

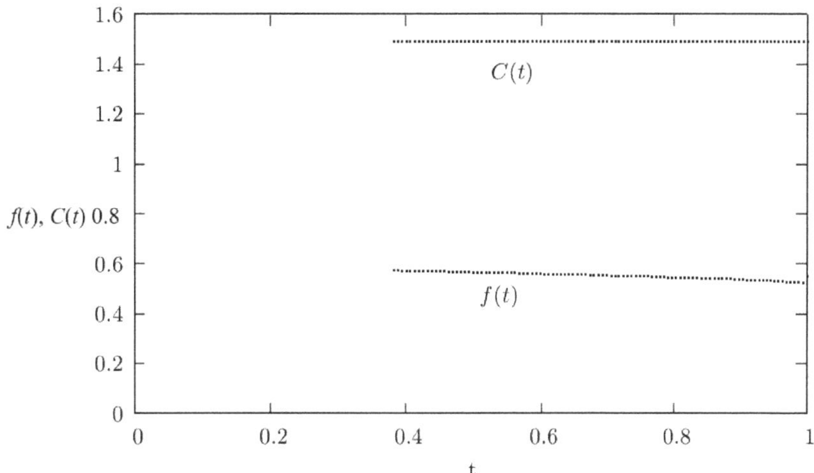

Figure 2. Equilibrium density $f(t)$ and cost $C(t)$ for $T = 1, \mu = 2, \gamma = 1$.

6. N Players

The same approach can be used when considering a service system with more than 3 players. Suppose that in the queueing system an arrival strategy $F(t)$ of the form (1) is used, and the (N)-th player is looking for a strategy that gives the best response. In this case, the found best answer must coincide with the strategy $F(t)$. Due to symmetry, it will be the equilibrium in the game.

By the state of the system at an instant t, we mean the set (k, s, i), where k is the number of claims entered into the system, i is the number of claims in the orbit, and the parameter s, which means for $s = 0$ the server free, and with $s = 1$ the server is busy at instant t. We denote by $W_i^j(t)$ the average time until the server seeks for a claim in orbit at the time t provided the system has not yet received j customers and there are i customers in the orbit. Our purpose is to find the average sojourn time of the (N)-th customer when they enter the system at the instant t. For $t = 0$, we obtain:

$$
\begin{aligned}
C(0) &= (1-p)^{N-1}\frac{1}{\mu} + \sum_{i=1}^{N-1} p^i (1-p)^{N-1-i}\left(\frac{1}{\mu} + W_i^{N-1-i}(0)\right) \\
&= \frac{1}{\mu} + \sum_{i=1}^{N-1} p^i (1-p)^{N-1-i} W_i^{N-1-i}(0).
\end{aligned}
$$

For $0 < t < t_e$, the average sojourn time of the (N)-th customer satisfies:

$$
\begin{aligned}
C(t) &= (1-p)^{N-1}\frac{1}{\mu} + \sum_{i=1}^{N-1} p^i (1-p)^{N-1-i}\left(\frac{1}{\mu} + W_i^{N-1-i}(t)\right) \\
&= \frac{1}{\mu} + \sum_{i=1}^{N-1} p^i (1-p)^{N-1-i} W_i^{N-1-i}(t).
\end{aligned}
$$

Finally, for $t \geq t_e$, the average sojourn time of the (N)-th customer equals:

$$C(t) = P(\text{idle server})\frac{1}{\mu}+$$
$$\sum_{k=1}^{N-1}\sum_{i=0}^{k-1} P(\text{busy server, } k \text{ arrived customers, } i \text{ orbital customers})\big(W_i^{N-1-k}(t) + \frac{1}{\mu}\big).$$

It gives:

$$C(t) = \sum_{k=1}^{N-1}\sum_{i=0}^{k-1}\Big(p_{k,0,i}(t)\frac{1}{\mu} + p_{k,1,i}(t)(W_i^{N-1-k}(t) + \frac{1}{\mu})\Big),$$

implying, after some algebra, the following expression:

$$C(t) = \frac{1}{\mu} + \sum_{k=1}^{N-1}\sum_{i=0}^{k-1} p_{k,1,i}(t)W_i^{N-1-k}(t). \quad (17)$$

In this expression $p_{k,s,i}(t)$ denotes the state probability at the instant t, which satisfy the following Kolmogorov backward equations:

$$\begin{aligned} p'_{k,1,i}(t) = & - (\mu + \lambda_k(t))p_{k,1,i}(t) + \lambda_{k-1}(t)p_{k-1,1,i-1}(t) \\ & + \lambda_{k-1}(t)p_{k-1,0,i}(t) + \gamma p_{k,0,i+1}(t), \end{aligned}$$

$$\begin{aligned} p'_{k,0,i}(t) = & - (\gamma + \lambda_k(t))p_{k,0,i}(t) + \mu p_{k,1,i}(t); \\ k = & \quad 1,2,\ldots,N-2; \quad i = 0,\ldots,k-1. \end{aligned}$$

We note that in this case,

$$\lambda_k(t) = (N-1-k)\frac{f(t)}{1-F(t)}, \quad k = 1,\ldots,N-2,$$

is the arrival rate, provided there are k customers in the system at the instant t.

Then we find the expression for $W_i^j(t)$, the average waiting time a customer spends in the orbit, starting with instant t, until they occupy the server. Then, substituting $W_i^j(t)$ in (17), we proceed as above to find the optimal strategy for the (N)-th player. To do this, we require that the following conditions:

$$C(0) = C(t_e) = C(t) = C^*, \quad t \in [t_e, T],$$

are satisfied. Finally, these conditions allow, in general, one to find the optimal strategy $F(t)$ along the same steps we described above for particular cases with 2 and 3 players.

7. Conclusions

We consider a single-server retrial queueing system in which the service of customers is handled in a strategic manner, i.e., unlike the conventional retrial queues, the customers are generated by players who choose the time to occupy the server. An equilibrium is sought in such a system when applications are distributed in such a way that the average service time is minimal. An equilibrium is sought in the class of mixed strategies, when players choose the login time randomly. It is shown that the optimal strategy is such that a player enters the system with a non-zero probability at the initial moment of time, otherwise, it takes a random pause, and then a distribution density is used, satisfying a system of the differential equations. This setting is described in detail for the case of two and three players and illustrated by a few numerical examples. Moreover the model for an arbitrary number of players is formulated. This scenario is assumed to be considered in detail in a future work.

Author Contributions: Conceptualization, J.C., V.M. and E.M.; methodology, J.C., V.M. and E.M.; software, J.C., V.M. and E.M.; validation, J.C., V.M. and E.M.; formal analysis, J.C., V.M. and E.M.; investigation, J.C., V.M. and E.M.; resources, J.C., V.M. and E.M.; data curation, J.C., V.M. and E.M.; writing—original draft preparation, J.C., V.M. and E.M; writing—review and editing, J.C., V.M. and E.M.; visualization, J.C., V.M. and E.M.; supervision, J.C., V.M. and E.M. All authors have read and agreed to the published version of the manuscript.

Funding: This research received no external funding.

Conflicts of Interest: The authors declare no conflict of interest.

References

1. Falin, G.I.; Templeton, J.G.D. *Retrial Queues*; Chapman & Hall: London, UK, 1997.
2. Artalejo, J.R.; Gomez-Corral, A. *Retrial Queueing Systems: A Computational Approach*; Springer: Berlin/Heidelberg, Germany, 2008.
3. Kim, J.; Kim, B. A survey of retrial queueing systems. *Ann. Oper. Res.* **2016**, *247*, 3–36. [CrossRef]
4. Altman, E.; Borovkov, A.A. On the stability of retrial queues. *Queueing Syst.* **1997**, *26*, 343–363. [CrossRef]
5. Morozov, E. A multiserver retrial queue: Regenerative stability analysis. *Queueing Syst.* **2007**, *56*, 157–168. [CrossRef]
6. Morozov, E.; Phung-Duc, T. Stability analysis of a multiclass retrial system with classical retrial policy. *Perform. Eval.* **2017**, *112*, 15–26. [CrossRef]
7. Morozov, E.; Steyaert, B. *Stability Analysis of Regenerative Queueing Models*; Springer: Berlin/Heidelberg, Germany, 2021. [CrossRef]
8. Fayolle, G. A simple telephone exchange with delayed feedback. In *Teletraffic Analysis and Computer Performance Evaluation*; Boxma, O.J., Cohen, J.W., Tijms, H.C., Eds.; Elsevier: Amsterdam, The Netherlands, 1986; Volume 7, pp. 245–253.
9. Choi, B.D.; Rhee, K.H.; Park, K.K. The M/G/1 retrial queue with retrial rate control policy. In *Probability in the Engineering and Informational Sciences*; Cambridge University Press: Cambridge, UK, 1993; Volume 7, pp. 29–46.
10. Avrachenkov, K.; Morozov, E.; Steyaert, B. Sufficient stability conditions for multi-class constant retrial rate systems. *Queueing Syst.* **2016**, *82*, 149–171. [CrossRef]
11. Morozov, E.; Zhukova, K. The Overflow Probability Asymptotics in a Single-Class Retrial System with General Retrieve Time. In Proceedings of the International Conference on Distributed Computer and Communication Networks, Moscow, Russia, 20–24 September 2021; DCCN 2021, LNCS 13144; Vishnevskiy, V.M., Samouylov, K.E., Kozyrev, D.V., Eds.; Springer: Cham, Switzerland, 2021; pp. 55–66.
12. Dimitriou, I. A queueing system for modeling cooperative wireless networks with coupled relay nodes and synchronized packet arrivals. *Perform. Eval.* **2017**, *114*, 16–31. [CrossRef]
13. Pappas, N.; Kountouris, M.; Ephremides, A.; Traganitis, A. Relay-assisted multiple access with full-duplex multi-packet reception. *IEEE Trans. Wirel. Commun.* **2015**, *14*, 3544–3558. [CrossRef]
14. Morozov, E.; Morozova, T. On the stationary remaining service time in the queuieng systems. *CEUR Workshop Proc.* **2020**, *2792*, 140–149.
15. Glazer, A.; Hassin, R. ?/M/1: On the equilibrium distribution of customer arrivals. *Eur. J. Oper. Res.* **1983**, *13*, 146–150. [CrossRef]
16. Glazer, A.; Hassin, R. Equilibrium arrivals in queues with bulk service at scheduled times. *Transp. Sci.* **1987**, *21*, 273–278. [CrossRef]
17. Altman, E.; Shimkin, N. Individually optimal dynamic routing in a processor sharing system. *Oper. Res.* **1998**, *46*, 776–784. [CrossRef]
18. Mazalov, V.V.; Chuiko, J.V. Nash equilibrium in the optimal arrival time problem. *Comput. Technol.* **2006**, *11*, 60–71.
19. Haviv, M.; Kella, O.; Kerner, Y. Equilibrium strategies in queues based on time or index of arrival. *Prob. Eng. Inform. Sci.* **2010**, *24*, 13–25. [CrossRef]
20. Hassin, R.; Kleiner, Y. Equilibrium and optimal arrival patterns to a server with opening and closing times. *IIE Trans.* **2011**, *43*, 164–175. [CrossRef]
21. Jane, R.; Juneja, S.; Shimkin, N. The concert queueing game: To wait or to be late. *Discret. Event Dyn. Syst.* **2011**, *21*, 103–138. [CrossRef]
22. Haviv, M. When to arrive at a queue with tardiness costs. *Perform. Eval.* **2013**, *70*, 387–399. [CrossRef]
23. Ravner, L.; Haviv, M. Strategic timing of arrivals to a finite queue multi-server loss system. *Queueing Syst.* **2015**, *81*, 71–96.
24. Mazalov, V.; Chirkova, J. *Networking Games. Network Forming Games and Games on Networks*; Academic Press: Cambridge, MA, USA, 2019; 322p.
25. Haviv, M.; Ravner, L. A survey of queueing systems with strategic timing of arrivals. *Queueing Syst.* **2021**, *99*, 163–198. [CrossRef]
26. Boon, M.A.A. A polling model with reneging at polling instants. *Ann. Oper. Res.* **2012**, *198*, 5–23. [CrossRef]

MDPI

Article
Self-Service System with Rating Dependent Arrivals

Alexander Dudin [1,2,*], Olga Dudina [1], Sergei Dudin [1] and Yulia Gaidamaka [2]

[1] Department of Applied Mathematics and Computer Science, Belarusian State University, 4, Nezavisimosti Ave., 220030 Minsk, Belarus; dudina@bsu.by (O.D.); dudins@bsu.by (S.D.)

[2] Applied Mathematics and Communications Technology Institute, Peoples' Friendship University of Russia (RUDN University), 6 Miklukho-Maklaya St., 117198 Moscow, Russia; gaydamaka-yuv@rudn.ru

* Correspondence: dudin-alexander@mail.ru

Abstract: A multi-server infinite buffer queueing system with additional servers (assistants) providing help to the main servers when they encounter problems is considered as the model of real-world systems with customers' self-service. Such systems are widely used in many areas of human activity. An arrival flow is assumed to be the novel essential generalization of the known Markov Arrival Process (MAP) to the case of the dynamic dependence of the parameters of the MAP on the rating of the system. The rating is the process defined at any moment by the quality of service of previously arrived customers. The possibilities of a customers immediate departure from the system at the entrance to the system and the buffer due to impatience are taken into account. The system is analyzed via the use of the results for multi-dimensional Markov chains with level-dependent behavior. The transparent stability condition is derived, as well as the expressions for the key performance indicators of the system in terms of the stationary probabilities of the Markov chain. Numerical results are provided.

Keywords: multi-server queueing model; rating; self-sufficient servers; self-checkout; assistants; multi-dimensional Markov chains

Citation: Dudin, A.; Dudina, O.; Dudin, S.; Gaidamaka, Y. Self-Service System with Rating Dependent Arrivals. *Mathematics* **2022**, *10*, 297. https://doi.org/10.3390/math10030297

Academic Editor: János Sztrik

Received: 11 December 2021
Accepted: 17 January 2022
Published: 19 January 2022

1. Introduction

Queueing theory is very useful for modeling various real-world systems, contact centers, airports, banks, telecommunication, and retail networks, in particular. The queueing model considered in this paper has two main novel features: (i) the mechanism of customer's arrival is dependent on the current rating of the system and (ii) consideration of self service of customers via so called self service devices (SSD) or self checkouts. Both these features are inherent for many real systems, e.g., entertainment systems, contact centers, food services, and retail networks. Thus, they have to be carefully taken into account in system design and management aiming to guarantee the effective operation of a system. The effective operation suggests earning the maximal profit received by the system via customers service and a high degree of customer satisfaction.

The standard queueing models considered in the literature suggest that the arrival flow of customers to the system does not depend on the system state. The flow entering the service may depend on this state via the mechanism of customers admission depending on the visibility of a queue, as well as its length and (or) the number of busy servers. In this paper, we assume the dependence of the arrival process not on the queue length but the rating of the system.

The rating is now the well-known notion that reflects the customers' satisfaction and has an influence on the choice among the competitive service systems in which the customer can obtain service. Ratings of various service systems are now very popular and easily available, e.g., on the Internet. Checking ratings/reviews before making consumption decisions has become a ritual for many of today's customers, see [1]. The rating dynamically evaluates the current quality of customer service in this system. The rating of the system may cause so-called word-of-mouth advertising and has quite a strong influence on the preferences of the customers and their arrival rate to any system. In turn, this has a high impact on the profit earned by this system, and the ratings have to be taken into account in the management of the system operation. Analysis of queueing systems with an arrival process depending on the rating of the system is of high practical importance.

The existing literature about the queues with servers rating is not extensive. In [1], a queueing model with two types of customers is considered. Sophisticated customers are well-informed of service-related information and make their joining-or-balking decisions strategically, whereas naive customers do not have such information and rely on online rating information to make such decisions. The problem of optimal pricing strategy is solved in [1]. The queueing model containing two competitive systems with customers' choice of the system to be joined via the comparison of the individual ratings of the systems was recently considered with the use of matrix analytic methods in [2].

In modern retail networks, hotels, banks, airports, etc., there is a robust trend for extending the use of self-service devices (SSD) or self-checkouts. SSDs have been defined by the new technological interfaces (e.g., the Quick Response (QR) codes, image and face recognition, radio frequency identification (RFID)) that allow customers to produce services without a service employee's involvement, see, e.g., [3]. The human operator (administrator, assistant, etc.) is involved in the service only upon request of a customer who asks for help in resolving certain problems that he/she met during a service or in the case of violation of the established rules by a customer. The use of SSD is becoming very wide nowadays, in particular, in many retail networks worldwide due to many reasons. The main reason is that it is profitable for both owners of services and customers.

The owners save money via non-payment of salaries to service employees and other operational costs. This creates better opportunities for successful competition with very popular now online shopping that is associated for many customers with the safest (from the perspective of health safety) and convenient way of shopping. Statistics show that one human operator (administrator, assistant, etc.) can easily control 6–10 SSDs. The SSDs take up less space than the regular cash registers, which allows optimizing the store space. With their help, it is possible to unload the cash register area and to increase its throughput. Among other things, SSDs encourage customers to make additional purchases. At one time, McDonald's found out during an experiment that visitors spend an average of 30 percent more on purchases when they are not worried that the person behind the cash register will evaluate their choice. The use of SSDs may allow the reduction of the actual and perceived waiting time that is strongly linked with customers satisfaction. In turn, this should imply higher loyalty of customers and the future profit of the owner.

The main profit gained by a customer consists of: (i) having a chance to avoid long waiting in the queue until the human server (cashier) will become available. Waiting generally is regarded as an undesirable activity that customers must undertake to complete the service. Waiting can lead to both emotional (anger, irritation, frustration, boredom, stress) and behavioral (e.g., abandonment or reneging) responses, especially when it is costly and limits the person's ability to engage in more productive or rewarding ways to spend their time; (ii) getting more control over his/her shopping experience; (iii) obtaining a possibility of better distancing from other buyers what is very important in the current era of the COVID-19 pandemic. With the continuous improvement in technology and the promotion of self-service retail stores in the market, their numbers will increase. Furthermore, the scales of users and transactions will rapidly increase in the future. For more existing literature about the perspectives and attractiveness of the use of SSDs, see, e.g., [4–8].

In the early beginning, the spread of the use of SSDs was fully justified by the huge investment of the companies to enforce the use of the new perspective technologies. After they are already implemented in life, it is necessary to effectively manage the operation of each concrete service system. To this end, besides many administrative problems, a whole bunch of pure mathematical problems has to be resolved. One of these problems is a traditional problem in modeling service systems. Namely, given the actual or expected characteristics of the arrival process, the distribution of service time of one customer, and the required values of the service level indicators, it is necessary to optimally choose the number of required servers (SSDs). Such indicators may be, e.g., the probability that the waiting time of an arbitrary customer will not exceed the given value with the fixed in advance value or the probability of customer abandonment.

It is known that, during the use of SSD for purchases, the customers can meet problems related, e.g., to the search of the necessary good on the shelves, readability of RFIDs, correct use of the scales, damage of the goods. To resolve the potentially arising problems, usually, the stores have some additional staff of administrators (assistants, helpers, etc.). They provide help to a customer if at least one of the assistants is not busy. Otherwise, the customer should wait until one of the assistants becomes available. Therefore, the problem of the optimal choice of the number of SSDs is supplemented by the problem of the optimal matching of the number of required assistants to the number of the SSDs. The redundant number of assistants implies higher, unjustified operational costs. The insufficient number of assistants causes long waiting times for help for a customer. That, in turn, implies longer total service time of this customer, longer waiting time of other customers, the higher probability of the abandonment and reneging by an arbitrary customer from the system, and the loss of potential profit that could be earned by service of customers. In this paper, we solve the problem of computation of the values of performance indicators under any fixed set of the numbers of SSDs and assistants. The problem is formulated and solved in borders of the matrix queueing theory. The usability of this result for the optimal choice of such a set is numerically illustrated.

Due to the practical importance of the effective use of SSDs, there are a lot of papers devoted to this topic. We mention only a few of them that operate with the notion of a customer waiting time. Analysis of the waiting time is one of the standard goals in queueing theory. In [6], the usefulness of queueing theory for the analysis of systems with SSD is noted. The question of the relation of the actual waiting time of a customer and the perceived waiting time, as well as their strong link with customer satisfaction, are discussed. Customer satisfaction is strongly associated with the loyalty of the customers, which is very important for service providers. Therefore, analysis of the ways to increase the loyalty of the customers strongly correlates with an analysis of the actual waiting time of a customer. Such a time is one of the key performance indicators of the majority of queueing systems. Thus, queueing analysis is an important part of solving the problem of the optimal design of the systems of SSDs. However, the analysis of many queueing systems is quite complicated. This explains why analysis of these systems is often implemented not via the use of the analytical and algorithmic methods of queueing theory but via the computer simulation methods. Namely, the method of computer simulation is used for the experimental study of the systems of SSDs. In [7], the correlation of the waiting time, customer experience, and satisfaction was discussed via the use of certain methods of sociology. The content of [8,9] is similar to [7], and the methods of sociology are also used.

In our paper, we provide an analysis of the queueing model with SSDs and assistants. The existing literature about the queues with service assistants is quite scarce. The recent paper [10] is devoted to the analysis of the set of SSDs described in terms of a tandem queueing model with a single-server first phase and a multi-server second phase. All distributions defining the system operation are exponential. The behavior of the system is described by a two-dimensional Markov chain that is the Quasi-Birth-and-Death-Process. This process is easily analyzed via the tools of the matrix-geometric method by M. Neuts, see [11].

The queueing model of the self-checkout (self-service) system considered in our paper assumes the existence of two multi-server sub-systems. Let us denote the number of servers in these two systems as N and M, correspondingly. The first sub-system defines the service process of customers by themselves. Any arriving customer that does not abandon the system (due to too long, in his/her opinion, queue) obtains service in this sub-system and successfully departs from the system if he/she does not encounter service problems.

If a problem occurs, the server from the second sub-system has to help in resolving this problem if they are available. If all servers of the second sub-system are busy, the customer that encountered the problem suspends their service until any server of the second sub-system becomes available. After the problem is resolved, the server in the first sub-system resumes the work while the corresponding assistant is released. A problem in service at the first sub-system can occur an arbitrary number of times. After service is completed, the customer departs from the system.

As follows from this brief description, our model does not belong to the class of tandem queues because the service of an arbitrary customer does not strictly consist of at most two sequential services at different queueing systems. This service may be the sequence of alternating services by servers from the first and the second sub-systems. Our model is more similar to the unreliable queue with repairmen. The first sub-system describes the service of customers, and the second sub-system describes the behavior of the pool of repairmen. However, the overwhelming majority of the papers devoted to this subject consider only the joint distribution of the number of non-broken servers and the number of busy repairmen. The duty of servers to provide service and a possible queue of customers are not taken into account; see the survey [12], paper [13], and references therein. In our model, namely, the characteristics of the customers' service quality are in the focus of the study.

Models similar to our queueing models (however, without the rating consideration) are considered in the following papers. In [14], the model with N servers and $M = 1$ assistants (called in [14] as the main servers and the consultant) is considered. Arrivals are defined by quite a general Markov Arrival Process (MAP); for definition, properties, and related research, see [15–18]. Other distributions characterizing the system are assumed to be exponential. The system is comprehensively analyzed using the matrix analytic methods. Extensive illustrative numerical examples to bring out the qualitative nature of the model are presented. In [19], the model with $N = 1$ servers and $M = 1$ assistants is analyzed. All parameters of the system depend on the state of a finite state random environment. All involved distributions are assumed to be exponential. The system is analyzed using the matrix analytic methods. In [20], the model with $N = 1$ servers and $M = 1$ assistants is analyzed. The input buffer is finite, and the number of opportunities to ask for help by each server is restricted. Service and help times have so-called phase-type (PH) distributions, see [11]. The system is analyzed using the matrix analytic methods. The model considered in [21] assumes an arbitrary number of servers and assistants (called in this paper, specialist servers); the possibility of providing help by the assistant is only after the main service. Service cannot be continued after receiving help. Practically, this means the consideration of a tandem queueing model. The arrival process is the MAP, and the help times have PH distributions. The system is analyzed using the matrix analytic methods. A little bit similar to our model under quite general assumptions about the arrival process (the Batch Markov Arrival process) and service times (phase-time distribution) was recently considered in [22]. In that model, if the server does not succeed in finishing the service of a customer within a certain time, then the so-called backup server joins with the server for the service of this customer. When both servers serve a customer, the service speed increases. Another difference is that in [22], after obtaining help from the backup server, the server mandatorily finishes the service. In our model, we suggest that the server can obtain help from the assistant many times, and the server and the assistant do not cooperate in service. During obtaining the help, service is not provided.

The additional two features of the model considered in this paper are the following.

- The model suggests the visible queue. This means that the arriving customer can see the number of customers in the queue and can use this information to decide to join a queue or abandon (balk) it, e.g., queueing systems managed by ticket technology is widely used in service industries, as well as government offices, see [23]. Upon arriving at a ticket queue, each customer is issued a numbered ticket. The number currently being served is displayed. An arriving customer balks if the difference between their ticket number and the displayed number exceeds their patience level. Analysis of a queueing model in [23] was implemented via the use of matrix analytic methods. Systems with a visible queue were also previously considered, e.g., in [24,25]. In [24], an empirical study of queue abandonment by the patients in an emergency department of a hospital is implemented by the methods of econometrics. In [25], the queueing/inventory model with a visible queue was analyzed via the use of matrix analytic methods.

- The model suggests the impatience of customers waiting in the buffer. Abandonment or balking means the refusal of a customer from joining the queue due to it being inappropriate for him/her in length. Another possible reason for customer loss is his/her impatience or reneging. Initially, a customer joins the queue. However, if his/her waiting time exceeds some critical (deterministic or random) level, the customer departs from the system. The phenomenon of impatience is important and received quite a lot of attention in the literature; for more references, see, e.g., [26–28]. In [27], in particular, the problem of wasting time by the server because the arriving customer decides to join a queue and receives a ticket but then departs from the system without canceling the ticket is considered. In [29], the authors analyzed the model in which customers' patience is exponentially distributed, and the system's waiting capacity is unlimited. Such a model is both rich and analyzable enough to provide information that is practically important for call center managers. The distinguishing feature of the paper [30] is that service time distributions in the considered multi-server queueing model are generally distributed. A simple and insightful solution is presented for the loss probability. The solution offered in [30] is exact for exponential services and is an excellent heuristic for general service times. In [31], a single server variant of the model from [30] is under study. In [32], the customer's loss probability in $M/M/c$ queue with impatient customers is expressed in a simple formula involving the waiting time probabilities in the $M/M/c$ queue with patient customers. In that paper, a probabilistic derivation of this formula is given, and possible use of this general formula in the $M/M/c$ retrial queue with impatient customers is outlined. In [33–39], the retrial models with impatient customers are also considered. The model considered in [40] assumes that the multi-server queue operates under the influence of the random environment, and the impatience rate depends on the current state of the random environment. In [41–43], multi-server queues with the MAP or marked MAP arrival flows and impatient customers as the models of call centers were analyzed via the matrix analytic methods.

The remainder of the text of this paper consists of the following. In Section 2, the mathematical model is completely described. In Section 3, the process describing the dynamics of the system is defined as the continuous-time multi-dimensional Markov chain with level-dependent transitions. The generator of this chain is given. Section 4 contains the ergodicity and non-ergodicity conditions for this Markov chain in the transparent form. Section 5 briefly touches on the question of computation of the stationary distribution of the Markov chain. In Section 6, expressions for computation of the key performance indicators of the system in terms of the computed stationary probabilities of the system states are presented. Section 7 contains the results of the numerical experiment, the aim of which is to give insights into the shape of dependencies of the performance indicators on the number of servers and assistants. It is worth noting that the numerical realization of the elaborated paper algorithm results, are implemented not for a toy example with a small number of servers and assistants but for the system with realistic numbers of SSDs and assistants. Section 8 concludes the paper.

2. Mathematical Model

We consider a multi-server queueing system with an infinite buffer having two types of servers. The structure of the system is presented in Figure 1.

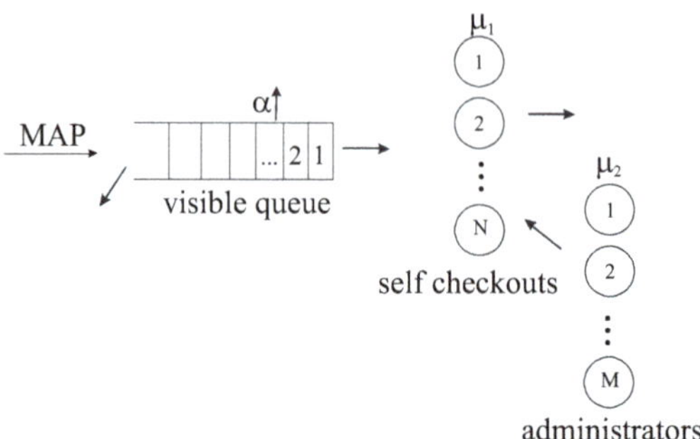

Figure 1. Structure of the system.

The servers of the first type correspond to self-checkouts (self-service devices). The total number of such servers is equal to N. The servers of the second type correspond to administrators (assistants). The total number of assistants is equal to M. If a customer is accepted for service, he/she firstly occupies one first-type server for service during an exponentially distributed time with the parameter μ_1. After this time expires, with a probability of $1-p$, the customer successfully finishes service and departs from the system. With the complementary probability, the customer meets a problem with service and requires some kind of help from an administrator. If at this moment, there is a free administrator, this administrator provides help to the customer to resolve the problem during an exponentially distributed with the parameter μ_2 time. After that, the customer continues his/her service during an exponentially distributed time with the parameter μ_1. The number of moments during service when a customer asks the help of an administrator (not necessarily the same as in the previous moments) is unlimited. If the help of an administrator is required while there is no free administrator, the customer suspends their service and waits until any administrator will become available.

The process of customers' arrival to the system is the generalization of the Markovian arrival process (MAP), see, e.g., [15–18], described below. Arrivals in the classical MAP are governed by the underlying Markov chain ν_t, $t \geq 0$, with the state space $\{1, 2, \ldots, W\}$. The generator D of this chain is represented in the additive form $D = D_0 + D_1$, where the components of the matrix D_1 define intensities of transitions of the chain that are accompanied by the customer's arrival. The non-diagonal components of the matrix D_0 define the intensities of transitions of the chain that are not accompanied by the customer's arrival. The diagonal components are negative. The moduli of these components define the rates of the exit from the corresponding state of the Markov chain.

In contrast to a classical MAP, we assume that the transition intensities of the underlying process ν_t additionally depend on the parameter r, $r = \overline{1, R}$, which defines the so-called current rating of the system. If the rating of the system is r, then the MAP is characterized by the square matrices $D_0^{(r)}$ and $D_1^{(r)}$ of size W. The average arrival rate of the MAP when the rating of the system is r is denoted as λ_r, which can be found as $\lambda_r = \boldsymbol{\theta}^{(r)} D_1^{(r)} \mathbf{e}$, $r = \overline{1, R}$, where $\boldsymbol{\theta}^{(r)}$ is an invariant vector of the MAP defined by the matrices $D_0^{(r)}$ and $D_1^{(r)}$, and $\mathbf{e} = (1, 1, \ldots, 1)^T$. We do not specify the concrete form of the matrices $D_0^{(r)}$ and $D_1^{(r)}$. We only suggest that the increase in the rating cannot imply the decrease in the average arrival rate, i.e., we require that the following inequalities hold true:

$$\lambda_1 \leq \lambda_2 \leq \cdots \leq \lambda_R.$$

The rating of the system can be dynamically changed during the system operation. We assume that if a customer is admitted to the system without waiting in the queue, its current rating r, $r = \overline{1, R-1}$, immediately increases by one with the fixed small probability r^+. For example, if we fix $r^+ = 0.001$, the service of 1000 customers without waiting in the queue leads, on average, to the increase in the rating by one. If a customer abandons the system without service, the current rating r, $r = \overline{2, R}$, decreases by one with the probability r^-. For example, if we fix $r^- = 0.01$, the loss of 100 customers leads, on average, to the decrease in the current rating by one. At the moment of the change of the rating, transition intensities of the underlying process ν_t immediately adjust their values to the new rating.

The queue is assumed to be visible. This means that an arriving customer can observe the number of customers in the queue and leave the system without service if he/she considers this number inappropriate. We assume that if during an arbitrary customer arrival epoch there is no free server and the number of customers in the buffer is i, the customer permanently leaves the system with the probability q_i, $i \geq 0$. With the complementary probability, the customer joins the queue. We suggest that the limit $q = \lim_{i \to \infty} q_i$ exists, $0 < q \leq 1$. Additionally, we assume that customers can be impatient and leave the buffer and depart from the system, independently of each other, after an exponentially distributed with the parameter $\alpha, \alpha \geq 0$, amount of time. In the case of $\alpha = 0$, the customers are patient.

Now, let us analyze the described queueing model.

3. Process of the System States

The behavior of the system under study can be described by the regular irreducible continuous-time Markov chain

$$\xi_t = \{i_t, n_t, r_t, \nu_t\},\ t \geq 0,$$

where, during the epoch t,

- i_t is the number of customers in the system, $i_t \geq 0$;
- n_t is the number of blocked (waiting for help from an assistant) servers, $n_t = \overline{0, \min\{i_t, N\}}$;
- r_t is the current rating of the system, $r_t = \overline{1, R}$;
- ν_t is the state of the underlying process of the MAP, $\nu_t = \overline{1, W}$.

Here and further, the notation $n = \overline{0, N}$ means that the parameter n admits values from the set $\{0, \ldots, N\}$.

To formally define the continuous-time Markov chain ξ_t, it is necessary to write down, for any pair of the states (i, n, r, ν) and (i', n', r', ν'), the intensity of the transitions between these states.

To avoid bulky denotations, following the standard methodology of investigation of multi-dimensional Markov chains having one denumerable component, we enumerate the states of the Markov chain $\xi_t = \{i_t, n_t, r_t, \nu_t\}$ in the direct lexicographic order of the components $\{n_t, r_t, \nu_t\}$ and combine the set of the states with the value i of the component i_t into the so-called level i, $i \geq 0$.

Let $Q_{i,j}$ be the matrix constituted by the transition intensities from level i to level j and let Q be the block matrix constituted by the blocks $Q_{i,j}$, $i \geq 0$, $j \geq 0$. It is clear that the matrix Q is the infinitesimal generator of the Markov chain ξ_t, $t \geq 0$.

Theorem 1. *The generator Q of the Markov chain ξ_t, $t \geq 0$, has the following block three-diagonal structure*

$$Q = \begin{pmatrix} Q_{0,0} & Q_{0,1} & O & O & \ldots & O & O & O & \ldots \\ Q_{1,0} & Q_{1,1} & Q_{1,2} & O & \ldots & O & O & O & \ldots \\ O & Q_{2,1} & Q_{2,2} & Q_{2,3} & \ldots & O & O & O & \ldots \\ \vdots & \vdots & \vdots & \vdots & \ddots & \vdots & \vdots & \vdots & \ddots \end{pmatrix}.$$

The non-zero blocks are defined as follows:

$$Q_{0,0} = \hat{D}_0,$$

$$Q_{i,i} = I_{i+1} \otimes \hat{D}_0 + (-\mu_2 C_i - \mu_1 \tilde{C}_i + \mu_2 C_i E_i^- + p\mu_1 \tilde{C}_i E_i^+) \otimes I_{RW},\ 0 < i < N,$$

$$Q_{i,i} = I_{N+1} \otimes \hat{D}_0 + q_{i-N} I_{N+1} \otimes \hat{D}_1((r_- \mathcal{R}^- + (1-r_-)I_R) \otimes I_W) + (-\mu_2 C_N - \mu_1 \tilde{C}_N + $$
$$+\mu_2 C_N E_N^- + p\mu_1 \tilde{C}_N E_N^+) \otimes I_{RW} - (i-N)\alpha I_{(N+1)RW},\ i \geq N,$$
$$Q_{i,i+1} = \tilde{E}_i \otimes \hat{D}_1((r_+ \mathcal{R}^+ + (1-r_+)I_{R+1}) \otimes I_W),\ 0 < i < N,$$
$$Q_{i,i+1} = (1-q_{i-N}) I_{N+1} \otimes \hat{D}_1,\ i \geq N,$$
$$Q_{i,i-1} = \mu_1(1-p)\tilde{C}_i \hat{E}_i \otimes I_{RW},\ 0 < i \leq N,$$
$$Q_{i,i-1} = (i-N)\alpha I_{(N+1)} \otimes ((r_- \mathcal{R}^- + (1-r_-)I_R) \otimes I_W) + \mu_1(1-p)\tilde{C}_N \otimes I_{RW},\ i > N,$$

where

$\otimes$ *indicates the symbol of the Kronecker product matrices, see [44];*

$\hat{D}_0 = \mathrm{diag}\{D_0^{(1)}, D_0^{(2)}, \ldots, D_0^{(R)}\}$, *where* $\mathrm{diag}\{\ldots\}$ *denotes the diagonal matrix with the diagonal entries listed in the brackets;*

$\hat{D}_1 = \mathrm{diag}\{D_1^{(1)}, D_1^{(2)}, \ldots, D_1^{(R)}\}$;

$C_i = \mathrm{diag}\{0, 1, \ldots, \min\{i-1, M\}, \min\{i, M\}\},\ i = \overline{1,N}$;

$\tilde{C}_i = \mathrm{diag}\{i, i-1, \ldots, 0\},\ i = \overline{1,N}$;

I_K *is the identity matrix having a size indicated in the suffix (if the size of the matrix is clear from the context, it can be omitted);*

O_K *is a zero matrix having a size indicated in the suffix (if the size of the matrix is clear from the context, it can be omitted);*

E_i^- *is a square matrix of size* $i+1$ *with all zero entries, except the entries* $(E_i^-)_{l,l-1},\ l = \overline{2,i+1},\ i = \overline{1,N}$, *which are equal to 1;*

E_i^+ *is a square matrix of size* $i+1$ *with all zero entries, except the entries* $(E_i^+)_{l,l+1},\ l = \overline{1,i},\ i = \overline{1,N}$, *which are equal to 1;*

$\tilde{E}_i$ *is a matrix of size* $(i+1) \times (i+2)$ *with all zero entries, except the entries* $(\tilde{E}_i)_{l,l},\ l = \overline{1,i+1},\ i = \overline{1,N}$, *which are equal to 1;*

$\hat{E}_i$ *is a matrix of size* $(i+1) \times i$ *with all zero entries, except the entries* $(\hat{E}_i)_{l,l},\ l = \overline{1,i},\ i = \overline{1,N}$, *which are equal to 1;*

$\mathcal{R}^+$ *is a matrix of size* $R \times R$ *with all zero entries, except the entries* $(\mathcal{R}^+)_{l,l+1},\ l = \overline{1,R-1}$, *and* $(\mathcal{R}^+)_{R,R}$, *which are equal to 1;*

$\mathcal{R}^-$ *is a matrix of size* $R \times R$ *with all zero entries, except the entries* $(\mathcal{R}^-)_{l,l-1},\ l = \overline{2,R}$, *and* $(\mathcal{R}^-)_{1,1}$, *which are equal to 1.*

Proof. The proof of Theorem 1 is implemented via careful analysis of all possible transitions of the Markov chain $\xi_t,\ t \geq 0$, and further combining the intensities of these transitions into the blocks of the generator.

The generator Q has all negative diagonal entries and non-negative non-diagonal entries. The diagonal entries of the generator Q define, up to the sign, the total intensity of leaving the corresponding state of the Markov chain $\xi_t,\ t \geq 0$. In the case $i = 0$, the Markov chain can leave the current state only if the underlying process of the arrival process makes a transition. The intensities of such transitions are defined as the modules of the diagonal entries of the matrix $\hat{D}_0$.

In the case when the system is not idle, but the buffer is empty, the Markov chain $\xi_t,\ t \geq 0$, can also change its state due to the finish of the service by a server or finish of help provided by an assistant. The intensities of such transitions are defined as the diagonal entries of the matrix $(\mu_2 C_i + \mu_1 \tilde{C}_i) \otimes I_{RW},\ i = \overline{1,N}$.

If the number of customers in the buffer is greater than zero, the Markov chain $\xi_t,\ t \geq 0$, can also change its state due to the departure of some customers from the buffer due to impatience. The intensities of such transitions are defined as the diagonal entries of the matrix $(i-N)\alpha I_{(N+1)RW},\ i > N$.

The non-diagonal entries of the matrices $Q_{i,i},\ i \geq 0$, define the intensities of transitions that do not lead to the change of the number of customers in the system i. Such transitions are the following:

(1) The underlying process of the arrival process transition, which does not imply the acceptance of a new customer to the system (the customer is not generated or lost). The intensities of such transitions are given as the non-diagonal entries of the matrix $I_{\min\{N,i\}+1} \otimes \hat{D}_0$ and the entries of the matrix $q_{i-N} I_{N+1} \otimes \hat{D}_1((r_- \mathcal{R}^- + (1-r_-)I_R) \otimes I_W)$ for $i \geq N$. Note, that here the matrix $(r_- \mathcal{R}^- + (1-r_-)I_R)$ defines the possible change of the system rating due to the loss of a customer;

(2) The number of blocked servers is increased by one. The intensities of such transitions are given as the entries of the matrix $(p\mu_1 \tilde{C}_{\min\{N,i\}} E^+_{\min\{N,i\}}) \otimes I_{RW}$;

(3) The number of blocked servers is decreased by one. The intensities of such transitions are given as the entries of the matrix $(\mu_2 C_{\min\{N,i\}} E^-_{\min\{N,i\}}) \otimes I_{RW}$.

The entries of the matrices $Q_{i,i+1}$, $i \geq 0$ define the intensities of transitions that lead to the increase in the number of customers i in the system by one. This can happen only in the case when the underlying arrival process makes a transition with the generation of a customer, and this customer is admitted to the system. The intensities of such transitions are given as the entries of the matrices $\tilde{E}_i \otimes \hat{D}_1((r_+ \mathcal{R}^+ + (1-r_+)I_{R+1}) \otimes I_W)$, in the case $i = \overline{1, N-1}$, and the entries of the matrices $(1 - q_{i-N}) I_{N+1} \otimes \hat{D}_1$, if $i \geq N$.

The entries of the matrices $Q_{i,i-1}$, $i \geq 1$, define the intensities of transitions that lead to the decrease in the number of customers i in the system by one. This can happen if a customer leaves the system successfully serviced (the intensities of such transitions are given as the entries of the matrices $\mu_1(1-p)\tilde{C}_{\min\{N,i\}} \hat{E}_{\min\{N,i\}} \otimes I_{RW}$) and if a customer leaves the non-empty buffer due to impatience (the intensities of such transitions are given as the entries of the matrices $(i-N)\alpha I_{(N+1)} \otimes ((r_- \mathcal{R}^- + (1-r_-)I_R) \otimes I_W)$).

The blocks $Q_{i,j}$, $i, j \geq 0$, $|i-j| > 1$, of the generator Q are zero matrices because the customers arrive and leave the system only one-by-one. □

4. Ergodicity Condition

An important step of analysis of any Markov chain with an infinite state space is establishing conditions for the ergodicity and non-ergodicity of this Markov chain (the stability condition). The ergodicity and non-ergodicity conditions for the Markov chain under study ξ_t are given by the following Theorem.

Theorem 2. *(a) If the impatience rate α is positive, the Markov chain ξ_t is ergodic under all finite values of other parameters of the considered queueing system;*

(b) If the limiting probability q is equal to 1, the Markov chain ξ_t is ergodic under all finite values of other parameters of the considered queueing system;

(c) If the customers staying in the buffer are patient, i.e., $\alpha = 0$, the Markov chain ξ_t is ergodic if the following inequality is fulfilled:

$$\lambda_1(1-q) < \sum_{n=0}^{N} \gamma_n (N-n)\mu_1, \tag{1}$$

where γ_n are the probabilities that at an arbitrary moment when the system is overloaded, the number of assistants providing help to the servers is equal to n, $n = \overline{0, N}$.

These probabilities are computed by the formula:

$$\gamma_n = \left(1 + \sum_{j=1}^{N} \prod_{l=1}^{j} \frac{p(N-l+1)\mu_1}{l\mu_2}\right)^{-1} \prod_{l=1}^{n} \frac{p(N-l+1)\mu_1}{l\mu_2}, \; n = \overline{0, N}; \tag{2}$$

(d) If the customers staying in the buffer are patient, i.e., $\alpha = 0$, the Markov chain ξ_t is non-ergodic if the following inequality is fulfilled:

$$\lambda_1(1-q) > \sum_{n=0}^{N} \gamma_n (N-n)\mu_1. \tag{3}$$

Proof. Let the generator of a Markov chain be the upper-Hessenbergian matrix, i.e., it has zero blocks below the first sub-diagonal and other blocks $Q_{i,i+k-1},\ k \geq 0$. Let the matrix T_i be the diagonal matrix, the diagonal entries of which also coincide with the modules of the diagonal entries of the matrix $Q_{i,i},\ i \geq 0$.

If the following limits exist:

$$Y^{(k)} = \lim_{i \to \infty} \mathcal{R}_i^{-1} Q_{i,i+k-1},\ k = 0,2,3,\ldots, \quad Y^{(1)} = \lim_{i \to \infty} \mathcal{R}_i^{-1} Q_{i,i} + I,$$

and the matrix $\sum_{k=0}^{\infty} Y^{(k)}$ is stochastic; then the Markov chain belongs to the class of Asymptotically Quasi-Toeplitz Markov chains ($AQTMC$), see [45].

The generator Q of the Markov chain describing the queueing system under study defined by Theorem 1 has the particular, with respect to the upper-Hessenbergian matrix, three-block diagonal structure.

If $\alpha > 0$, then it can be verified that for this Markov chain the matrices $Y_k,\ k = 0,1,2$ exist and are defined by: $Y^{(0)} = I,\ Y^{(k)} = O,\ k = 1,2$.

If $\alpha = 0$, then it can be verified that the matrices $Y^{(k)}$ exist and are defined by

$$Y^{(0)} = T^{-1}\tilde{Q}^{-},$$

$$Y^{(1)} = T^{-1}\tilde{Q}^{0} + I,$$

$$Y^{(2)} = T^{-1}\tilde{Q}^{+},$$

where

$$\tilde{Q}^{-} = \mu_1(1-p)\tilde{C}_N \otimes I_{RW},$$

$$\tilde{Q}^{0} = I_{N+1} \otimes \hat{D}_0 + qI_{N+1} \otimes \hat{D}_1((r_{-}\mathcal{R}^{-} + (1-r_{-})I_R) \otimes I_W) + (-\mu_2 C_N - \mu_1 \tilde{C}_N + \mu_2 C_N E_N^{-} + p\mu_1 \tilde{C}_N E_N^{+}) \otimes I_{RW},$$

$$\tilde{Q}^{+} = (1-q)I_{N+1} \otimes \hat{D}_1,$$

and T is the diagonal matrix the diagonal entries of which coincide with the modules of the diagonal entries of the matrix $\tilde{Q}^0$.

Therefore, in both cases, $\alpha > 0$ and $\alpha = 0$, the limits $Y^{(k)},\ k = 0,1,2$, exist and it is easy to check that their sum is the stochastic matrix. This implies that the considered Markov chain belongs to the class of $AQTMC$. The sufficient condition for the ergodicity of $AQTMC$, see [45], rewritten for the three-block diagonal generator is the fulfillment of the inequality:

$$\boldsymbol{\psi} Y^{(0)}\mathbf{e} > \boldsymbol{\psi} Y^{(2)}\mathbf{e} \tag{4}$$

where the vector $\boldsymbol{\psi}$ is the unique solution of equations

$$\boldsymbol{\psi}(Y^{(0)} + Y^{(1)} + Y^{(2)}) = \boldsymbol{\psi},\ \boldsymbol{\psi}\mathbf{e} = 1. \tag{5}$$

The sufficient condition for the non-ergodicity is the fulfillment of the inequality

$$\boldsymbol{\psi} Y^{(0)}\mathbf{e} < \boldsymbol{\psi} Y^{(2)}\mathbf{e}.$$

Because for $\alpha > 0$ we have $Y^{(0)} = I,\ Y^{(k)} = O,\ k = 1,2$, inequality (4) takes a trivial form: $1 > 0$. Thus, the chain is ergodic for all finite values of other parameters of the considered queueing system. Statement (a) of the theorem is proven.

Consider now the case $\alpha = 0$. It can be verified that in this case system (4) and inequality (5) are equivalent to the system

$$\boldsymbol{\varphi}(\tilde{Q}^{-} + \tilde{Q}^{0} + \tilde{Q}^{+}) = \mathbf{0},\ \boldsymbol{\varphi}\mathbf{e} = 1 \tag{6}$$

and the inequality

$$\boldsymbol{\varphi}\tilde{Q}^{-}\mathbf{e} > \boldsymbol{\varphi}\tilde{Q}^{+}\mathbf{e}. \tag{7}$$

It can be verified that

$$\tilde{Q}^{-} + \tilde{Q}^{0} + \tilde{Q}^{+} = I_{N+1} \otimes ((1-q)\hat{D}_1 + \hat{D}_0 + q\hat{D}_1((r_{-}\mathcal{R}^{-} + (1-r_{-})I_R) \otimes I_W)) + H$$

where

$$H = \begin{pmatrix} -pN\mu_1 & pN\mu_1 & O & \dots & O & O \\ \mu_2 & -\mu_2 - p(N-1)\mu_1 & p(N-1)\mu_1 & \dots & O & O \\ O & 2\mu_2 & -2\mu_2 - p(N-2)\mu_1 & \dots & O & O \\ \vdots & \vdots & \vdots & \ddots & \vdots & \vdots \\ O & O & O & \dots & N\mu_2 & -N\mu_2 \end{pmatrix} \otimes I_{RW}.$$

Using the so-called mixed product rule for Kronecker product of matrices, see [44], it is possible to show that a solution of (6) has the representation

$$\boldsymbol{\varphi} = \boldsymbol{\Delta}_1 \otimes \boldsymbol{\Delta}_2 \tag{8}$$

where the row vector $\boldsymbol{\Delta}_1$ is a solution to the system

$$\boldsymbol{\Delta}_1 \left[I_{N+1} \otimes ((1-q)\hat{D}_1 + \hat{D}_0 + q\hat{D}_1((r_{-}\mathcal{R}^{-} + (1-r_{-})I_R) \otimes I_W)) \right] = \mathbf{0}, \tag{9}$$

$$\boldsymbol{\Delta}_1\mathbf{e} = 1,$$

and the row vector $\boldsymbol{\Delta}_2$ is a solution to the system

$$\boldsymbol{\Delta}_2 H = \mathbf{0},\ \boldsymbol{\Delta}_2\mathbf{e} = 1. \tag{10}$$

By the direct substitution into (9), it is possible to check that vector $\boldsymbol{\Delta}_1$ defined by

$$\boldsymbol{\Delta}_1 = (\boldsymbol{\theta}^{(1)}, \mathbf{0}, \dots, \mathbf{0}) \tag{11}$$

is the solution of equation (9).

By the direct substitution into (10), it is possible to check that vector $\boldsymbol{\Delta}_2$ is defined by

$$\boldsymbol{\Delta}_2 = (\gamma_0, \dots, \gamma_N) \tag{12}$$

where probabilities $\gamma_n,\ n = \overline{0,N}$, are given by Formula (2). Taking into account (8), (11), and (12) in (4), we obtain Formula (1).

It is clear that if $q = 1$, inequality (1) is always true. This proves statement (b) of the theorem. Statement (c) is also proven. The proof of statement (d) is easily made analogously to the proof of (b). □

Remark 1. *Inequality (1) is intuitively transparent. Usually, the ergodicity condition is equivalent to the requirement that in the situation when the system is very overloaded, the rate of customers admission is less than the rate of customers service. The left-hand side of (1) is the rate of customers' admission to the system. As it is seen from (1), this rate here depends only on the arrival rate λ_1 and does not depend on the rates $\lambda_r,\ r = \overline{2,R}$. This stems from the fact that the rating of the system, when it is very overloaded, is equal to 1 due to the abandonment (balking) of many customers. The right-hand side of (1) is the rate of customers' departure from the service when the system is overloaded. The values $\gamma_n,\ n = \overline{0,N}$, are the probabilities that n assistants provide help to the servers and these servers do not serve customers. Correspondingly, the number of servers providing service with the rate μ_1 is equal to $N-n$. It follows from the formula of total probability that the* *right-hand side of (1) indeed is the rate of customers' departure from the service when the system is overloaded.*

5. Computation of the Stationary Distribution of the Markov Chain

If the ergodicity condition is fulfilled, the stationary probabilities of the Markov chain $\xi_t,\ t \geq 0,$

$$\pi(i,n,r,v) = \lim_{t\to\infty} P\{i_t = i, n_t = n, r_t = r, v_t = v\},$$

$$i \geq 0,\ n = \overline{0,\min\{i,N\}},\ r = \overline{1,R},\ v = \overline{1,W};$$

exist.

We form the row vectors $\boldsymbol{\pi}_i,\ i \geq 0$, of these stationary probabilities enumerated in the lexicographic order of the components $(n,r,v),\ n = \overline{0,\min\{i,N\}},\ r = \overline{1,R},\ v = \overline{1,W}$.

It is a well-known fact that these stationary probabilities can be found as the solution of the following system:

$$(\boldsymbol{\pi}_0, \boldsymbol{\pi}_1, \ldots, \boldsymbol{\pi}_N, \ldots)Q = \mathbf{0},$$

$$(\boldsymbol{\pi}_0, \boldsymbol{\pi}_1, \ldots, \boldsymbol{\pi}_N, \ldots)\mathbf{e} = 1.$$

Because the Markov chain $\xi_t,\ t \geq 0$, is a level-dependent quasi-birth-and-death process having one countable component, this system cannot be solved using the standard matrix analytic methods. To solve this system, we recommend using the algorithms from papers [46,47].

6. Performance Indicators

The probability $p_r,\ r = \overline{1,R}$, that the value of the rating of the system at an arbitrary epoch is equal to r can be found as

$$p_r = \sum_{i=0}^{\infty} \sum_{n=0}^{\min\{i,N\}} \pi(i,n,r)\mathbf{e}.$$

The average rating $\bar{R}$ of the system can be found as

$$\bar{R} = \sum_{r=1}^{R} r p_r.$$

The average customer's arrival rate λ can be found as

$$\lambda = \sum_{r=1}^{R} p_r \lambda_r.$$

The average output rate λ_{out} of successfully serviced customers can be found as

$$\lambda_{out} = \sum_{i=1}^{\infty} \sum_{n=0}^{\min\{i,N\}-1} (\min\{i,N\} - n)(1-p)\mu_1 \pi(i,n)\mathbf{e}.$$

The average number N_{cust} of customers in the system is calculated as

$$N_{cust} = \sum_{i=1}^{\infty} i\boldsymbol{\pi}_i\mathbf{e}.$$

The average number N_{buf} of customers in the buffer is calculated as

$$N_{buf} = \sum_{i=N+1}^{\infty} (i-N)\boldsymbol{\pi}_i\mathbf{e}.$$

The average number N_{serv-1} of occupied servers is calculated as

$$N_{serv-1} = \sum_{i=1}^{\infty} \min\{i, N\} \boldsymbol{\pi}_i \mathbf{e}.$$

The average number $N_{blocked}$ of blocked servers is calculated as

$$N_{blocked} = \sum_{i=1}^{\infty} \sum_{n=1}^{\min\{i,N\}} n \boldsymbol{\pi}(i, n) \mathbf{e}.$$

The average number $N_{blocked-1}$ of blocked servers that are currently obtaining the help of an assistant is calculated as

$$N_{blocked-1} = \sum_{i=1}^{\infty} \sum_{n=1}^{\min\{i,N\}} \min\{n, M\} \boldsymbol{\pi}(i, n) \mathbf{e}.$$

The average number $N_{blocked-2}$ of blocked servers that are waiting until any assistant will become available is calculated as

$$N_{blocked-2} = \sum_{i=M+1}^{\infty} \sum_{n=M+1}^{\min\{i,N\}} (n - M) \boldsymbol{\pi}(i, n) \mathbf{e} = N_{blocked} - N_{blocked-1}.$$

The average number N_{serv-2} of busy assistants is equal to $N_{blocked-1}$.

The loss probability P_{ent} of an arbitrary customer at the entrance to the system due to the unwillingness to wait in a long queue can be found as

$$P_{ent} = \frac{1}{\lambda} \sum_{i=N}^{\infty} \sum_{n=0}^{N} \sum_{r=1}^{R} q_{i-N} \boldsymbol{\pi}(i, n, r) D_1^{(r)} \mathbf{e}.$$

The loss probability P_{imp} of an arbitrary customer due to impatience can be found as

$$P_{imp} = \frac{\alpha N_{buf}}{\lambda}.$$

The loss probability P_{loss} of an arbitrary customer can be found as

$$P_{loss} = 1 - \frac{\lambda_{out}}{\lambda} = P_{ent} + P_{imp}.$$

7. Numerical Example

The purpose of this example is to illustrate the dependencies of the main performance measures of the system on the number N of servers and the number M of assistants. Let us assume that the system can have up to 50 servers and up to 10 assistants. Thus, below we vary the parameter N over the interval $(1, 50)$ and the parameter M over the interval $(1, 10)$ with the same step 1.

We assume the following values of the parameters of the system:

- The service intensity of an arbitrary customer by a server is $\mu_1 = 0.5$.
- The probability p that a customer meets a problem with service and requires the help of an assistant is $p = 0.25$.
- The rate of help provided by an assistant is $\mu_2 = 1.5$.
- The parameter R that defines the maximum possible rating of the system is $R = 10$.
- The probability of rating increasing is $r^+ = 0.001$.
- The probability of rating decreasing is $r^- = 0.005$.
- The parameter α describing the impatience rate of customers is equal to 0.06.

- The probabilities $q_i = q_{i,N}$ that a customer will leave the system having N servers during the arrival epoch when the number of customers in the buffer is i, are defined as:

$$q_{i,N} = \begin{cases} \frac{i}{i+100N}, & if\, 0 \leq i \leq N; \\ \frac{i}{i+40N}, & if\, N < i \leq \max\{10, 2N\}; \\ \frac{i}{i+10N}, & if\, \max\{10, 2N\} < i \leq \max\{20, 5N\}; \\ \frac{i}{i+N}, & if\, \max\{20, 5N\} < i \leq \max\{100, 10N\}; \\ \frac{i}{i+0.1N}, & otherwise. \end{cases}$$

- The MAP arrival flow of customers when the system has the rating r is defined by the matrices $D_0^{(r)} = rD_0^{base}$ and $D_1^{(r)} = rD_1^{base}$, where the matrices D_0^{base} and D_1^{base} are defined by

$$D_0^{base} = \begin{pmatrix} -2.5 & 0.02 \\ 0.001 & -0.8 \end{pmatrix}; D_1^{base} = \begin{pmatrix} 2.46 & 0.02 \\ 0.001 & 0.798 \end{pmatrix}.$$

The base arrival flow has the average intensity $\lambda = 0.879048$, the coefficient of correlation of successive inter-arrival times $c_{cor} = 0.0557495$ and the coefficient of variation of inter-arrival times $c_{var} = 1.12815$.

The dependence of the average rating of the system $\bar{R}$ on the parameters N and M is presented in Figure 2.

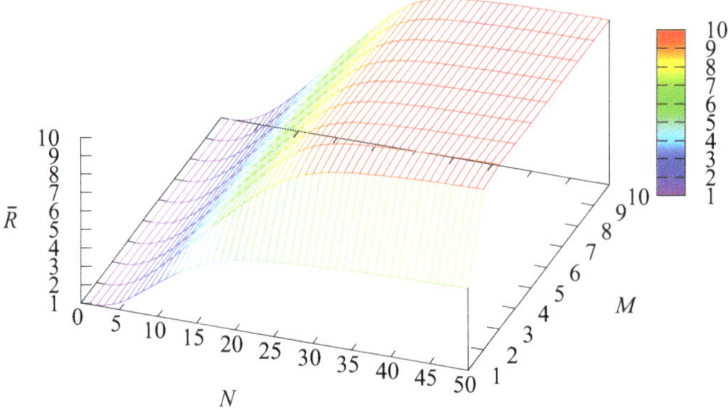

Figure 2. Dependence of the average rating of the system $\bar{R}$ on N and M.

As is seen in Figure 2, the average rating increases with the increase in the number of servers and assistants. Here, we do not consider the situations when $M \geq N$. It is evident that in such a situation, adding a new assistant does not improve the system performance, and the rating is not changed. In other cases, adding new servers and (or) assistants leads to a better quality of customer service. As the result, the average rating increases.

Figure 3 illustrates the dependence of the average customer's arrival rate λ on the parameters N and M.

The average arrival rate of customers also increases with the increase in the number of servers and (or) assistants. This is explained by the evident fact that a higher rating of the system leads to a higher arrival rate.

The dependence of the average number N_{buf} of customers in the buffer on the parameters N and M is presented in Figure 4.

The dependence of the average number N_{buf} of customers in the buffer on the parameters N and M is complicated and hardly predictable intuitively. On the one hand, as for a classical system, the increase in the number of servers leads to a decrease in the waiting time. However, for the considered system, the increase in the number of servers leads to an increase in the arrival rate that may imply an increase in the waiting time. As one can see from the figure, sometimes the first factor prevails over the second one, and the average number N_{buf} of customers in the buffer decreases with the growth of N and M; sometimes the situation changes oppositely.

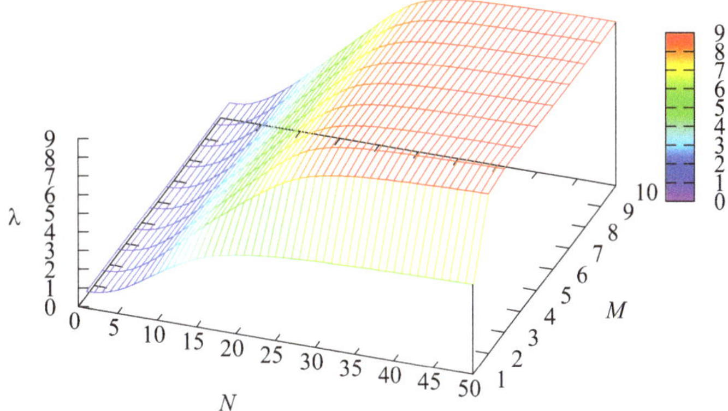

Figure 3. Dependence of the average customer's arrival rate λ on N and M.

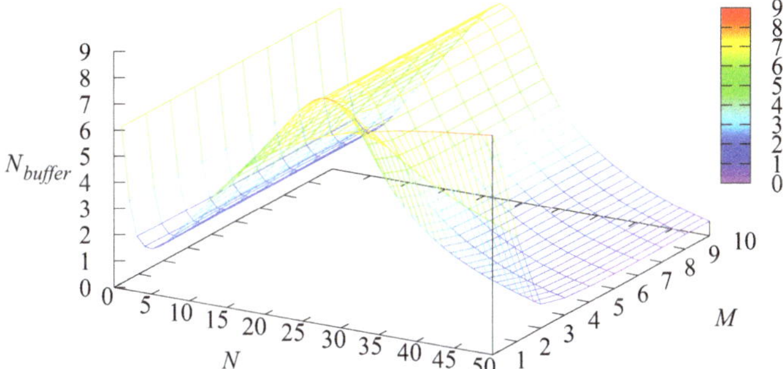

Figure 4. Dependence of the average number N_{buf} of customers in the buffer on N and M.

Figures 5–7 illustrate the dependencies of the average number N_{serv-1} of occupied servers, the average number $N_{blocked-1}$ of blocked servers that are under unblocking by an assistant, and the average number $N_{blocked-2}$ of blocked servers that are waiting until an assistant will become available on the parameters N and M.

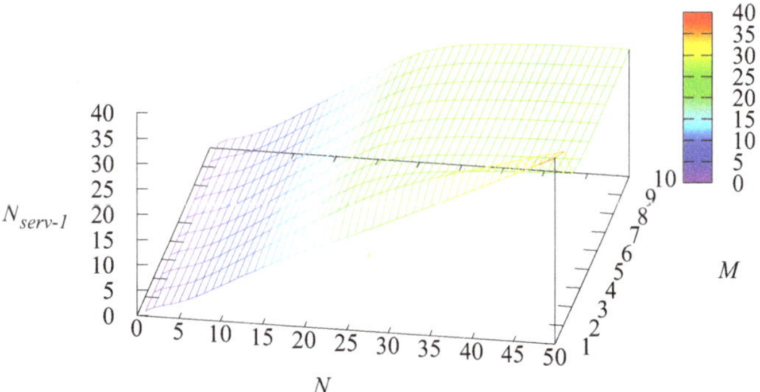

Figure 5. Dependence of the average number N_{serv-1} of occupied type-1 servers on N and M.

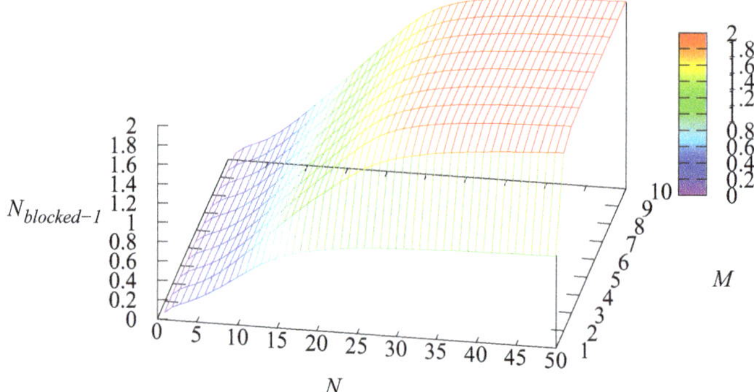

Figure 6. Dependence of the average number $N_{blocked-1}$ on N and M.

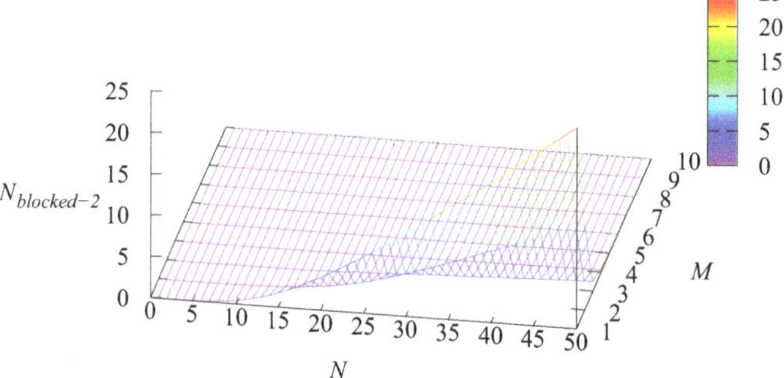

Figure 7. Dependence of the average number $N_{blocked-2}$ on N and M.

The average number N_{serv-1} of occupied servers essentially increases with the increase in the total number of servers N. When the number of assistants M increases, the average number of blocked servers decreases if the average arrival rate is constant. Thus, the increase in the number of assistants M may lead to the decrease in the number of busy servers. However, with an increasing value of M, the rating and the intensity of the arrival flow can also increase, which may cause an increase in the number of busy servers. Therefore, the dependence of the number N_{serv-1} on the parameter M is hardly predictable.

The average number $N_{blocked-1}$ of blocked servers currently receiving the help of assistants increases with the growth of N and M. This growth is mainly caused by the increase in the average arrival rate. In the considered example, the average number $N_{blocked-2}$ of blocked servers that are waiting until an assistant will become available grows when the number of servers N grows and decreases with the increasing value of M.

The dependence of the loss probability P_{ent} of an arbitrary customer at the entrance to the system due to the unwillingness to wait in a long queue on the parameters N and M is presented in Figure 8.

As it is seen from Figure 8, the loss probability of an arbitrary customer at the entrance to the system decreases when the number N of servers grows. The dependence of P_{ent} on M is less essential, and P_{ent} may decrease or increase with the growth of M.

The dependence of the loss probability P_{imp} of an arbitrary customer due to impatience on the parameters N and M is presented in Figure 9.

The behavior of the dependence of the loss probability P_{imp} of an arbitrary customer due to impatience on the parameters N and M is also complicated. Note that the loss probability P_{imp} essentially depends on the average number of customers in the buffer (see Figure 4) and the average arrival rate λ (see Figure 3).

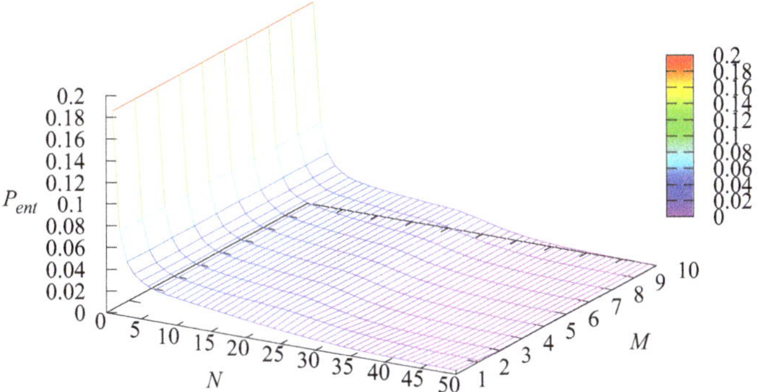

Figure 8. Dependence of the loss probability P_{ent} on N and M.

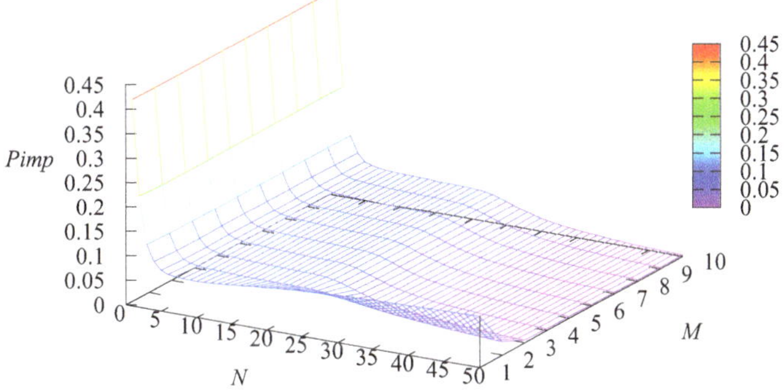

Figure 9. Dependence of the loss probability P_{imp} on N and M.

The dependence of the loss probability P_{loss} of an arbitrary customer on the parameters N and M is presented in Figure 10.

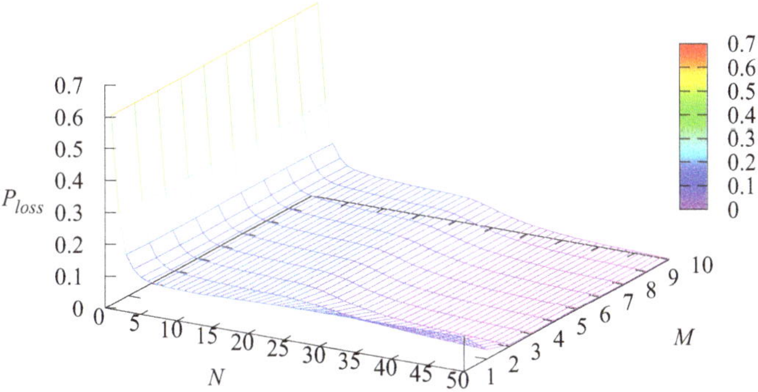

Figure 10. Dependence of the loss probability P_{loss} on N and M.

The loss probability P_{loss} is the sum of loss probabilities P_{ent} and P_{imp}. Because in the considered case, the main losses occur due to customers' impatience, the behavior of P_{loss} is similar to the behavior of the probability P_{imp}.

We considered the dependencies of the main performance measures on the number of servers N and the number of assistants M and can conclude that these dependencies can be quite hardly predictable. Therefore, mathematical modeling with the use of the results presented in this paper is required for the exact estimation of the system performance characteristics. It is worth noting that, from the point of view of potential practical applications, the more important problem is to define the optimal number N of servers and the number M of assistants. To find these optimal values, first of all, it is necessary to choose an appropriate cost criterion. In this paper, we assume that the quality of the system operation is characterized by the following economic criterion:

$$E = E(M, N) = a_1 \lambda_{out} - b_1 \lambda P_{ent} - b_2 \lambda P_{imp} - d_1 N - d_2 M.$$

Here, the cost coefficients a_1, b_1, b_2, d_1, and d_2 have the following meaning:

a_1 is the profit obtained by the system for the successful service of one customer;

b_1 and b_2 are the charges paid by the system for the loss of a customer at the entrance of the system and due to impatience, correspondingly;

d_1 and d_2 are the charges paid by the system for maintaining one server and one assistant per unit time, correspondingly;

Thus, the economic criterion E has the meaning of the average profit obtained by the system per unit of time.

In this numerical example, we fix the following cost coefficients:

$$a_1 = 1,\ b_1 = 2,\ b_2 = 3,\ d_1 = 0.05,\ d_2 = 0.1.$$

We aim to maximize the average profit of the system.

Figure 11 illustrates the dependence of the economic criterion E on the parameters N and M.

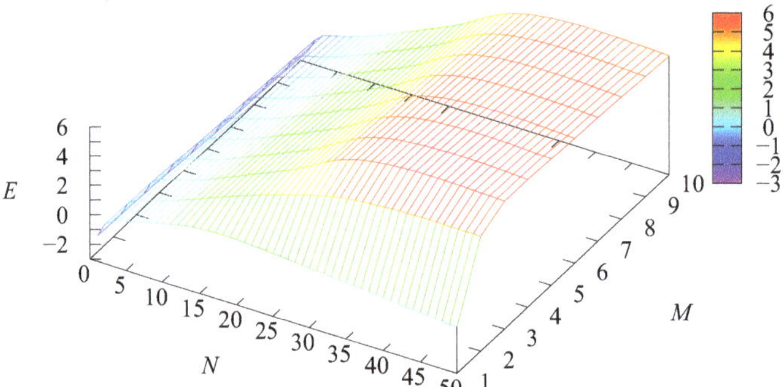

Figure 11. Dependence of the values of economic criterion E on N and M.

As is seen from Figure 11, the profit of the system essentially increases when the number of the exploited servers grows from 1 to about 30. When the number of servers grows from 30 to 40, the profit still increases, but not so essentially. A further increase in the number of servers N leads to a slight decrease in the system profit because the cost of adding a new server exceeds the obtained additional profit from the service of customers. The same situation occurs when the number M of assistants grows. Thus, it can be verified that in the considered example, the optimal value E^* of the economic criterion E is $E^* = 5.87082$. This optimal value is achieved when the number of servers is $N = 40$, and the number of assistants is $M = 4$.

It is worth noting that the maximal size of the blocks of the generator is defined by the number $(N+1)RW$. One of the goals of this numerical experiment was to demonstrate the feasibility of the elaborated algorithms for more or less realistic values of the number N of SSDs and the number M of assistants. In this experiment, we have fixed $N = 50$ and $M = 10$ and the range of rating as the set $\{1, \dots, 10\}$, which is quite enough for the modeling of even a quite large real hypermarket. Therefore, the maximal size of the block in this example is 510 W. Because the total number of points (N, M) for which computations were performed to show the shape of the considered dependencies and solve the optimization problem is $NM = 500$, we restricted ourselves by the base MAP of order 2, just to avoid long computations. For $W = 2$, the maximal size of the block is more than 1000, and computation for 500 points takes several minutes. The increase in W does not create any essential problem in computations except the increase in computation time with the increase in W. Note that the use of the MAP of order 2 is often enough for good matching the main characteristics of the MAP flow to the corresponding characteristics of even quite bursty real flows.

8. Conclusions

In this paper, we have considered a queueing model having a finite number N of servers and M assistants that help the servers when certain service problems occur. This model fits a description of the operation of a huge variety of real-world systems with so-called self-service of customers. We assume the novel description of an arrival process as the generalization of the MAP to the case of rating-dependent arrival rates. Rating dependent arrivals are typical in many real systems with competing service providers. The effects of possible customers abandonment (balking) and impatience (reneging) are accepted for consideration. Algorithmic analysis of this queueing model based on the use of level-dependent multi-dimensional Markov chains is implemented. This analysis includes the derivation of conditions for stability and non-stability of the model, computation of the steady-state distribution of the chain, numerical illustration of the dependence of the main performance measures of the system on N and M, and the solution of the optimization problem.

The obtained results can be extended to the models with other possible mechanisms for calculating the value of the rating, more complicated distributions of service and help times, unreliable servers or (and) assistants, heterogeneous (experienced and non-experienced) customers, etc.

Author Contributions: Conceptualization, S.D., Y.G. and A.D.; methodology, S.D., O.D. and Y.G.; software, S.D. and O.D.; validation, S.D. and O.D.; formal analysis, S.D., Y.G. and A.D.; investigation, A.D.; writing, original draft preparation, Y.G. and A.D.; writing, review and editing A.D. and S.D.; supervision A.D. and Y.G.; project administration O.D. and A.D. All authors read and agreed to the published version of the manuscript.

Funding: This paper has been supported by the RUDN University Strategic Academic Leadership Program.

Institutional Review Board Statement: Not applicable.

Informed Consent Statement: Not applicable.

Data Availability Statement: Not applicable.

Conflicts of Interest: The authors declare no conflict of interest.

References

1. Huang, F.; Guo, P.; Wang, Y. Cyclic Pricing When Customers Queue with Rating Information. *Prod. Oper. Manag.* **2019**, *28*, 2471–2485. [CrossRef]
2. Dudin, S.; Dudin, A.; Dudina, O.; Samouylov, K. Competitive queueing systems with comparative rating dependent arrivals. *Oper. Res. Perspect.* **2020**, *7*, 100139. [CrossRef]
3. Meuter, M.L.; Ostrom, A.L.; Roundtree, R.I.; Bitner, M.J. SSTs: Understanding customer satisfaction with technology-based service encounters. *J. Mark.* **2000**, *54*, 50–64. [CrossRef]
4. Lyu, F.; Lim, H.A.; Choi, J. Customer Acceptance of Self-service Technologies in Retail: A Case of Convenience Stores in China. *Asia Pac. J. Inf. Syst.* **2019**, *29*, 428–447. [CrossRef]
5. Li, M.; Choi, T.Y.; Rabinovich, E.; Crawford, A. Self-service operations at retail stores: The role of inter-customer interactions. *Prod. Oper. Manag.* **2013**, *22*, 888–914. [CrossRef]
6. Kokkinou, A.; Cranage, D.A. Using self-service technology to reduce customer waiting times. *Int. J. Hosp. Manag.* **2013**, *33*, 435–445. [CrossRef]
7. Djelassi, S.; Diallo, M.F.; Zielke, S. How self-service technology experience evaluation affects waiting time and customer satisfaction? A moderated mediation model. *Decis. Support Syst.* **2018**, *111*, 38–47. [CrossRef]

8. Hwang, Y.; Kim, D.J. Customer self-service systems: The effects of perceived Web quality with service contents on enjoyment, anxiety, and e-trust. *Decis. Support Syst.* **2007**, *43*, 746–760. [CrossRef]
9. Morimura, F.; Nishioka, K. Waiting in exit-stage operations: expectation for self-checkout systems and overall satisfaction. *J. Mark. Channels* **2016**, *23*, 241–254. [CrossRef]
10. Wu, C.H.; Yang, D.Y. Bi-objective optimization of a queueing model with two-phase heterogeneous service. *Comput. Oper. Res.* **2021**, *130*, 105230. [CrossRef]
11. Neuts, M. *Matrix-Geometric Solutions in Stochastic Models*; John Hopkins University Press: Baltimore, MD, USA, 1981.
12. Haque, L.; Armstrong, M.J. A survey of the machine interference problem. *Eur. J. Oper. Res.* **2007**, *179*, 469–482. [CrossRef]
13. Singh, P.R.; Sharma, G.C.; Jain, M. Some perspectives of machine repair problems. *Int. J. Eng.* **2010**, *23*, 253–268.
14. Chakravarthy, S.R. A multi-server queueing model with server consultations. *Eur. J. Oper. Res.* **2014**, *233*, 625–639. [CrossRef]
15. Lucantoni, D.M. New Results on the Single Server Queue with a Batch Markovian Arrival Process. *Commun. Stat. Stoch. Model.* **1991**, *7*, 1–46. [CrossRef]
16. Chakravarthy, S.R. The batch Markovian arrival process: A review and future work. *Adv. Probab. Theory Stoch. Process.* **2001**, *1*, 21–49.
17. Dudin, A.N.; Klimenok, V.I.; Vishnevsky, V.M. *The Theory of Queuing Systems with Correlated Flows*; Springer Nature: Cham, Switzerland, 2019; ISBN 978-3-030-32072-0.
18. Vishnevskii, V.M.; Dudin, A.N. Queueing systems with correlated arrival flows and their applications to modeling telecommunication networks. *Autom. Remote Control* **2017**, *78*, 1361–1403. [CrossRef]
19. Krishnamoorthy, A.; Thekkiniyedath, R.; Lakshmy, B. On a Two-Server Queue with Consultation in Random Environment. In *International Conference on Information Technologies and Mathematical Modelling*; Springer: Cham, Switzerland, 2019; pp. 217–229.
20. Resmi, T.; Ravikumar, K. Three-Server Queue with Consultations by Main Server with a Buffer at the Main Server. In *International Conference on Information Technologies and Mathematical Modelling*; Springer: Cham, Switzerland, 2020; pp. 131–142.
21. Lal, T.S.; Krishnamoorthy, A.; Joshua, V.C.; Vishnevsky, V. A Two-Stage Tandem Queue with Specialist Servers. In *Applied Probability and Stochastic Processes*; Springer: Berlin/Heidelberg, Germany, 2020; pp. 335–353.
22. Klimenok, V.; Dudin, A.; Samouylov, K. Analysis of the $BMAP/PH/N$ queueing system with backup servers. *Appl. Math. Model.* **2018**, *57*, 64–84. [CrossRef]
23. Xu, S.H.; Gao, L.; Ou, J. Service performance analysis and improvement for a ticket queue with balking customers. *Manag. Sci.* **2007**, *53*, 971–990. [CrossRef]
24. Batt, R.J.; Terwiesch, C. Waiting patiently: An empirical study of queue abandonment in an emergency department. *Manag. Sci.* **2015**, *61*, 39–59. [CrossRef]
25. Sun, B.; Dudin, A.; Dudin, S. Queueing system with impatient customers, visible queue and replenishable inventory. *Appl. Comput. Math.* **2018**, *17*, 161–174.
26. Wang, K.; Li, N.; Jiang, Z. Queueing system with impatient customers: A review. In Proceedings of the 2010 IEEE International Conference on Service Operations and Logistics, and Informatics, Qingdao, China, 15–17 July 2010; pp. 82–87.
27. Hanukov, G.; Hassoun, M.; Musicant, O. On the Benefits of Providing Timely Information in Ticket Queues with Balking and Calling Times. *Mathematics* **2021**, *9*, 2753. [CrossRef]
28. Jouini, O.; Koole, G.; Roubos, A. Performance indicators for call centers with impatient customers. *IIE Trans.* **2013**, *45*, 341–354. [CrossRef]
29. Garnett, O.; Mandelbaum, A.; Reiman, M. Designing a call center with impatient customers. *Manuf. Serv. Oper. Manag.* **2002**, *4*, 208–227. [CrossRef]
30. Boots, N.K.; Tijms, H. A multiserver queueing system with impatient customers. *Manag. Sci.* **1999**, *45*, 444–448. [CrossRef]
31. De Kok, A.G.; Tijms, H.C. A queueing system with impatient customers. *J. Appl. Probab.* **1985**, *22*, 688–696. [CrossRef]
32. Boots, N.K.; Tijms, H. An $M/M/c$ queue with impatient customers. *Top* **1999**, *7*, 213–220. [CrossRef]
33. Shin, Y.W.; Choo, T.S. $M/M/s$ queue with impatient customers and retrials. *Appl. Math. Model.* **2009**, *33*, 2596–2606. [CrossRef]
34. Kuki, A.; Berczes, T.; Sztrik, J.; Toth, A. Reliability analysis of a retrial queueing systems with collisions, impatient customers, and catastrophic breakdowns. In Proceedings of the 2021 International Conference on Information and Digital Technologies (IDT), Zilina, Slovakia, 22–24 June 2021; pp. 254–259.
35. Wüchner, P.; Sztrik, J.; De Meer, H. Finite-source $M/M/S$ retrial queue with search for balking and impatient customers from the orbit. *Comput. Netw.* **2009**, *53*, 1264–1273. [CrossRef]
36. Kuki, A.; Berczes, T.; Toth, A.; Sztrik, J. Numerical analysis of finite source Markov retrial system with non-reliable server, collision, and impatient customers. *Ann. Math. Inform.* **2020**, *51*, 53–63. [CrossRef]
37. Klimenok, V.I.; Orlovsky, D.S.; Dudin, A.N. A $MAP/PH/N$ system with impatient repeated calls. *Asia-Pac. J. Oper. Res.* **2007**, *24*, 293–312. [CrossRef]
38. Van Do, T.; Do, N.H.; Zhang, J. An enhanced algorithm to solve multiserver retrial queueing systems with impatient customers. *Comput. Ind. Eng.* **2013**, *65*, 719–728.
39. Danilyuk, E.; Vygoskaya, O.; Moiseeva, S. Retrial queue $M/M/N$ with impatient customer in the orbit. In Proceedings of the International Conference on Distributed Computer and Communication Networks, Moscow, Russia, 17–21 September 2018; pp. 493–504.
40. Perel, N.; Yechiali, U. Queues with slow servers and impatient customers. *Eur. J. Oper. Res.* **2010**, *201*, 247–258. [CrossRef]

41. Dudin, S.A.; Dudina, O.S. Call center operation model as a $MAP/PH/N/R-N$ system with impatient customers. *Probl. Inf. Transm.* **2011**, *47*, 364–377. [CrossRef]
42. Kim, C.; Dudin, A.; Dudin, S.; Dudina, O. Tandem queueing system with impatient customers as a model of call center with Interactive Voice Response. *Perform. Eval.* **2013**, *70*, 440–453. [CrossRef]
43. Dudin, S.; Kim, C.; Dudina, O. $MMAP/M/N$ queueing system with impatient heterogeneous customers as a model of a contact center. *Comput. Oper. Res.* **2013**, *40*, 1790–1803. [CrossRef]
44. Graham, A. *Kronecker Products and Matrix Calculus with Applications*; Courier Dover Publications: New York, NY, USA, 2018.
45. Klimenok, V.I.; Dudin, A.N. Multi-dimensional asymptotically quasi-Toeplitz Markov chains and their application in queueing theory. *Queueing Syst.* **2006**, *54*, 245–259. [CrossRef]
46. Dudin, S.; Dudina, O. Retrial multi-server queuing system with PHF service time distribution as a model of a channel with unreliable transmission of information. *Appl. Math. Model.* **2019**, *65*, 676–695. [CrossRef]
47. Dudin, S.; Dudin, A.; Kostyukova, O.; Dudina, O. Effective algorithm for computation of the stationary distribution of multi-dimensional level-dependent Markov chains with upper block-Hessenberg structure of the generator. *J. Comput. Appl. Math.* **2020**, *366*, 112425. [CrossRef]

Article

A Continuous-Time Network Evolution Model Describing 2- and 3-Interactions

István Fazekas * and Attila Barta

Faculty of Informatics, University of Debrecen, Kassai Street 26, 4028 Debrecen, Hungary; barta.attila@inf.unideb.hu
* Correspondence: fazekas.istvan@inf.unideb.hu

Abstract: A continuous-time network evolution model is considered. The evolution of the network is based on 2- and 3-interactions. 2-interactions are described by edges, and 3-interactions are described by triangles. The evolution of the edges and triangles is governed by a multi-type continuous-time branching process. The limiting behaviour of the network is studied by mathematical methods. We prove that the number of triangles and edges have the same magnitude on the event of non-extinction, and it is $e^{\alpha t}$, where α is the Malthusian parameter. The probability of the extinction and the degree process of a fixed vertex are also studied. The results are illustrated by simulations.

Keywords: network evolution; random graph; multi-type branching process; continuous-time branching process; 2- and 3-interactions; Malthusian parameter; Poisson process; life-length; extinction

MSC: 05C80; 60J85

Citation: Fazekas, I.; Barta, A. A Continuous-Time Network Evolution Model Describing 2- and 3-Interactions. *Mathematics* **2021**, *9*, 3143. https://doi.org/10.3390/math9233143

Academic Editors: Alexander Zeifman, Victor Korolev and Alexander Sipin

Received: 18 October 2021
Accepted: 28 November 2021
Published: 6 December 2021

1. Introduction

Network theory has a vast literature. In the book of Barabási [1], the general aspects can be found, while the book of van der Hofstad [2] is devoted to the mathematical models. Any network can be considered as a graph. The nodes of the network are the vertices, and the connections are the edges of the graph. A most famous model is the preferential attachment model proposed by Albert and Barabási [3]. It is a discrete time network evolution model and it describes connections of two nodes. In real life, the meaning of connection can be any interaction or any cooperation.

There are models for cooperation more than two units. For example, Backhausz and Móri studied three-interactions in [4]. Their model is generalised for N-interactions by Fazekas and Porvázsnyik in [5]. Both of these papers consider cliques where, inside a team, all members cooperate. In some sense, the opposite of the cliques, i.e., star-like connections were considered by Fazekas and Perecsényi [6]. In [6], there is no cooperation between two peripheral members of the team but all of them cooperate with the central member of the team. Despite [3], in papers [4–6] the preferential attachment rule is used for certain subgraphs and not for vertices.

We mention that in [7] the Erdős–Rényi graph, the configuration model and the preferential attachment graph were studied when the population was split into two types. The mathematical tool of the analysis in [7] is the theory of multi-type branching processes.

There are several continuous-time network evolution models. Here, we list only some papers using continuous time branching processes. Early works in this direction are [8,9]. Recently, in [10], multi-type preferential attachment trees were studied. In [10], the results of [11] on multi-type continuous time branching processes were applied to describe the evolution of the network.

In this paper, we study a new network evolution model. The structure and the rules of the evolution of our model were inspired both by some everyday experiences and deep scientific results on motifs. On the one hand, we had in our mind activities and structures based on personal connections of the actors and where teams of some persons are important. Thus, we considered the friendship, the recruitment of party members and cooperation among party members, the recruitment and cooperation of volunteers, cooperation among scientists, informal connections among the employees of a company, etc. In these cases, the network consists of relatively small teams, a person can be a member of several teams at the same time, new teams can be born, and they can die, a newcomer can join the network if he/she joins an existing team.

On the second hand, our model is supported by the theory of motifs and their applications for real life networks. Here, we list only a few papers on this topic.

In [12], the authors used network motifs: 'patterns of interconnections occurring in complex networks at numbers that are significantly higher than those in randomized networks'. They developed an algorithm for detecting network motifs and found motifs with three or four vertices in biological and technological networks.

In [13], the authors analyse the local structure of several networks such as protein signaling, developmental genetic networks, power grids, protein-structure networks, World Wide Web links, social networks, and word-adjacency networks. For the study, they used motifs on three or our vertices. In [14], the authors found the numbers of all 3- and 4-node subgraphs, in both directed and non-directed geometric networks. In [15], a method for the identification of all ordered 3-node substructures and the visualization of their significance profile are offered.

Therefore, we wanted to study a network that consists of small substructures, a node can be a member of several substructures at the same time, new substructures can be born and they can die, a new node can join to the network if it joins to an existing substructure.

Concerning the mathematical tools, we follow the line of Móri and Rokob [16], where connections of two units were described by edges and the evolution of the edges was governed by a continuous-time branching process. In [17] we extended the model of [16] to 3-interactions. In this paper, our aim is to study networks containing groups of different sizes. For the sake of simplicity, we consider only groups of sizes 2 and 3. In this case, we can obtain explicit formulae for some quantities and our implicit formulae will be also transparent. The extension of our results to larger groups is possible. We emphasize that in our model all the nodes have the same type. Thus, despite [7,10], the theory of multi-type branching processes will be applied for groups of nodes and not just for the nodes.

In our model, when a new member joins to the network, it joins directly to an existing team. If that team consists of two members, then either a new team of two members or a new team of three members is produced. Similarly, if the new member joins to a team of three members, then new teams having two or three members can emerge. Thus, we obtain a two-type continuous time branching process, in which an individual can be either an edge or a triangle of the network.

The starting individual (that is the ancestor) can be either an edge or a triangle. It produces offspring at each time given by the driving branching process. These offspring can be edges or triangles, and after their birth they also start their own reproduction processes. An evolution step of the generic triangle is the following. Always one vertex is born, but it is connected to the triangle with random number of edges. The new vertex can be connected to $1, 2$ or 3 vertices of the triangle. Therefore, the offspring of a triangle can be a new edge, or one new triangle or three new triangles. The lifetime of a triangle is determined by the number of its offspring. The reproduction process of an edge is similar.

The structure of this paper is the following. In Section 2, a detailed description of our model is given. In Section 3, the general results are presented. These are the survival functions of an edge and of a triangle (Theorem 1), the mean offspring number of an edge and of a triangle (Corollary 1), the Perron root and the Malthusian parameter. As usual, we obtain only implicit expression for the Malthusian parameter, but our expression is simple and numerically tractable.

In Section 4, asymptotic theorems on the number of edges and triangles (Theorem 2) are proved. Both of them have magnitude $e^{\alpha t}$ on the event of non-extinction, where α is the Malthusian parameter. To prove Theorem 2, we use the underlying branching process counted with certain random characteristics and apply the asymptotic theorems of [11].

In Section 5, the generating functions are calculated. Using the generating functions, the probability of extinction are studied. In Section 6, the asymptotic behaviour of the degree of a fixed vertex is considered. Here, we again apply the asymptotic theorems of [11] but with other characteristics than in Section 4. In Section 7, we present some simulation results supporting our theorems. Our figures and tables show that the values obtained by simulation fit well to the theoretical results.

The proofs are based on known general results of multi-type continuous-time branching processes. Therefore, for the reader's convenience, in Section 8, we list several results on multi-type Crump–Mode–Jagers processes.

We mention that our model was presented in our conference paper [18]. In that paper some preliminary theoretical results were announced together with some numerical evidence but without mathematical proofs.

2. The Model

We study the following network evolution model. At the initial time $t = 0$ the network consists of one single object, this object can be either an edge or a triangle. This object is called the ancestor. During the evolution, this ancestor object produces offspring objects, which can be either edges or triangles. Then, these offspring objects produce their offspring objects and so on. The reproduction times of any fixed object, including the ancestor, are the occurrences in its own Poisson process with rate 1.

From the theory of branching processes, we apply the following usual assumptions. That is we suppose that the reproduction processes of different objects are independent. Moreover, we assume that the reproduction processes of the edges are independent copies of the reproduction process of the generic edge. Similarly, the reproduction processes of the triangles are independent copies of the reproduction process of the generic triangle.

First, we explain the evolution of the generic edge. A Poisson process $\Pi_2(t)$ with parameter 1 gives its reproduction times. At any jumping time of this Poisson process, a new vertex appears and it is connected to the generic edge with one or two edges. The probability that this new vertex is connected to the generic edge by one new edge is r_1, where $0 \leq r_1 \leq 1$. The other end point of this new edge is chosen from the two vertices of the generic edge uniformly at random. We see that in this case the generic edge produces always one new edge. The other case is that when the new vertex is connected to both vertices of the generic edge. Its probability is $r_2 = 1 - r_1$. In this second case the offspring of the generic edge is a triangle consisting of the generic edge and the two new edges. We emphasize that in this last case the generic edge itself and the new triangle will produce offspring, but the two new edges are not substantive parts of the reproduction process, so they alone will not produce offspring.

The reproduction process of the generic triangle is similar. The Poisson process with rate 1 corresponding to the generic triangle is denoted by $\Pi_3(t), t \geq 0$. The jumping times of $\Pi_3(t)$ are the birth times of the generic triangle. At every birth time a new vertex is born and it joins to the existing graph so that it is connected to our generic triangle with 1, 2 or 3 edges. Denote by p_j $(j = 1, 2, 3)$ the probability that the new vertex is connected to j vertices of our generic triangle. The vertices of the generic triangle to be connected to the new vertex are chosen uniformly at random.

By the above definition of the evolution process, at each birth step we add precisely 1 new vertex. When the new vertex is connected to one vertex of the generic triangle, the generic triangle gives birth to one new edge. This event has probability p_1. However, in the remaining two cases we count only the new triangles and not the new edges. When the new edge is connected to the generic triangle by two edges, these two edges and one edge of the generic triangle form a new triangle. Therefore, with probability p_2, the generic triangle produces one child triangle. When the new edge is connected to the generic triangle by three edges, these edges and the edges of the generic triangle form three new triangles. Thus, with probability p_3, the generic triangle produces three children triangles.

Any edge is called a type 2 object, and any triangle is called a type 3 object. We use subscript 2 for edges and subscript 3 for triangles. Thus, we denote by $\xi_{i,j}(t)$ the number of type j offspring of the type i generic object up to time t ($i, j = 2, 3$). Recall that $\xi_{i,j}$, $i, j = 2, 3$, are point processes. Then

$$\xi_2(t) = \xi_{2,2}(t) + \xi_{2,3}(t) \tag{1}$$

gives the total number of offspring (that is both edges and triangles) of the generic edge up to time t. We can also see that

$$\xi_3(t) = \xi_{3,2}(t) + \xi_{3,3}(t) \tag{2}$$

is the number of all offspring (edges or triangles) of the generic triangle up to time t.

We denote by $\tau_3(1), \tau_3(2), \ldots$ the birth times of the generic triangle, and we denote by $\varepsilon_3(1), \varepsilon_3(2), \ldots$ the corresponding total litter sizes. That is, at the ith birth event, the generic triangle bears $\varepsilon_3(i)$ children being either triangles or edges. The discrete random variables $\varepsilon_3(1), \varepsilon_3(2), \ldots$ are independent and identically distributed having distribution $\mathbb{P}(\varepsilon_3(i) = j) = q_j, j \geq 1$. By the above evolution process, we have

$$\mathbb{P}(\varepsilon_3(i) = 1) = q_1 = p_1 + p_2, \ \mathbb{P}(\varepsilon_3(i) = 3) = q_3 = p_3,$$

$$\mathbb{P}(\varepsilon_3(i) = j) = q_j = 0, \ \text{ if } \ j \notin \{1, 3\}.$$

We assume that the litter sizes are independent of the birth times.

Let λ_3 be the life-length of the generic triangle. It is a finite, non-negative random variable. We assume that the reproduction terminates at the death of the individual. Therefore, $\xi_3(t) = \xi_3(\lambda_3)$ for $t > \lambda_3$. Then, the reproduction process of a triangle can be formulated as

$$\xi_3(t) = \sum_{\tau_3(i) \leq t \wedge \lambda_3} \varepsilon_3(i) = S_3(\Pi_3(t \wedge \lambda_3)), \tag{3}$$

where $\Pi_3(t)$ is the Poisson process, $S_3(n) = \varepsilon_3(1) + \cdots + \varepsilon_3(n)$ gives the total number of offspring of the generic triangle before the $(n+1)$th birth event and by $x \wedge y$ we denote the minimum of $\{x, y\}$.

The survival function of the life-length. Let $L_3(t)$ denote the distribution function of the triangle's life-length λ_3. Then, the survival function of λ_3 is

$$1 - L_3(t) = \mathbb{P}(\lambda_3 > t) = \exp\left(-\int_0^t l_3(u)du\right), \tag{4}$$

where $l_3(t)$ is the hazard rate of the life-length λ_3. We suppose that the hazard rate depends on the total number of offspring, so that

$$l_3(t) = b + c\xi_3(t) \tag{5}$$

with fixed positive constants b and c.

Let λ_2 be the life-length of the generic edge. Then, $\xi_2(t) = \xi_2(\lambda_2)$ for $t > \lambda_2$. As the edge always gives birth to one offspring (which can be an edge or a triangle); therefore,

$$\xi_2(t) = \Pi_2(t \wedge \lambda_2) \tag{6}$$

is the total number of offspring of the generic edge, where $\Pi_2(t)$ is the Poisson process.

We denote by $L_2(t)$ the distribution function of λ_2. Then, the survival function of the life-length of an edge is

$$1 - L_2(t) = \exp\left(-\int_0^t l_2(u)du\right), \tag{7}$$

where l_2 is the hazard rate of the life-length λ_2. We suppose that l_2 is of the form $l_2(t) = b + c\xi_2(t)$.

We emphasize that we do not delete any edge or any triangle when it dies, because its ingredients can belong to other triangles or edges, too. Thus, dead triangles and edges will be considered as inactive objects not producing new offspring.

In Figure 1, an example is shown for our graph evolution model. For a clear view it contains only three birth steps after the initial time $t = 0$. The nodes of the ancestor are highlighted by red. The edges are labelled with the birth times t. The following objects appear in Figure 1, which are described by the labels of their nodes:

- (1-2-3): is a triangle, the ancestor with birth time $t = 0$,
- (1-2-3-4): represents three triangles, i.e., the offspring of (1-2-3) at its first reproduction time $t = 0.571$,
- (1-5): an edge, offspring of (1-2-3) with birth time $t = 0.847$,
- (1-5-6): a triangle, offspring of (1-5) with birth time $t = 1.06$.

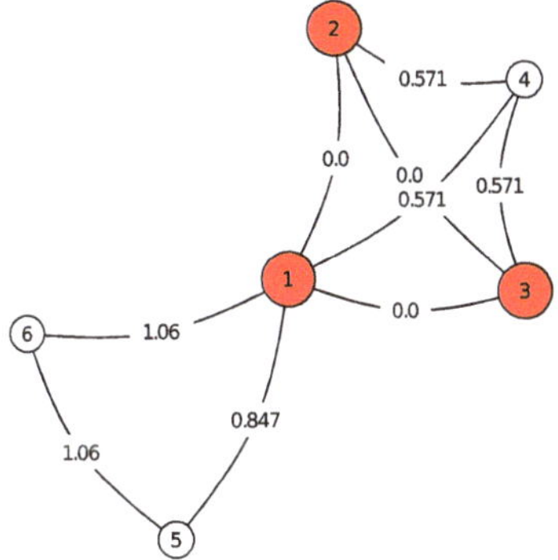

Figure 1. Example of the graph evolution model with parameter set: $r_1 = 0.1,\ p_1 = 0.4,\ p_2 = 0.2,\ b = 0.1,\ c = 0.1$.

Two more examples are shown in Figure 2 with different parameters. In Figure 2a the ancestor is an edge, while in Figure 2b the ancestor is a triangle.

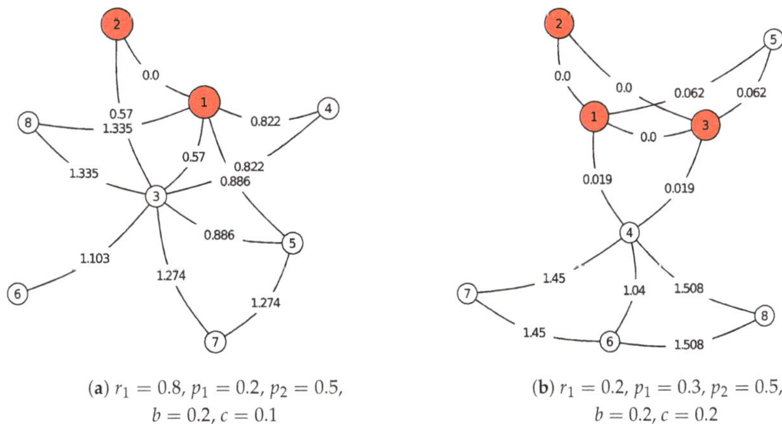

(**a**) $r_1 = 0.8$, $p_1 = 0.2$, $p_2 = 0.5$, $b = 0.2$, $c = 0.1$

(**b**) $r_1 = 0.2$, $p_1 = 0.3$, $p_2 = 0.5$, $b = 0.2$, $c = 0.2$

Figure 2. Examples of the graph evolution model with two different parameter sets.

3. General Results

The survival functions.

Theorem 1. *The survival function for a triangle is*

$$\mathbb{P}(\lambda_3 > t) = e^{-t(b+1)} e^{\frac{3(p_1+p_2)\left(1-e^{-ct}\right)+p_3\left(1-e^{-3ct}\right)}{3c}}. \tag{8}$$

The survival function for an edge is

$$\mathbb{P}(\lambda_2 > t) = e^{-t(b+1)} e^{\frac{1-e^{-ct}}{c}}. \tag{9}$$

Proof. At the first part of the proof we omit subscripts 2 and 3, because the calculations are the same for edges and triangles. Let $t > 0$ and assume that $\Pi(t) = k$. Then, the first k birth events happened before time t. Thus, the birth times $\tau(1), \tau(2), \ldots, \tau(k)$ and the corresponding litter sizes $\varepsilon(1), \varepsilon(2), \ldots, \varepsilon(k)$ are known. Therefore, the reproduction process $\xi(u)$ is also known for $u < t$. By (5), a simple calculation shows that the survival function of an object is

$$1 - L(t) = \exp\left(-\int_0^t l(u)du\right) = \exp\left(-\left(bt + c\int_0^t \xi(u)du\right)\right) =$$
$$= \exp(-(bt + ctS(k) - c(\varepsilon(1)\tau(1) + \cdots + \varepsilon(k)\tau(k)))).$$

Then

$$\mathbb{P}(\lambda > t | \Pi(t) = k, \tau(1), \ldots, \tau(k), \varepsilon(1), \ldots, \varepsilon(k)) =$$
$$= \exp(-(bt + ctS(k) - c(\varepsilon(1)\tau(1) + \cdots + \varepsilon(k)\tau(k)))).$$

Let $\left(U_1^*, \ldots, U_k^*\right)$ be an ordered sample of size k from uniform distribution on $[0,1]$. Then, the joint conditional distribution of the birth times $\tau(1), \ldots, \tau(k)$ given $\Pi(t) = k$, coincides with the distribution of $\left(tU_1^*, \ldots, tU_k^*\right)$. Therefore

$$\mathbb{P}(\lambda > t | \Pi(t) = k) = \mathbb{E}\exp\left(-\left(bt + ct\sum_{i=1}^{k} \varepsilon(i)\left(1 - \frac{\tau(i)}{t}\right)\right)\right) =$$
$$= \mathbb{E}\exp\left(-bt + ct\sum_{i=1}^{k} \varepsilon(i)(U_i^* - 1)\right),$$

because $\tau(i) = tU_i^*$. The litter sizes $\varepsilon(1), \ldots, \varepsilon(k)$ are independent identically distributed random variables, which are independent also of $U_1^*, \ldots, U_k^*$. Hence

$$\mathbb{P}(\lambda > t | \Pi(t) = k) = \mathbb{E}\exp\left(-bt + ct\sum_{i=1}^{k} \varepsilon(i)(U_i - 1)\right) =$$
$$= e^{-bt}\mathbb{E}\prod_{i=1}^{k} e^{ct\varepsilon(i)(U_i-1)} = e^{-bt}\left(\mathbb{E}_{\varepsilon(i)}\left(\mathbb{E}_{U_i}\left(e^{ct\varepsilon(i)U_i}\right)e^{-ct\varepsilon(i)}\right)\right)^k =$$
$$= e^{-bt}\left(\sum_{j=1}^{\infty} q_j \frac{e^{ctj} - 1}{ctj} e^{-ctj}\right)^k = e^{-bt}\left(\sum_{j=1}^{\infty} q_j \frac{1 - e^{-ctj}}{ctj}\right)^k,$$

where we applied that U_i is uniformly distributed. Using this and the total probability theorem, we find

$$\mathbb{P}(\lambda > t) = \sum_{k=0}^{\infty} \mathbb{P}(\Pi(t) = k)\mathbb{P}(\lambda > t | \Pi(t) = k) =$$

$$= \sum_{k=0}^{\infty} \frac{t^k}{k!} e^{-t} e^{-bt} \left(\sum_{j=1}^{\infty} q_j \frac{1 - e^{-ctj}}{ctj} \right)^k =$$

$$= e^{-(b+1)t} \sum_{k=0}^{\infty} \frac{1}{k!} \left(\sum_{j=1}^{\infty} q_j \frac{1 - e^{-ctj}}{cj} \right)^k =$$

$$= e^{-(b+1)t} e^{\sum_{j=1}^{\infty} q_j \frac{1-e^{-ctj}}{cj}}.$$

Therefore, the survival function for a triangle is

$$\mathbb{P}(\lambda_3 > t) = e^{-t(b+1)} e^{\frac{3(p_1+p_2)\left(1-e^{-ct}\right)+p_3\left(1-e^{-3ct}\right)}{3c}}.$$

Finally, the survival function for an edge is

$$\mathbb{P}(\lambda_2 > t) = e^{-t(b+1)} e^{\frac{1-e^{-ct}}{c}}.$$

□

The mean offspring number. Let us denote by $m_{i,j}(t) = \mathbb{E}\xi_{i,j}(t)$ the expectation of the number of type j offspring of a type i mother until time t.

Corollary 1. *For any $t \geq 0$, we have*

$$m_{2,2}(t) = r_1 F(t), \qquad m_{2,3}(t) = r_2 F(t), \tag{10}$$

where

$$F(t) = \int_0^t (1 - L_2(s))ds = \int_0^t e^{-(b+1)s} e^{\frac{1-e^{-cs}}{c}} ds = \frac{1}{c} \int_0^{1-e^{-ct}} (1-u)^{\frac{b+1}{c}-1} e^{\frac{u}{c}} du.$$

$$\mathbb{E}\lambda_2 = \frac{1}{c} \int_0^1 (1-u)^{\frac{b+1}{c}-1} e^{\frac{u}{c}} du. \tag{11}$$

For any $t \geq 0$, we have

$$m_{3,2}(t) = p_1 G(t), \qquad m_{3,3}(t) = (p_2 + 3p_3) G(t), \tag{12}$$

where

$$G(t) = \int_0^t (1 - L_3(s))ds = \int_0^t e^{-s(b+1)} e^{\frac{3(p_1+p_2)(1-e^{-cs})+p_3\left(1-e^{-3cs}\right)}{3c}} ds =$$

$$= \frac{1}{c} \int_0^{1-e^{-ct}} (1-u)^{\frac{b+1}{c}-1} e^{\frac{u}{3c}\left(p_3 u^2 - 3p_3 u + 3\right)} du.$$

$$\mathbb{E}\lambda_3 = \frac{1}{c} \int_0^1 (1-u)^{\frac{b+1}{c}-1} e^{\frac{u}{3c}\left(p_3 u^2 - 3p_3 u + 3\right)} du. \tag{13}$$

$0 < \mathbb{E}\lambda_2, \mathbb{E}\lambda_3 < \infty$ *because $b \geq 0$.*

Proof. We have

$$m_{i,j}(t) = \mathbb{E}\xi_{i,j}(t) = \mathbb{E}\big(\varepsilon_{i,j}(1) + \varepsilon_{i,j}(2) + \cdots + \varepsilon_{i,j}(\Pi(t \wedge \lambda_i))\big),$$

where $\varepsilon_{i,j}(k)$ is the number of type j offspring of a type i mother at her kth birth event. Using Wald's identity, the average number of children is

$$m_{i,j}(t) = \mathbb{E}\big(\varepsilon_{i,j}(1)\big)\mathbb{E}(\Pi(t \wedge \lambda_i)). \tag{14}$$

Using that Π is a Poisson process with rate 1, and $t \wedge \lambda$ is bounded for any t, from (14), we obtain that the average number of children is

$$m_{i,j}(t) = \mathbb{E}(\varepsilon_{i,j}(1))\mathbb{E}(t \wedge \lambda_i) = \mathbb{E}(\varepsilon_{i,j}(1)) \int_0^t (1 - L_i(s))ds. \tag{15}$$

Now, consider $m_{2,2}(t)$. Applying (9) and using the substitution $u = 1 - e^{-cs}$, we obtain

$$m_{2,2}(t) = r_1 \int_0^t e^{-(b+1)s} e^{\frac{1-e^{-cs}}{c}} ds = \frac{r_1}{c} \int_0^{1-e^{-ct}} (1-u)^{\frac{b+1}{c}-1} e^{\frac{u}{c}} du. \tag{16}$$

If we write r_2 instead of r_1, then we obtain $m_{2,3}(t)$. Thus, we obtained (10). Moreover, with $t \to \infty$, we have $\mathbb{E}\lambda_2 = \int_0^\infty \mathbb{P}(\lambda_2 > t)dt$. Thus, (11) follows from (16).

Now, we turn to $m_{3,3}(t)$. Applying (8), and using the substitution $u = 1 - e^{-cs}$, we obtain ,

$$\int_0^t \mathbb{P}(\lambda_3 > s)ds = \int_0^t e^{-s(b+1)} e^{\frac{3(p_1+p_2)(1-e^{-cs})+p_3\left(1-e^{-3cs}\right)}{3c}} ds =$$

$$= \frac{1}{c} \int_0^{1-e^{-ct}} (1-u)^{\frac{b+1}{c}-1} e^{\frac{u}{3c}\left(p_3u^2 - 3p_3u + 3(p_1+p_2+p_3)\right)} du. \tag{17}$$

As $\mathbb{E}(\varepsilon_{3,3}(1)) = p_2 + 3p_3$, so from (15) we obtain $m_{3,3}(t)$. Using that $\mathbb{E}(\varepsilon_{3,2}(1)) = p_1$, we obtain $m_{3,2}(t)$. Thus, we obtained (12). Moreover, we have $\mathbb{E}\lambda_3 = \int_0^\infty \mathbb{P}(\lambda_3 > t)dt$. Thus, (13) follows from (17) with $t \to \infty$. □

Let

$$m^*_{i,j}(\kappa) = \int_0^\infty e^{-\kappa t} m_{i,j}(dt), \qquad i,j = 2,3,$$

be the Laplace transform of $m_{i,j}$.

Proposition 1. *For any $\kappa \geq 0$, we have*

$$m^*_{2,2}(\kappa) = r_1 A(\kappa), \qquad m^*_{2,3}(\kappa) = r_2 A(\kappa), \tag{18}$$

where

$$A(\kappa) = \int_0^\infty e^{-\kappa s} e^{-(b+1)s} e^{\frac{1-e^{-cs}}{c}} ds = \frac{1}{c} \int_0^1 (1-u)^{\frac{\kappa+b+1}{c}-1} e^{\frac{u}{c}} du. \tag{19}$$

For any $\kappa \geq 0$, we have

$$m^*_{3,2}(\kappa) = p_1 B(\kappa), \qquad m^*_{3,3}(\kappa) = (p_2 + 3p_3)B(\kappa), \tag{20}$$

where

$$B(\kappa) = \int_0^\infty e^{-\kappa s} e^{-s(b+1)} e^{\frac{3(p_1+p_2)(1-e^{-cs})+p_3\left(1-e^{-3cs}\right)}{3c}} ds =$$

$$= \frac{1}{c} \int_0^1 (1-u)^{\frac{\kappa+b+1}{c}-1} e^{\frac{u}{3c}\left(p_3u^2 - 3p_3u + 3\right)} du.$$

Proof. Apply the definition of $m^*_{i,j}(\kappa)$, Corollary 1 and substitution $u = 1 - e^{-cs}$. □

The Perron root and the Malthusian parameter. Let

$$M(\kappa) = \begin{pmatrix} m^*_{2,2}(\kappa) & m^*_{2,3}(\kappa) \\ m^*_{3,2}(\kappa) & m^*_{3,3}(\kappa) \end{pmatrix} \tag{21}$$

be the matrix of the Laplace transforms. Direct calculation gives that the characteristic roots of $M(\kappa)$ are

$$\varrho_{1,2}(\kappa) = \frac{(p_2 + 3p_3)B(\kappa) + r_1 A(\kappa) \pm \sqrt{((p_2 + 3p_3)B(\kappa) - r_1 A(\kappa))^2 + 4p_1 B(\kappa) r_2 A(\kappa)}}{2}. \tag{22}$$

The greater of the values $\varrho_1(\kappa)$ and $\varrho_2(\kappa)$ is called the Perron root, so

$$\varrho(\kappa) = \varrho_1(\kappa) = \frac{(p_2 + 3p_3)B(\kappa) + r_1 A(\kappa) + \sqrt{((p_2 + 3p_3)B(\kappa) - r_1 A(\kappa))^2 + 4p_1 B(\kappa) r_2 A(\kappa)}}{2} \tag{23}$$

is the Perron root.

We assume that our process is supercritical; that is,

$$\varrho(0) > 1. \tag{24}$$

For supercriticality, condition

$$\max\{(p_2 + 3p_3)B(0), r_1 A(0)\} > 1$$

is sufficient.

That value of κ for which the Perron root is equal to 1 is called the Malthusian parameter. Thus, using the usual notation in the theory of branching processes, α is the Malthusian parameter if $\varrho(\alpha) = 1$. In this paper, we assume the existence of the Malthusian parameter. From relation $\varrho(\alpha) = 1$ and (23), we obtain that the Malthusian α satisfies the equation

$$r_1 A(\alpha)(p_2 + 3p_3)B(\alpha) - (r_1 A(\alpha) + (p_2 + 3p_3)B(\alpha)) = r_2 A(\alpha) p_1 B(\alpha) - 1. \tag{25}$$

Later, we use the eigenvectors of $M(\alpha)$. To this end, let α be the Malthusian parameter, and let $(v_2, v_3)^\top$ be the right eigenvector of $M(\alpha)$ corresponding to eigenvalue 1 and satisfying condition $v_2 + v_3 = 1$. Then, direct calculation shows that

$$v_2 = \frac{(r_1 - 1)A(\alpha)}{(2r_1 - 1)A(\alpha) - 1}, \qquad v_3 = \frac{r_1 A(\alpha) - 1}{(2r_1 - 1)A(\alpha) - 1}. \tag{26}$$

Again, let α be the Malthusian parameter and let $(u_2, u_3)^\top$ be the left eigenvector of $M(\alpha)$ satisfying condition $u_2 v_2 + u_3 v_3 = 1$. Direct calculation shows that

$$u_2 = \frac{p_1 B(\alpha)((2r_1 - 1)A(\alpha) - 1)}{p_1 B(\alpha)(r_1 - 1)A(\alpha) - (r_1 A(\alpha) - 1)^2}, \qquad u_3 = \frac{(1 - r_1 A(\alpha))((2r_1 - 1)A(\alpha) - 1)}{p_1 B(\alpha)(r_1 - 1)A(\alpha) - (r_1 A(\alpha) - 1)^2}. \tag{27}$$

4. Asymptotic Theorems on the Number of Triangles and Edges

In this section, we use Proposition 4 from Section 8. So we should check the conditions given in Section 8. For condition (a) from Section 8, we should guarantee that not all measures $m_{i,j}$ are concentrated on a lattice. By Corollary 1, these measures are absolutely continuous, and thus it is satisfied.

Concerning condition (b1), we underline that we suppose the existence of a positive Malthusian parameter α. To this end, in this section, we assume that (25) has a finite positive solution α. We can check numerically the existence of this value. For (b2), we assume (24). Condition (c) from Section 8 will be checked later in the proofs of the results together with other conditions related to it.

Now, we analyse condition (d). We can see from Corollary 1 that $F(\infty)$ and $G(\infty)$ are positive. Thus, we can concentrate on parameters r_i and p_i. If $r_2 = p_1 = 0$, then (d) is not satisfied; however, in this case, one can study separately the process of edges (it grows at any birth time by 1), and the process of triangles (this is described in [17]). If $r_1 = 0$ and $p_2 + p_3 = 0$, then (d) is not satisfied, and the evolution process is an alternating one. If either $r_2 = 0$ or $p_1 = 0$, then (d) is not satisfied.

To guarantee condition (d), in this section, we assume that $0 \le r_1 < 1$, $0 < p_1 \le 1$, and it is excluded that both $r_1 = 0$ and $p_1 = 1$ are satisfied at the same time. In this case, condition (d) from Section 8 is satisfied.

The denominator in the limit theorem. In the following theorem, we need the next formulae. In Section 8, we see that the denominator of m_∞^Φ in the limiting expression is independent of Φ, and it is

$$\sum_{l,j=1}^{p} u_l v_j \int_0^\infty t e^{-\alpha t} m_{l,j}(dt).$$

It can be written in the form (and considering our two-dimensional case)

$$D(\alpha) = \sum_{l,j=2}^{3} u_l v_j \Big(-m^*_{l,j}(\alpha)\Big)'. \tag{28}$$

Here, u_i and v_i are from Equations (26) and (27). Moreover, by Corollary 1 or by Proposition 1, we have that

$$\Big(-m^*_{2,2}(\alpha)\Big)' = r_1(-A'(\alpha)), \qquad (-m^*_{2,3}(\alpha))' = r_2(-A'(\alpha)), \tag{29}$$

$$\Big(-m^*_{3,2}(\alpha)\Big)' = p_1(-B'(\alpha)), \qquad (-m^*_{3,3}(\alpha))' = (p_2 + 3p_3)(-B'(\alpha)), \tag{30}$$

where

$$-A'(\alpha) = \int_0^\infty s e^{-\alpha s} e^{-(b+1)s} e^{\frac{1-e^{-cs}}{c}} ds = -\frac{1}{c^2}\int_0^1 \ln(1-u)(1-u)^{\frac{\alpha+b+1}{c}-1} e^{\frac{u}{c}} du, \tag{31}$$

$$\begin{aligned} -B'(\alpha) &= \int_0^\infty s e^{-\alpha s} e^{-s(b+1)} e^{\frac{3(p_1+p_2)(1-e^{-cs})+p_3\left(1-e^{-3cs}\right)}{3c}} ds = \\ &= -\frac{1}{c^2}\int_0^1 \ln(1-u)(1-u)^{\frac{\alpha+b+1}{c}-1} e^{\frac{u}{3c}(p_3u^2-3p_3u+3)} du. \end{aligned} \tag{32}$$

Now, we turn to the number of edges and triangles. Recall that an edge is a type 2, and a triangle is a type 3 object.

Theorem 2. *Assume that (24) is satisfied and (25) has a finite positive solution α. Assume that $0 \le r_1 < 1$, $0 < p_1 \le 1$ and it is excluded that both $r_1 = 0$ and $p_1 = 1$ are satisfied at the same time.*

Let ${}_iE(t)$ denote the number of all edges being born up to time t if the ancestor of the population was a type i object, $i = 2, 3$. Then

$$\lim_{t\to\infty} e^{-\alpha t}\,{}_iE(t) = {}_iW\frac{v_i u_2}{\alpha D(\alpha)} \tag{33}$$

almost surely for $i = 2, 3$.

Let ${}_i\hat{E}(t)$ denote the number of all edges present at time t if the ancestor of the population was a type i object, $i = 2, 3$. Then

$$\lim_{t\to\infty} e^{-\alpha t}\,{}_i\hat{E}(t) = {}_iW\frac{v_i u_2 A(\alpha)}{D(\alpha)} \tag{34}$$

almost surely for $i = 2, 3$.

Let ${}_iT(t)$ denote the number of all triangles being born up to time t if the ancestor of the population was a type i object, $i = 2, 3$. Then

$$\lim_{t\to\infty} e^{-\alpha t}\,{}_iT(t) = {}_iW\frac{v_i u_3}{\alpha D(\alpha)} \tag{35}$$

almost surely for $i = 2, 3$.

Let ${}_i\hat{T}(t)$ denote the number of all triangles present at time t if the ancestor of the population was a type i object, $i = 2, 3$. Then,

$$\lim_{t\to\infty} e^{-\alpha t}\,{}_i\hat{T}(t) = {}_iW\frac{v_i u_3 B(\alpha)}{D(\alpha)} \tag{36}$$

almost surely for $i = 2, 3$.

The quantities ${}_2W$ and ${}_3W$ are a.s. non-negative, $\mathbb{E}({}_2W) = \mathbb{E}({}_3W) = 1$, ${}_2W$ and ${}_3W$ are a.s. positive on the event of survival.

Proof. We apply Proposition 4. To obtain condition (71), it is enough to show that

$$\mathbb{E}\left[{}_{\alpha}\xi_i(\infty)\log^{+}{}_{\alpha}\xi_i(\infty)\right] < \infty, \qquad i = 2,3, \tag{37}$$

where

$${}_{\alpha}\xi_i(\infty) = \int_0^{\infty} e^{-\alpha t}\xi_i(dt), \qquad i = 2,3, \tag{38}$$

and

$$\xi_i(t) = \xi_{i,2}(t) + \xi_{i,3}(t), \qquad i = 2,3. \tag{39}$$

If $i = 2$, then $\xi_2(t)$ is the birth process of an edge, and the children can be both edges and triangles. Therefore, at each birth, there is one child. Therefore,

$${}_{\alpha}\xi_2(\infty) = \int_0^{\infty} e^{-\alpha t}\xi_2(dt) = \sum_{\tau(i)\leq\lambda_2} 1e^{-\alpha\tau(i)} \leq \sum_{i=1}^{\infty} 1e^{-\alpha\tau(i)} = M,$$

where $\tau(1), \tau(2), \ldots$ are the jumps of the Poisson process Π_2. In the Poisson process $\Pi_2(t)$ the distribution of the interarrival time $(\tau(i) - \tau(i-1))$ is exponential with rate 1. Therefore, $\tau(i)$ has Γ-distribution $\Gamma(i,1)$. Using this, we have

$$\mathbb{E}(M) = \sum_{i=1}^{\infty} \mathbb{E}\left(e^{-\alpha\tau(i)}\right) = \sum_{i=1}^{\infty} \frac{1}{(1+\alpha)^i} = \frac{1}{\alpha}. \tag{40}$$

Let us denote by η_i the interarrival time $\tau(i) - \tau(i-1)$. Let η_0 be an exponentially distributed random variable with rate 1 that is independent of M. Then,

$$e^{-\alpha\eta_0}(1+M) = e^{-\alpha\eta_0} + e^{-\alpha\eta_0}\sum_{i=1}^{\infty} e^{-\alpha(\eta_1+\cdots+\eta_i)} = \sum_{i=0}^{\infty} e^{-\alpha(\eta_0+\eta_1+\cdots+\eta_i)}.$$

Therefore, the distribution of $e^{-\alpha\eta_0}(1+M)$ coincides with the distribution of M. Therefore, using (40), we have

$$\mathbb{E}M^2 = \mathbb{E}\left(e^{-\alpha\eta_0}(1+M)\right)^2 = \frac{1}{1+2\alpha}\left(1 + \frac{2}{\alpha} + \mathbb{E}M^2\right).$$

From this, we find

$$\mathbb{E}M^2 = \frac{\alpha+2}{2\alpha^2} < \infty.$$

Thus, (37) is true for $i = 2$.

If $i = 3$, then $\xi_3(t)$ is the birth process of a triangle and the children can be both edges and triangles. Therefore, at each birth there are at most three children. Therefore,

$${}_{\alpha}\xi_3(\infty) = \int_0^{\infty} e^{-\alpha t}\xi_3(dt) = \sum_{\tau(i)\leq\lambda_3} \varepsilon(i)e^{-\alpha\tau(i)} \leq 3\sum_{i=1}^{\infty} 1e^{-\alpha\tau(i)} = 3M,$$

where $\tau(1), \tau(2), \ldots$ are the jumps of the Poisson process Π_3. By the above calculation $\mathbb{E}M^2 < \infty$, so (37) is true for $i = 3$.

If we show that $\int_0^{\infty} t^2 e^{-\alpha t} m_{i,j}(dt) < \infty$, for $i,j = 2,3$, then conditions (c) and (iv) of Section 8 will be proved. Now, for $i = 2$ and $j = 2,3$, we have from Corollary 1

$$\int_0^{\infty} t^2 e^{-\alpha t} m_{2,j}(dt) \leq \max\{r_1, r_2\}\int_0^{\infty} t^2 e^{-\alpha t} e^{-t(b+1)} e^{\frac{1-e^{-ct}}{c}} dt \leq \int_0^{\infty} t^2 e^{-t(\alpha+b+1-1)} dt < \infty$$

because $\alpha + b > 0$.

For $i = 3$ and $j = 2,3$, we have from Corollary 1

$$\int_0^{\infty} t^2 e^{-\alpha t} m_{3,j}(dt) \leq \max\{p_1, p_2 + 3p_3\}\int_0^{\infty} t^2 e^{-\alpha t} e^{-t(b+1)} e^{(p_1+p_2)\frac{1-e^{-ct}}{c} + p_3\frac{1-e^{-3ct}}{3c}} dt \leq$$

$$\leq \int_0^\infty t^2 e^{-t(\alpha+b+1-1)} dt < \infty$$

Thus, conditions (c) and (iv) of Section 8 are proved.

Now, turn to the number of edges.

To obtain (33), let $\Phi_x(t) = 1$ if x is an edge, and $\Phi_x(t) = 0$ if x is a triangle. Therefore, $\mathbb{E}\Phi_2(t) = 1$ and $\mathbb{E}\Phi_3(t) = 0$. Conditions $(i) - (ii) - (iii)$ and (v) of Section 8 are satisfied. Thus, (69) and (70) imply (33).

To obtain (34), let $\Phi_x(t) = 1$ if x is an edge and it is present at t, and $\Phi_x(t) = 0$ if x is a triangle. Therefore, $\mathbb{E}\Phi_2(t) = 1 - L_2(t)$ and $\mathbb{E}\Phi_3(t) = 0$. Conditions $(i) - (ii) - (iii)$ and (v) of Section 8 are satisfied. Now,

$$\int_0^\infty e^{-\alpha t}\mathbb{E}\Phi_2(t)dt = \int_0^\infty e^{-\alpha t}(1 - L_2(t))dt = A(\alpha).$$

Thus, (69) and (70) imply (34).

Now, we turn to the number of triangles.

To obtain (35), let $\Phi_x(t) = 0$ if x is an edge, and $\Phi_x(t) = 1$ if x is a triangle. Therefore, $\mathbb{E}\Phi_2(t) = 0$ and $\mathbb{E}\Phi_3(t) = 1$. Conditions $(i) - (ii) - (iii)$ and (v) of Section 8 are satisfied. Thus, (69) and (70) imply (35).

To obtain (36), let $\Phi_x(t) = 0$ if x is an edge, and $\Phi_x(t) = 1$ if x is a triangle, and it is present at t. Therefore, $\mathbb{E}\Phi_2(t) = 0$ and $\mathbb{E}\Phi_3(t) = 1 - L_3(t)$. Conditions $(i) - (ii) - (iii)$ and (v) of Section 8 are satisfied. Now,

$$\int_0^\infty e^{-\alpha t}\mathbb{E}\Phi_3(t)dt = \int_0^\infty e^{-\alpha t}(1 - L_3(t))dt = B(\alpha).$$

Thus, (69) and (70) imply (36). □

5. Generating Functions and the Probability of Extinction

The joint generating function of $\Pi_2(\lambda_2)$, $\xi_{22}(\lambda_2)$ and $\xi_{23}(\lambda_2)$. Recall that Π_2 is the Poisson process describing the reproduction times of the generic edge and λ_2 is its life length. Thus,

$$w_{i,j,k} = \mathbb{P}(\Pi_2(\lambda_2) = i, \xi_{22}(\lambda_2) = j, \xi_{23}(\lambda_2) = k)$$

is the joint distribution of the offspring size of the generic edge during its whole life and its last reproduction time. We have

$$w_{i,j,k} = \mathbb{P}(\tau_i \leq \lambda_2 < \tau_{i+1}, \xi_{22}(\tau_i) = j, \xi_{23}(\tau_i) = k),$$

where τ_i is the ith jumping time of the Poisson process Π_2. Thus, it again shows that $w_{i,j,k}$ is the probability that the ith birth event is the last one that occurred before death, and the total numbers of the two types of offspring up to time τ_i are equal to j and k, respectively.

Now, consider the sequence

$$u_{i,j,k} = \mathbb{P}(\tau_i \leq \lambda_2, \xi_{22}(\tau_i) = j, \xi_{23}(\tau_i) = k).$$

Let $\xi_2(\tau_{i-1}) = m$ and assume for a while that τ_i and τ_{i-1} are fixed. Then, using (4) and (5) for the hazard rate, we can calculate that, for fixed τ_i and τ_{i-1},

$$\mathbb{P}(\lambda_2 \geq \tau_i | \lambda_2 \geq \tau_{i-1}) = \exp(-(b + cm)(\tau_i - \tau_{i-1})).$$

We know that the increment $(\tau_i - \tau_{i-1})$ is exponential with parameter 1; therefore,

$$\mathbb{P}(\lambda_2 \geq \tau_i | \lambda_2 \geq \tau_{i-1}) = \mathbb{E}_{\tau_i - \tau_{i-1}} \exp(-(b + cm)(\tau_i - \tau_{i-1})) = \frac{1}{1 + b + cm}. \quad (41)$$

At each birth step, the new individual can be either an edge or a triangle. Therefore, using the above calculations, the total probability theorem, and the independence of the type of the newly born individual and (Π_2, λ_2), we have the following recursion for $u_{i,j,k}$.

$$u_{i,j,k} = u_{i-1,j-1,k}\frac{r_1}{1+b+c(j+k-1)} + u_{i-1,j,k-1}\frac{r_2}{1+b+c(j+k-1)}. \tag{42}$$

Now, by the definition of $w_{i,j,k}$, we can see that

$$\begin{aligned} w_{i,j,k} &= \mathbb{P}(\tau_i \le \lambda_2 < \tau_{i+1}, \xi_{22}(\tau_i) = j, \xi_{23}(\tau_i) = k) = \\ &= \mathbb{P}(\lambda_2 < \tau_{i+1} | \tau_i \le \lambda_2, \xi_{22}(\tau_i) = j, \xi_{23}(\tau_i) = k)\mathbb{P}(\tau_i \le \lambda_2, \xi_{22}(\tau_i) = j, \xi_{23}(\tau_i) = k) = \\ &= \frac{b+c(j+k)}{1+b+c(j+k)} u_{i,j,k}, \end{aligned}$$

where by (41), $\frac{b+c(j+k)}{1+b+c(j+k)}$ is the probability that the generic individual dies before the next birth event.

Let $v_{i,j,k} = \dfrac{w_{i,j,k}}{b+c(j+k)} = \dfrac{u_{i,j,k}}{1+b+c(j+k)}$. Then, from (42), we obtain the following recursion for the sequence $v_{i,j,k}$

$$(1+b+c(j+k))v_{i,j,k} = v_{i-1,j-1,k}r_1 + v_{i-1,j,k-1}r_2, \tag{43}$$

where the initial values are

$$v_{0,0,0} = \frac{1}{1+b} \text{ and } v_{0,j,k} = 0 \text{ for } j \ne 0 \text{ or } k \ne 0. \tag{44}$$

Now, we calculate the generating function $G(x,y,z)$ of the sequence $v_{i,j,k}$. We have

$$G(x,y,z) = \sum_{i=0}^{\infty}\sum_{j=0}^{\infty}\sum_{k=0}^{\infty} v_{i,j,k}x^i y^j z^k.$$

First, multiplying with $x^i y^j z^k$ and then taking the sum of both sides of (43), we obtain

$$\begin{aligned} &\sum_{i=1}^{\infty}\sum_{j=0}^{\infty}\sum_{k=0}^{\infty} v_{i,j,k}x^i y^j z^k (1+b+cj+ck) = \\ &= r_1 xy \sum_{i=1}^{\infty}\sum_{j=0}^{\infty}\sum_{k=0}^{\infty} v_{i-1,j-1,k}x^{i-1}y^{j-1}z^k + r_2 xz \sum_{i=1}^{\infty}\sum_{j=0}^{\infty}\sum_{k=0}^{\infty} v_{i-1,j,k-1}x^{i-1}y^j z^{k-1}, \end{aligned}$$

where $v_{0,j,k}, j = 0,1,\dots$ is given by (44), and we define $v_{i,j,k} = 0$ if $j < 0$ or $k < 0$. From this equation, we find

$$\begin{aligned} (1+b)\left(G(x,y,z) - \frac{1}{1+b}\right) + ycG_y'(x,y,z) + zcG_z'(x,y,z) = \\ = r_1 xyG(x,y,z) + r_2 xzG(x,y,z). \end{aligned} \tag{45}$$

Let $h(t) = G(x,ty,tz)$. Now, substituting y with ty, z with tz in (45), we can obtain the following linear differential equation.

$$h'(t) + h(t)\left(\frac{1+b}{ct} - \frac{r_1 xy + r_2 xz}{c}\right) = \frac{1}{ct} \tag{46}$$

with the initial value condition

$$h(0) = \frac{1}{1+b}. \tag{47}$$

Now, we can use the well-known method for linear differential equations. We obtain that the solution of the initial value problem (46) and (47) is

$$h(t) = t^{-\frac{1+b}{c}} e^{\frac{r_1 xy + r_2 xz}{c} t} \frac{1}{c} \int_0^t s^{\frac{1+b-c}{c}} e^{-\frac{r_1 xy + r_2 xz}{c} s} ds.$$

With $t = 1$, we obtain that

$$G(x, y, z) = h(1) = \frac{1}{c} \int_0^1 s^{\frac{1+b-c}{c}} e^{\frac{r_1 xy + r_2 xz}{c}(1-s)} ds.$$

We need the generating function of $w_{i,j,k} = v_{i,j,k}(b + c(j + k))$. It is

$$H(x, y, z) = \sum_{i=0}^{\infty} \sum_{j=0}^{\infty} \sum_{k=0}^{\infty} v_{i,j,k}(b + c(j + k)) x^i y^j z^k =$$

$$= bG(x, y, z) + cyG'_y(x, y, z) + czG'_z(x, y, z). \tag{48}$$

From here, we obtain

Proposition 2. *The joint generating function of $\Pi_2(\lambda_2)$, $\xi_{22}(\lambda_2)$ and $\xi_{23}(\lambda_2)$ is*

$$H(x, y, z) =$$

$$e^{\frac{r_1 xy + r_2 xz}{c}} \frac{1}{c} \int_0^1 s^{\frac{1+b-c}{c}} e^{-\frac{r_1 xy + r_2 xz}{c} s} [b + (r_1 xy + r_2 xz)(1 - s)] ds, \tag{49}$$

where $-1 \le x, y, z \le 1$.

Corollary 2. *The generating function of the total offspring distribution of the generic edge is*

$$f_2(y, z) = H(1, y, z) = e^{\frac{r_1 y + r_2 z}{c}} \frac{1}{c} \int_0^1 s^{\frac{1+b-c}{c}} e^{-\frac{r_1 y + r_2 z}{c} s} [b + (r_1 y + r_2 z)(1 - s)] ds. \tag{50}$$

The joint generating function of $\Pi_3(\lambda_3)$, $\xi_{32}(\lambda_3)$ and $\xi_{33}(\lambda_3)$. Here, we study the offspring of a triangle. To distinguish the notation of this subsection and the previous subsection, but avoid too many subscripts, we use bar. Thus, here $\overline{w}_{i,j,k}$, $\overline{u}_{i,j,k}$, $\overline{v}_{i,j,k}$, $\overline{G}(x, y, z)$ and $\overline{H}(x, y, z)$ denote quantities relating offspring of the generic triangle. Recall that Π_3 is the Poisson process describing the reproduction times of the generic triangle and λ_3 is the life length of the triangle. Thus,

$$\overline{w}_{i,j,k} = \mathbb{P}(\Pi_3(\lambda_3) = i, \xi_{32}(\lambda_3) = j, \xi_{33}(\lambda_3) = k)$$

is the joint distribution of the offspring size of the generic triangle during its whole life and its last reproduction time. We have

$$\overline{w}_{i,j,k} = \mathbb{P}(\tau_i \le \lambda_3 < \tau_{i+1}, \xi_{32}(\tau_i) = j, \xi_{33}(\tau_i) = k),$$

where τ_i is the ith jumping time of the Poisson process Π_3. Thus, we again show that $\overline{w}_{i,j,k}$ is the probability that the ith birth event is the last one that happened before death, and the total numbers of the two types of offspring up to time τ_i are equal to j and k, respectively.

Let

$$\overline{u}_{i,j,k} = \mathbb{P}(\tau_i \le \lambda_3, \xi_{32}(\tau_i) = j, \xi_{33}(\tau_i) = k).$$

Let $\xi_3(\tau_{i-1}) = m$, and assume for a while that τ_i and τ_{i-1} are fixed. Then, using (4) and (5) for the hazard rate, we can calculate that, for fixed τ_i and τ_{i-1},

$$\mathbb{P}(\lambda_3 \ge \tau_i | \lambda_3 \ge \tau_{i-1}) = \exp(-(b + cm)(\tau_i - \tau_{i-1})).$$

We know that the increment $(\tau_i - \tau_{i-1})$ is exponential with parameter 1; therefore,

$$\mathbb{P}(\lambda_3 \geq \tau_i | \lambda_3 \geq \tau_{i-1}) = \mathbb{E}_{\tau_i - \tau_{i-1}} \exp(-(b+cm)(\tau_i - \tau_{i-1})) = \frac{1}{1+b+cm}. \quad (51)$$

At each birth step, the new individual can be either an edge or a triangle. Therefore, using the above calculations, the total probability theorem, and the independence of the type of the newly born individual and (Π_3, λ_3), we have the following recursion for $\overline{u}_{i,j,k}$.

$$\overline{u}_{i,j,k} = \overline{u}_{i-1,j-1,k} \frac{p_1}{1+b+c(j+k-1)} + \\ + \overline{u}_{i-1,j,k-1} \frac{p_2}{1+b+c(j+k-1)} + \overline{u}_{i-1,j,k-3} \frac{p_3}{1+b+c(j+k-3)}. \quad (52)$$

Now, by the definition of $\overline{w}_{i,j,k}$, we can see that

$$\overline{w}_{i,j,k} = \mathbb{P}(\tau_i \leq \lambda_3 < \tau_{i+1}, \xi_{32}(\tau_i) = j, \xi_{33}(\tau_i) = k) = \\ = \mathbb{P}(\lambda_3 < \tau_{i+1} | \tau_i \leq \lambda_3, \xi_{32}(\tau_i) = j, \xi_{33}(\tau_i) = k) \mathbb{P}(\tau_i \leq \lambda_3, \xi_{32}(\tau_i) = j, \xi_{33}(\tau_i) = k) = \\ = \frac{b+c(j+k)}{1+b+c(j+k)} \overline{u}_{i,j,k},$$

where by (51), $\frac{b+c(j+k)}{1+b+c(j+k)}$ is the probability that the generic individual dies before the next birth event.

Now, let $\overline{v}_{i,j,k} = \dfrac{\overline{w}_{i,j,k}}{b+c(j+k)} = \dfrac{\overline{u}_{i,j,k}}{1+b+c(j+k)}$. Then, from (52), we obtain the following recursion for the sequence $\overline{v}_{i,j,k}$

$$(1+b+c(j+k))\overline{v}_{i,j,k} = \overline{v}_{i-1,j-1,k} p_1 + \overline{v}_{i-1,j,k-1} p_2 + \overline{v}_{i-1,j,k-3} p_3, \quad (53)$$

where the initial values are

$$\overline{v}_{0,0,0} = \frac{1}{1+b} \text{ and } \overline{v}_{0,j,k} = 0 \text{ for } j \neq 0 \text{ or } k \neq 0. \quad (54)$$

Now, we calculate the generating function $\overline{G}(x,y,z)$ of the sequence $\overline{v}_{i,j,k}$. We have

$$\overline{G}(x,y,z) = \sum_{i=0}^{\infty} \sum_{j=0}^{\infty} \sum_{k=0}^{\infty} \overline{v}_{i,j,k} x^i y^j z^k.$$

First, multiplying with $x^i y^j z^k$ and then taking the sum of both sides of (53), we obtain

$$\sum_{i=1}^{\infty} \sum_{j=0}^{\infty} \sum_{k=0}^{\infty} \overline{v}_{i,j,k} x^i y^j z^k (1+b+cj+ck) = p_1 xy \sum_{i=1}^{\infty} \sum_{j=0}^{\infty} \sum_{k=0}^{\infty} \overline{v}_{i-1,j-1,k} x^{i-1} y^{j-1} z^k + \\ + p_2 xz \sum_{i=1}^{\infty} \sum_{j=0}^{\infty} \sum_{k=0}^{\infty} \overline{v}_{i-1,j,k-1} x^{i-1} y^j z^{k-1} + p_3 xz^3 \sum_{i=1}^{\infty} \sum_{j=0}^{\infty} \sum_{k=0}^{\infty} \overline{v}_{i-1,j,k-3} x^{i-1} y^j z^{k-3},$$

where $\overline{v}_{0,j,k}, j = 0, 1, \ldots$ is given by (54) and we define $\overline{v}_{i,j,k} = 0$ if $j < 0$ or $k < 0$. From this equation, we find

$$(1+b)\left(\overline{G}(x,y,z) - \frac{1}{1+b}\right) + yc\overline{G}_y'(x,y,z) + zc\overline{G}_z'(x,y,z) = \\ = p_1 xy \overline{G}(x,y,z) + p_2 xz \overline{G}(x,y,z) + p_3 xz^3 \overline{G}(x,y,z). \quad (55)$$

Let $\overline{h}(t) = \overline{G}(x, ty, tz)$. Now, substituting y with ty, z with tz in (55), we can obtain the following linear differential equation.

$$\overline{h}'(t) + \overline{h}(t)\left(\frac{1+b}{ct} - \frac{p_1xy + p_2xz + p_3xz^3t^2}{c}\right) = \frac{1}{ct} \tag{56}$$

with the initial value condition

$$\overline{h}(0) = \frac{1}{1+b}. \tag{57}$$

One can see that the solution of the initial value problem (56) and (57) is

$$\overline{h}(t) = t^{-\frac{1+b}{c}} e^{\frac{p_1xy+p_2xz}{c}t + \frac{p_3xz^3}{3c}t^3} \frac{1}{c}\int_0^t s^{\frac{1+b-c}{c}} e^{-\frac{p_1xy+p_2xz}{c}s - \frac{p_3xz^3}{3c}s^3} ds.$$

With $t = 1$, we obtain that

$$\overline{G}(x,y,z) = \overline{h}(1) = \frac{1}{c}\int_0^1 s^{\frac{1+b-c}{c}} e^{\frac{p_1xy+p_2xz}{c}(1-s) + \frac{p_3xz^3}{3c}(1-s^3)} ds.$$

Therefore, the generating function of $\overline{w}_{i,j,k} = \overline{v}_{i,j,k}(b + c(j+k))$ is

$$\overline{H}(x,y,z) = \sum_{i=0}^{\infty}\sum_{j=0}^{\infty}\sum_{k=0}^{\infty} \overline{v}_{i,j,k}(b + c(j+k))x^iy^jz^k =$$

$$= b\overline{G}(x,y,z) + cy\overline{G}_y'(x,y,z) + cz\overline{G}_z'(x,y,z). \tag{58}$$

From here, we obtain

Proposition 3. *The joint generating function of $\Pi_3(\lambda_3)$, $\xi_{32}(\lambda_3)$ and $\xi_{33}(\lambda_3)$ is*

$$\overline{H}(x,y,z) = \tag{59}$$

$$= \frac{1}{c}\int_0^1 s^{\frac{1+b-c}{c}} e^{\frac{p_1xy+p_2xz}{c}(1-s) + \frac{p_3xz^3}{3c}(1-s^3)} \left[b + (p_1xy + p_2xz)(1-s) + p_3xz^3(1-s^3)\right] ds,$$

where $-1 \le x, y, z \le 1$.

Corollary 3. *The generating function of the total offspring distribution of the generic triangle is*

$$f_3(y,z) = \overline{H}(1,y,z) = \tag{60}$$

$$= e^{\frac{p_1y+p_2z}{c} + \frac{p_3z^3}{3c}} \frac{1}{c}\int_0^1 s^{\frac{1+b-c}{c}} e^{-\frac{p_1y+p_2z}{c}s - \frac{p_3z^3}{3c}s^3} \left[b + (p_1y + p_2z)(1-s) + p_3z^3(1-s^3)\right] ds.$$

The probability of extinction. In Theorem 3, we give the probability of extinction. To determine the extinction probability of the process, we consider the well-known embedded multi type Galton–Watson process. At time $t = 0$, the 0th generation of the Galton–Watson process consists of a single individual, i.e., the ancestor. The first generation consists of all offspring of the ancestor. The offspring of the individuals of the nth generation form the $(n+1)$th generation. Under some assumptions, the extinction of our original process has the same probability as the extinction of this embedded Galton–Watson process. The reproduction process $\xi_{i,j}(t)$ gives the number of type j offspring of an ancestor of type i up to time t. With $t \to \infty$, we obtain that the total number of offspring is $\xi_{i,j}(\infty)$. Therefore, Corollary 1 gives us the 2×2 matrix of the expected total offspring number as

$$\mathbb{M} = \left(m_{i,j}(\infty)\right)_{i,j=2}^{3}.$$

Actually, $m_{i,j}(\infty)$ is the expected offspring number of the embedded Galton–Watson process.

Let s_2 and s_3 denote the probability of extinction of our process when the ancestor is an edge, resp. triangle.

Theorem 3. *Assume that $0 \le r_1 < 1$, $0 < p_1 \le 1$ and it is excluded that both $r_1 = 0$ and $p_1 = 1$ are satisfied at the same time. Let ϱ be the Perron–Frobenius root of $\mathbb{M}$. If $\varrho \le 1$, then $s_2 = s_3 = 1$. If $\varrho > 1$, then $s_2 < 1$ and $s_3 < 1$. In any case, (s_2, s_3) is the smallest non-negative solution of the vector equation*

$$(s_2, s_3) = (f_2(s_2, s_3), f_3(s_2, s_3)),$$

where f_2 and f_3 are given in Corollaries 2 and 3.

Proof. We apply Theorem 7.1 in Chapter 1 of [19]. By Corollary 1, $m_{i,j}(0) = 0$ and $m_{i,j}(t)$ is finite for any i, j. Therefore, by Theorem 7.1 in Chapter 3 of [19], the extinction of our original process has the same probability as the extinction of the embedded Galton–Watson process. Thus, we can apply Theorem 7.1 in Chapter 1 of [19]. Here, $\mathbb{M}$ is the matrix of the expected offspring numbers of the embedded Galton–Watson process. Now, $\mathbb{M}$ is positively regular because we assume that $0 \le r_1 < 1$, $0 < p_1 \le 1$ and it is excluded, that both $r_1 = 0$ and $p_1 = 1$ are satisfied at the same time. Thus, our result follows from Theorem 7.1 in Chapter 1 of [19]. □

6. The Asymptotic Behaviour of the Degree of a Fixed Vertex

The process of the 'good children'. To describe the degree of a fixed vertex, we introduce a new branching process that we call the process of 'good children'. This process contains those objects that contribute to the degree of the fixed vertex. We can see that a newly born vertex can have 1 or 2 edges if its parent is an edge object and 1, 2 or 3 edges if its parent is a triangle object.

First, we consider the case when the newly born vertex has one edge, and thus, at the beginning, it belongs to an edge object. In this paragraph, we call this edge the 'parent' edge. We fix the newly born vertex. Then, we distinguish those children objects of the 'parent' edge, which contribute to the degree of our fixed vertex. We call a child object of the 'parent' edge a 'good child' if it contains our fixed vertex. We can see that only the 'good children' and their 'good children' offspring can contribute to the degree of the fixed vertex. Then, the distribution of the number of 'good children' at a reproduction event of the 'parent' edge is

$$\mathbb{P}(\tilde{\varepsilon}_{22} = 0) = 1 - \frac{1}{2}r_1, \quad \mathbb{P}(\tilde{\varepsilon}_{22} = 1) = \frac{1}{2}r_1, \quad \mathbb{P}(\tilde{\varepsilon}_{23} = 0) = 1 - r_2, \quad \mathbb{P}(\tilde{\varepsilon}_{23} = 1) = r_2,$$

where $\tilde{\varepsilon}_{22}$ denotes the number of edge type 'good children' and $\tilde{\varepsilon}_{23}$ denotes the triangle type 'good children'. We have to consider the reproduction process of the 'good child', which is the following

$$\tilde{\xi}_{2,2}(t) = \tilde{\varepsilon}_{22}(1) + \tilde{\varepsilon}_{22}(2) + \cdots + \tilde{\varepsilon}_{22}(\Pi(t \wedge \lambda_2)), \tag{61}$$

$$\tilde{\xi}_{2,3}(t) = \tilde{\varepsilon}_{23}(1) + \tilde{\varepsilon}_{23}(2) + \cdots + \tilde{\varepsilon}_{23}(\Pi(t \wedge \lambda_2)), \tag{62}$$

where $\tilde{\xi}_{2,2}(t)$ denotes the number of all edge type 'good children', and $\tilde{\xi}_{2,3}(t)$ denotes the number of all triangle type 'good children' born by the 'parent' edge, $\tilde{\varepsilon}_{22}(1), \tilde{\varepsilon}_{22}(2), \ldots$ are i.i.d. copies of $\tilde{\varepsilon}_{22}$ and $\tilde{\varepsilon}_{23}(1), \tilde{\varepsilon}_{23}(2), \ldots$ are i.i.d. copies of $\tilde{\varepsilon}_{23}$. Using Corollary 1, we see that the mean values of the number of edge type and triangle type 'good children' are

$$\tilde{m}_{2,2}(t) = \mathbb{E}\tilde{\xi}_{2,2}(t) = \mathbb{E}(\tilde{\varepsilon}_{22})\mathbb{E}(\Pi(t \wedge \lambda_2)) = \frac{1}{2}r_1 F(t) = \frac{1}{2}m_{2,2}(t),$$

$$\tilde{m}_{2,3}(t) = \mathbb{E}\tilde{\xi}_{2,3}(t) = \mathbb{E}(\tilde{\varepsilon}_{23})\mathbb{E}(\Pi(t \wedge \lambda_2)) = r_2 F(t) = m_{2,3}(t).$$

Now, consider the second case where the newly born vertex has two edges, and thus the 'parent' object is a single triangle. Let $\tilde{\varepsilon}_{32}$ and $\tilde{\varepsilon}_{33}$ denote the number of edge, resp. triangle type 'good children' of the 'parent' triangle. The distribution of the number of 'good children' will be the following

$$\mathbb{P}(\tilde{\varepsilon}_{32}=0)=1-\frac{1}{3}p_1,\quad \mathbb{P}(\tilde{\varepsilon}_{32}=1)=\frac{1}{3}p_1,$$

$$\mathbb{P}(\tilde{\varepsilon}_{33}=0)=1-\frac{2}{3}p_2-p_3,\quad \mathbb{P}(\tilde{\varepsilon}_{33}=1)=\frac{2}{3}p_2,\quad \mathbb{P}(\tilde{\varepsilon}_{33}=2)=p_3.$$

Let $\tilde{\xi}_{3,2}(t)$ denote the number of all edge type 'good children', and $\tilde{\xi}_{3,3}(t)$ denote the number of all triangle type 'good children' born by the 'parent' triangle. We obtain from Corollary 1 that

$$\tilde{m}_{3,2}(t)=\mathbb{E}\tilde{\xi}_{3,2}(t)=\mathbb{E}(\tilde{\varepsilon}_{32})\mathbb{E}(\Pi(t\wedge\lambda_3))=\frac{1}{3}p_1G(t)=\frac{1}{3}m_{3,2}(t),$$

$$\tilde{m}_{3,3}(t)=\mathbb{E}\tilde{\xi}_{3,3}(t)=\mathbb{E}(\tilde{\varepsilon}_{33})\mathbb{E}(\Pi(t\wedge\lambda_3))=\frac{2}{3}(p_2+3p_3)G(t)=\frac{2}{3}m_{3,3}(t).$$

Therefore, from Proposition 1, it is easily seen that the Laplace transforms of the average number of offspring are

$$\tilde{m}^*_{2,2}(\kappa)=\frac{1}{2}r_1A(\kappa),\quad \tilde{m}^*_{2,3}(\kappa)=r_2A(\kappa),\quad \tilde{m}^*_{3,2}(\kappa)=\frac{1}{3}p_1B(\kappa),\quad \tilde{m}^*_{3,3}(\kappa)=\frac{2}{3}(p_2+3p_3)B(\kappa).$$

Let

$$\tilde{M}(\kappa)=\begin{pmatrix}\tilde{m}^*_{2,2}(\kappa) & \tilde{m}^*_{2,3}(\kappa)\\ \tilde{m}^*_{3,2}(\kappa) & \tilde{m}^*_{3,3}(\kappa)\end{pmatrix}$$

be the matrix of the previous Laplace transforms. The Perron root that is the largest eigenvalue of $\tilde{M}(\kappa)$ is

$$\tilde{\varrho}(\kappa)=\frac{\frac{2}{3}(p_2+3p_3)B(\kappa)+\frac{1}{2}r_1A(\kappa)+\sqrt{\left(\frac{2}{3}(p_2+3p_3)B(\kappa)-\frac{1}{2}r_1A(\kappa)\right)^2+\frac{4}{3}p_1B(\kappa)r_2A(\kappa)}}{2}. \tag{63}$$

In the following, we assume supercriticality of the 'good children' process; that is, we suppose that $\tilde{\varrho}(0)>1$. We can see that the reproduction process of the 'good children' is supercritical if

$$\max\left\{\frac{1}{2}r_1A(0),\frac{2}{3}(p_2+3p_3)B(0)\right\}>1.$$

We assume the existence of finite and positive Malthusian parameter of the 'good children' process. Thus, let $\tilde{\alpha}$ be the Malthusian parameter; it satisfies equation $\tilde{\varrho}(\tilde{\alpha})=1$. From this equation and from (63), we see that $\tilde{\alpha}$ is the solution of

$$\frac{1}{3}(r_1(p_2+3p_3)-r_2p_1)A(\tilde{\alpha})B(\tilde{\alpha})-\frac{1}{2}r_1A(\tilde{\alpha})-\frac{2}{3}(p_2+3p_3)B(\tilde{\alpha})+1=0. \tag{64}$$

Let $(\tilde{v}_2,\tilde{v}_3)^\top$ denote the right eigenvector of $\tilde{M}(\tilde{\alpha})$ corresponding to the eigenvalue 1, and let $(\tilde{u}_2,\tilde{u}_3)^\top$ be the left eigenvector with the conditions $\tilde{v}_2+\tilde{v}_3=1$ and $\tilde{v}_2\tilde{u}_2+\tilde{v}_3\tilde{u}_3=1$. Direct calculations show that

$$\tilde{v}_2=\frac{(1-r_1)A(\tilde{\alpha})}{(1-\frac{3}{2}r_1)A(\tilde{\alpha})+1},\qquad \tilde{v}_3=\frac{1-\frac{1}{2}r_1A(\tilde{\alpha})}{(1-\frac{3}{2}r_1)A(\tilde{\alpha})+1},$$

$$\tilde{u}_2=\frac{((1-\frac{3}{2}r_1)A(\tilde{\alpha})+1)\frac{1}{3}p_1B(\tilde{\alpha})}{\frac{1}{3}r_2A(\tilde{\alpha})p_1B(\tilde{\alpha})+\left(\frac{1}{2}r_1A(\tilde{\alpha})-1\right)^2},$$

$$\tilde{u}_3 = \frac{((\frac{3}{2}r_1 - 1)A(\tilde{\alpha}) - 1)\left(\frac{1}{2}r_1 A(\tilde{\alpha}) - 1\right)}{\frac{1}{3}r_2 A(\tilde{\alpha})p_1 B(\tilde{\alpha}) + \left(\frac{1}{2}r_1 A(\tilde{\alpha}) - 1\right)^2}.$$

Limit results for the degree. We have already mentioned that the 'good children' and only they can contribute to the degree of the fixed vertex. Thus, its degree is equal to the initial degree plus the number of 'good children'. Let ${}_2\tilde{C}(t)$ be the degree of a fixed vertex at time t after its birth in the case when the vertex belongs to an edge at its birth. Similarly, ${}_3\tilde{C}(t)$ is its degree in the case when the vertex belongs to triangle at its birth. Up to an additive constant, ${}_i\tilde{C}(t)$ is the number of 'good children' offspring of an i type 'parent' object at time t. It is the sum of the number of edge type 'good children' ${}_i\tilde{E}(t)$ and the triangle type 'good children' ${}_i\tilde{T}(t)$. To apply Proposition 4, we can use the same method as in Theorem 2. Thus, for the edges, we can again use the random characteristic $\Phi_x(t) = 1$ if x is an edge and $\Phi_x(t) = 0$ if x is a triangle, but the underlying process is the process of 'good children'. This is similar for triangles.

Therefore, we have almost surely

$$\lim_{t\to\infty} e^{-\tilde{\alpha}t}{}_i\tilde{C}(t) = \lim_{t\to\infty} e^{-\tilde{\alpha}t}\left({}_i\tilde{E}(t) + {}_i\tilde{T}(t)\right) = {}_i\tilde{W}\frac{\tilde{v}_i(\tilde{u}_2 + \tilde{u}_3)}{\tilde{\alpha}\tilde{D}(\tilde{\alpha})},$$

for $i = 2, 3$, where ${}_2\tilde{W}$ and ${}_3\tilde{W}$ are positive on the event of non-extinction of the 'good children'.

The last case is when the newly born vertex has three edges. Then, three triangles contribute to the degree of that vertex. Let ${}_3\tilde{\tilde{C}}(t)$ be the degree of this vertex. Then, ${}_3\tilde{\tilde{C}}(t)$ is the sum if 'good' offspring of three triangles. Thus, almost surely,

$$\lim_{t\to\infty} e^{-\tilde{\alpha}t}{}_3\tilde{\tilde{C}}(t) = ({}_3\tilde{W}_1 + {}_3\tilde{W}_2 + {}_3\tilde{W}_3)\frac{\tilde{v}_3(\tilde{u}_2 + \tilde{u}_3)}{\tilde{\alpha}\tilde{D}(\tilde{\alpha})},$$

where ${}_3\tilde{W}_1, {}_3\tilde{W}_2, {}_3\tilde{W}_3$ are independent copies of ${}_3\tilde{W}$.

Checking the conditions of Proposition 4 for the 'good children' process. To complete the previous reasoning, we should check the conditions of Proposition 4. First, we find the the denominator in the limit theorem that is we calculate $\tilde{D}$. By Section 8, we see that

$$\tilde{D}(\tilde{\alpha}) = \sum_{l,j=2}^{3} \tilde{u}_l \tilde{v}_j \left(-\tilde{m}^*_{l,j}(\tilde{\alpha})\right)'. \tag{65}$$

Here, $\tilde{u}_i$ and $\tilde{v}_i$ are the eigenvectors. Moreover,

$$\left(-\tilde{m}^*_{2,2}(\tilde{\alpha})\right)' = \tfrac{r_1}{2}(-A'(\tilde{\alpha})), \qquad (-\tilde{m}^*_{2,3}(\tilde{\alpha}))' = r_2(-A'(\tilde{\alpha})), \tag{66}$$

$$\left(-\tilde{m}^*_{3,2}(\tilde{\alpha})\right)' = \tfrac{p_1}{3}(-B'(\tilde{\alpha})), \qquad (-\tilde{m}^*_{3,3}(\tilde{\alpha}))' = \frac{2}{3}(p_2 + 3p_3)(-B'(\tilde{\alpha})), \tag{67}$$

where $\tilde{\alpha}$ is the Malthusian parameter in the process of 'good children' and A', B' denotes the derivatives given in (31) and (32).

Condition (a) of Proposition 4 is true because the measures $\tilde{m}_{i,j}$ are non-lattice as they are absolutely continuous. For condition (b1), we assume the existence of a positive Malthusian parameter. That is, we assume that (64) has a finite and positive solution $\tilde{\alpha}$. Condition (b2) is true, because we assume that $\tilde{\varrho}(0) > 1$. Condition (c) is a consequence of Section 4, because $\tilde{m}_{i,j}(t)$ has shape $cm_{i,j}$, where c is positive number.

To guarantee condition (d), in this section, we assume that $0 \le r_1 < 1, 0 < p_1 \le 1$, and it is excluded that both $r_1 = 0$ and $p_1 = 1$ are satisfied at the same time. Conditions (i)-(ii)-(iii) and (v) are true because of the shape of Φ. Conditions (iv) and (vi) are consequences of $\tilde{\xi}_{i,j}(t) \le \xi_{i,j}(t)$ as one can see from the proof of Theorem 2.

The extinction of the degree process. The extinction of the degree process means that the degree of the vertex does not increase after a certain time, that is, the reproduction process of the 'good children' dies out. The probability of this kind of extinction is the smallest non-negative root $(\tilde{s}_2, \tilde{s}_3)$ of the equation

$$(\tilde{s}_2, \tilde{s}_3) = \left(\tilde{f}_2(\tilde{s}_2, \tilde{s}_3), \tilde{f}_3(\tilde{s}_2, \tilde{s}_3)\right),$$

where $\tilde{f}_2$ and $\tilde{f}_3$ are the generating functions of the total 'good children' distribution of an edge, resp. a triangle. Now, by (61) and (62),

$$\tilde{f}_2(y,z) = h_{\Pi_2(\lambda_2)}\left(h_{\tilde{\varepsilon}_{2,2},\tilde{\varepsilon}_{2,3}}(y,z)\right),$$

where $h_{\Pi_2(\lambda_2)}$ is the generating function of $\Pi_2(\lambda_2)$, and $h_{\tilde{\varepsilon}_{2,2},\tilde{\varepsilon}_{2,3}}$ is the joint generating function of $\tilde{\varepsilon}_{2,2}$ and $\tilde{\varepsilon}_{2,3}$. Here, by (49),

$$h_{\Pi_2(\lambda_2)}(x) = H(x,1,1) = \frac{1}{c}\int_0^1 s^{\frac{1+b-c}{c}} e^{\frac{(r_1+r_2)x}{c}(1-s)}[b + (r_1+r_2)x(1-s)]ds.$$

By direct calculation,

$$h_{\tilde{\varepsilon}_{2,2},\tilde{\varepsilon}_{2,3}}(y,z) = \frac{1}{2}r_1 + \frac{1}{2}r_1 y + r_2 z.$$

Similarly,

$$\tilde{f}_3(y,z) = h_{\Pi_3(\lambda_3)}\left(h_{\tilde{\varepsilon}_{3,2},\tilde{\varepsilon}_{3,3}}(y,z)\right),$$

where by (59), the generating function of $\Pi_3(\lambda_3)$ is

$$h_{\Pi_3(\lambda_3)}(x) = \overline{H}(x,1,1) = \frac{1}{c}\int_0^1 s^{\frac{1+b-c}{c}} e^{\frac{(p_1+p_2)x}{c}(1-s)+\frac{p_3 x}{3c}(1-s^3)}\left[b + (p_1 x + p_2 x)(1-s) + p_3 x(1-s^3)\right]ds.$$

Moreover, the joint generating function of $\tilde{\varepsilon}_{3,2}$ and $\tilde{\varepsilon}_{3,3}$ is

$$h_{\tilde{\varepsilon}_{3,2},\tilde{\varepsilon}_{3,3}}(y,z) = \frac{2}{3}p_1 + \frac{1}{3}p_2 + \frac{1}{3}p_1 y + \frac{2}{3}p_2 z + p_3 z^2.$$

7. Simulations

In this section, we provide some empirical results for our asymptotic theorems. We generated our process in the programming language Julia. We needed an environment, where the priority queues were highly applicable. Using this structure, the running time was reasonable. A more detailed explanation of the algorithm can be found in [20].

According to Theorem 2, for large t, the graphs of the numbers of edges and triangles are approximately straight lines on the logarithmic scale. To obtain empirical evidence of our Theorem 2, we investigated the slope of the simulated number of edges and triangles being born and being present up to time t on the logarithmic scale. The initial instability of the single processes (Figure 3) motivated us to exclude the first few observations from the calculations, but the lack of them was not relevant, because the asymptotic properties can be observed in the later stage of the processes.

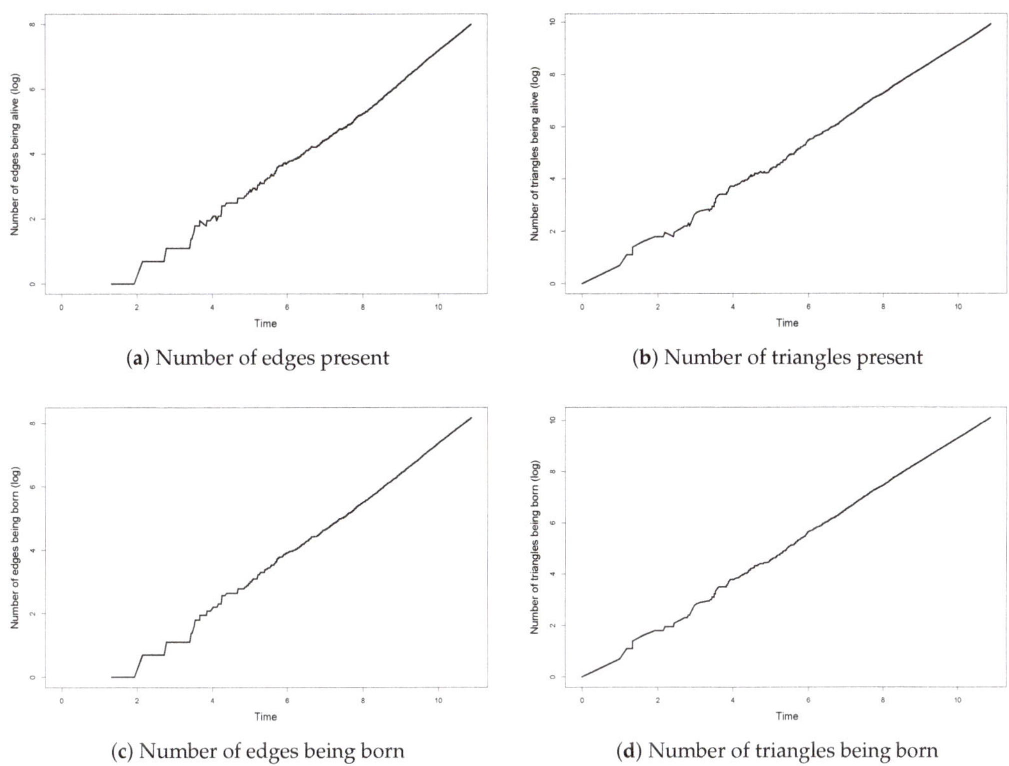

(**a**) Number of edges present

(**b**) Number of triangles present

(**c**) Number of edges being born

(**d**) Number of triangles being born

Figure 3. Measurements of a single process on a logarithmic scale.

For each parameter set, we stored the mentioned measurements only in integer time steps, and then we took the average of 100 simulated processes. In Figure 4, an example is shown for a specific parameter set ($r_1 = 0.1$, $p_1 = 0.2$, $p_3 = 0.6$, $b = 0.25$, $c = 0.25$). The values of the averages are plotted by dots. In each case, we fitted a regression line (plotted by continuous red line) to the last 9 values. We can see that the fit is perfect, thus, supporting our theorem.

Our main goal was to obtain a 95% confidence interval for the slope of the linear regression line, as that was our simulated approximation of the Malthusian parameter α. Table 1 contains the boundaries of the 95% confidence intervals for α. The columns labelled with 2.5% and 97.5% refer to the lower and the upper bounds obtained from simulations, while the column of $\hat{\alpha}$ refers to the numerical solution of Equation (25).

For each fixed parameter set $\{r_1, p_1, p_2, b, c\}$, we present the confidence intervals calculated from the number of edges being born (E) resp. being present ($\tilde{E}$) and from the number of triangles being born (T) resp. being present ($\tilde{T}$) up to time $t = 14$. The confidence intervals containing the numerical Malthusian parameter $\hat{\alpha}$ are highlighted with the $*$ symbol. We see that any confidence interval is narrow, and it either contains $\hat{\alpha}$, or $\hat{\alpha}$ is very close to the interval. These results show that the approximation is good for moderate values of t.

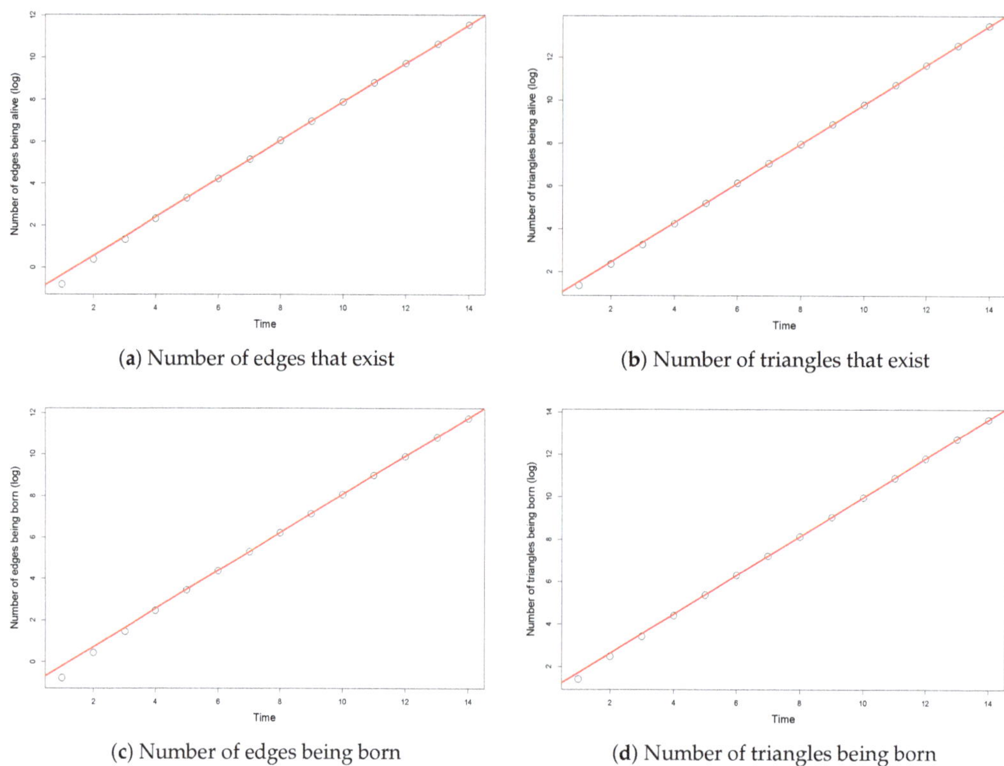

(**a**) Number of edges that exist (**b**) Number of triangles that exist

(**c**) Number of edges being born (**d**) Number of triangles being born

Figure 4. The average of 100 processes generated by the same parameter set and the regression line.

Finally, we present some simulation results for Theorem 3, that is, for the probability of extinction of the evolution process. We made the following computer experiment for any fixed parameter set $\{r_1, p_1, p_2, b, c\}$ and for type 2 and type 3 ancestors. We started to generate the process. If this process reached 2^{10} birth steps, then we stopped it and considered it as a non-extinct process. Otherwise, when the process did not reach 2^{10} birth steps, then the process died out. Applying the above method, we generated 10^5 processes for each parameter sets and counted the relative frequencies of the processes being extinct.

In Table 2, we show some of the results. Column Ancestor contains the type of the ancestor. In the column Numeric we show the numeric solution of the non-linear equation in Theorem 3. We used Julia's trust region method. Column Simulation contains the relative frequencies extracted from the simulations. The simulation results slightly underestimate the numeric values. This is reasonable because we stopped all processes at a fixed time.

Table 1. The 95% confidence intervals for α.

	r_1	p_1	p_2	b	c	$\hat{\alpha}$	2.5%	97.5%
E	0.1	0.5	0.5	0.2	0.4	**0.5394**	**0.5393 ***	**0.5443 ***
T							**0.5390 ***	**0.5440 ***
$\tilde{E}$							0.5410	0.5453
$\tilde{T}$							0.5395	0.5444
E	0.1	0.2	0.6	0.25	0.25	**0.9133**	**0.9130 ***	**0.9141 ***
T							0.9134	0.9142
$\tilde{E}$							**0.9133 ***	**0.9141 ***
$\tilde{T}$							**0.9133 ***	**0.9148 ***
E	0.1	0.2	0.6	0.45	0.35	**0.6622**	**0.6585 ***	**0.6659 ***
T							**0.6606 ***	**0.6648 ***
$\tilde{E}$							**0.6608 ***	**0.6647 ***
$\tilde{T}$							**0.6597 ***	**0.6638 ***

Table 2. Comparison of the numeric values of the extinction probabilities and their relative frequencies from 10^5 repetitions.

r_1	p_1	p_2	b	c	Ancestor	Numeric	Simulation
0.1	0.2	0.6	0.8	0.8	2	0.9095	0.9053
					3	0.8855	0.8805
0.2	0.3	0.6	0.7	0.7	2	0.9247	0.9184
					3	0.9141	0.9070
0.3	0.3	0.5	0.6	0.6	2	0.7371	0.7207
					3	0.6896	0.6834

8. Basic Facts on Branching Processes

In our paper, we use known results of the theory of continuous-time branching processes. The single type general Crump–Mode–Jagers branching processes have been described e.g., in [21–23]. The general multi-type branching processes have been studied, e.g., in [11,19,24].

Here, we give a short description of the general multi-type branching processes based on [11]. The individuals of this process can be of p different types, which we denote by $1, 2, \ldots, p$. Any individual x is described by the quantities $\lambda_x, \xi_x, \Phi_x, \Psi_x, \ldots$. The quantities $\lambda_x, \xi_x, \Phi_x, \Psi_x, \ldots$ are independent copies of the quantities $\lambda, \xi, \Phi, \Psi, \ldots$. Thus, we should give the definition of $\lambda, \xi, \Phi, \Psi, \ldots$, which we consider as the quantities corresponding to the generic individual.

The lifetime λ is a non-negative random variable which is not necessarily independent from the reproduction. The lifetime distribution is $L(t) = \mathbb{P}(\lambda \leq t)$. The reproduction process is $\xi_i(t) = (\xi_{i,1}(t), \ldots, \xi_{i,p}(t))$, $t \geq 0$. Here, the random point process $\xi_{i,j}$ describes the births of type j offspring of a type i mother. $\xi_{i,j}(t)$ gives the number of type j offspring of a type i mother up to time t. $\xi_{i,j}$ is determined by the birth events and the numbers of offspring. The process starts at time $t = 0$ with one individual called the ancestor and denoted by x_0. When a child is born, it starts its own reproduction process and so on. The birth time of the individual x is denoted by σ_x.

Let $\Phi(t)$ be a non-negative random function that describes a certain aspect of the life history of the individual. It is usually assumed that $\Phi(t) = 0$ for $t \leq 0$. Then, $\Phi(t)$ is called a random characteristic. Let $\Psi(t)$ be another random characteristic. Thus, the behaviour of the individual x is described by $\xi_x, \lambda_x, \Phi_x, \Psi_x, \ldots$.

Let us define the branching process ${}_{x_0}Z^{\Phi}(t)$ counted by the characteristic Φ as

$$ {}_{x_0}Z^{\Phi}(t) = \sum_x \Phi_x(t - {}_{x_0}\sigma_x), $$

where we summarize for all individuals x. Here, the left subscript x_0 of Z and of the birth time σ_x is important, because it denotes that the process starts with ancestor x_0 and the type of x_0 has influence for the evolution of the population.

Let us denote by $m_{i,j}(t)$ the reproduction function, which is the expected reproduction number $m_{i,j}(t) = \mathbb{E}\xi_{i,j}(t)$.

The following facts are well-known (see [11] or [24]).

We assume the following basic conditions in this section.

(a) Not all of the measures $m_{i,j}$ are concentrated on a lattice.

Let

$$ m^*_{i,j}(\kappa) = \int_0^\infty e^{-\kappa t} m_{i,j}(dt), \qquad i,j = 1,\dots,p, $$

be the Laplace transform of $m_{i,j}$. Let $M(\kappa)$ be the matrix

$$ M(\kappa) = \left(m^*_{i,j}(\kappa) \right)_{i,j=1}^p. $$

(b1) There exists a positive Malthusian parameter α that is a finite positive value so that $M(\alpha)$ has finite entries only, and the Perron–Frobenius root of $M(\alpha)$ is equal to 1. Here, the Perron–Frobenius root is the largest eigenvalue of the matrix. Let $(v_1,\dots,v_p)^\top$ be the right positive eigenvector and $(u_1,\dots,u_p)^\top$ the left positive eigenvector of $M(\alpha)$ corresponding to the Perron–Frobenius root. We normalize them as $\sum_{i=1}^p v_i = 1$ and $\sum_{i=1}^p u_i v_i = 1$.

(b2) The matrix $\left(m_{i,j}(\infty)\right)_{i,j=1}^p$ has an infinite entry, or all of them are finite, and its Perron–Frobenius root is greater than 1.

(c) The first moment of $e^{-\alpha t} m_{i,j}(dt)$ is finite and positive; that is,

$$ 0 < \int_0^\infty t e^{-\alpha t} m_{i,j}(dt) < \infty, \qquad i,j = 1,\dots,p. $$

(d) There exists a finite positive integer K so that all elements of the Kth power of the matrix $\left(m_{i,j}(\infty)\right)_{i,j=1}^p$ are positive.

Let

$$ {}_\alpha\xi_{i,j}(\infty) = \int_0^\infty e^{-\alpha t} \xi_{i,j}(dt). \tag{68} $$

Proposition 4. *Let α be the Malthusian parameter. Assume that the random characteristic Φ satisfies the following conditions:*

(i) $\Phi(t) \geq 0$,

(ii) The trajectories of Φ belong to the Skorohod space D, i.e., they do not have discontinuities of the second kind,

(iii) $\mathbb{E}(\sup_t \Phi(t)) < \infty$.

Assume also

(iv) for some $\varepsilon > 0$

$$ \int_0^\infty t(\log(1+t))^{1+\varepsilon} e^{-\alpha t} m_{i,j}(dt) < \infty, \qquad i,j = 1,\dots,p $$

and

(v) for some $\varepsilon > 0$

$$ \mathbb{E} \sup_{t \geq 0} \left\{ \max\left\{ t(\log(1+t))^{1+\varepsilon}, 1 \right\} e^{-\alpha t} \Phi(t) \right\} < \infty $$

for any ancestor.

Then,

$$\lim_{t \to \infty} e^{-\alpha t} {}_{x_0}Z^{\Phi}(t) = {}_{x_0}Y_{\infty} v_i m_{\infty}^{\Phi} \tag{69}$$

is likely, where i is the type of x_0,

$$m_{\infty}^{\Phi} = \frac{\sum_{j=1}^{p} u_j \int_0^{\infty} e^{-\alpha t} \mathbb{E}\Phi_j(t) dt}{\sum_{l,j=1}^{p} u_l v_j \int_0^{\infty} t e^{-\alpha t} m_{l,j}(dt)}, \tag{70}$$

${}_{x_0}Y_{\infty}$ is an a.s. non-negative random variable depending on the type of the ancestor x_0 but not depending on the choice of Φ.

If, in addition, we assume that

(vi)

$$\mathbb{E}\left[{}_{\alpha}\xi_{i,j}(\infty) \log^{+} {}_{\alpha}\xi_{i,j}(\infty) \right] < \infty, \qquad i,j = 1,\ldots,p, \tag{71}$$

then $\mathbb{E}({}_{x_0}Y_{\infty}) = 1$, ${}_{x_0}Y_{\infty}$ is positive with positive probability, and ${}_{x_0}Y_{\infty}$ is a.s. positive on the survival set.

The proof is a simple consequence of Theorem 2.4 and Proposition 4.1 of [11].

9. Discussion

In this paper, a new network evolution model was introduced. This model was inspired by those networks where small substructures play important role. In social life, such substructures could be a group of friends. In the theory of networks, these substructures are called motifs. In this paper, for the sake of simplicity, we consider only two types of substructures, the edges and the triangles. The novelty of the paper is the usage of a two-type continuous time branching process to describe these two types of interactions. Thus, despite [7,10], the theory of multi-type branching processes was applied for certain substructures of the network and not just for the nodes. Our paper extends the former studies of [16,17], where only one type of interaction was considered.

In this paper, we proved that the magnitude of the number of triangles on the event of non-extinction is $e^{\alpha t}$, where α is the Malthusian parameter. We obtained similar results for the number of edges. We also studied the degree process of a fixed vertex and the probability of the extinction. Our results are similar to the ones obtained for the simpler models in [16,17]. In addition to mathematical proofs, the results were illustrated by simulations.

In future extensions of the model, more than two types of substructures can be studied using the theory of multi-type branching processes.

Author Contributions: Conceptualization and methodology, I.F.; software, A.B.; writing the original draft, I.F.; editing, A.B.; visualization, A.B.; supervision, I.F.; Section 4 is due to I.F., Section 6 is due to A.B. All authors have read and agreed to the published version of the manuscript.

Funding: The publication is supported by the EFOP-3.6.1-16-2016-00022 project. The project is co-financed by the European Union and the European Social Fund.

Institutional Review Board Statement: Not applicable.

Informed Consent Statement: Not applicable.

Data Availability Statement: Not applicable.

Acknowledgments: The authors would like to thank the referees for their helpful remarks.

Conflicts of Interest: The authors declare no conflict of interest.

References

1. Barabási, A.-L. *Network Science*; Cambridge University Press: Cambridge, UK, 2018.
2. van der Hofstad, R. *Random Graphs and Complex Networks. Vol. 1.*; Cambridge Series in Statistical and Probabilistic Mathematics; Cambridge University Press: Cambridge, UK, 2017.
3. Barabási, A.-L.; Albert, R. Emergence of scaling in random networks. *Science* **1999**, *286*, 509–512. [CrossRef] [PubMed]
4. Backhausz, Á.; Móri, T.F. A random graph model based on 3-interactions. *Ann. Univ. Sci. Budapest. Sect. Comput.* **2012**, *36*, 41–52.
5. Fazekas, I.; Porvázsnyik, B. Scale-free property for degrees and weights in an N-interactions random graph model. *J. Math. Sci. (N. Y.)* **2016**, *214*, 69–82. [CrossRef]
6. Fazekas, I.; Noszály, C.; Perecsényi, A. The N-star network evolution model. *J. Appl. Probab.* **2019**, *56*, 416–440. [CrossRef]
7. Deijfen, M.; Fitzner, R. Birds of a feather or opposites attract - effects in network modelling. *Internet Math.* **2017**, *1*, 24.
8. Rudas, A.; Tóth, B.; Valkó, B. Random trees and general branching processes. *Random Struct. Algorithms* **2007**, *31*, 186–202. [CrossRef]
9. Athreya, K.B.; Ghosh, A.P.; Sethuraman, S. Growth of preferential attachment random graphs via continuous-time branching processes. *Proc. Indian Acad. Sci. Math. Sci.* **2008**, *118*, 473–494. [CrossRef]
10. Rosengren, S. A Multi-type Preferential Attachment Tree. *Internet Math.* **2018**, *1*, 16.
11. Iksanov, A.; Meiners, M. Rate of convergence in the law of large numbers for supercritical general multi-type branching processes. *Stoch. Proc. Appl.* **2015**, *125*, 708–738. [CrossRef]
12. Milo, R.; Shen-Orr, S.; Itzkovitz, S.; Kashtan, N.; Chklovskii, D.; Alon, U. Network Motifs: Simple Building Blocks of Complex Networks. *Science* **2002**, *298*, 824–827. [CrossRef]
13. Milo, R.; Itzkovitz, S.; Kashtan, N.; Levitt, R.; Shen-Orr, S.; Ayzenshtat, I.; Sheffer, M.; Alon, U. Superfamilies of Evolved and Designed Networks. *Science* **2004**, *303*, 1538–1542. [CrossRef]
14. Itzkovitz, S.; Alon, U. Subgraphs and network motifs in geometric networks. *Phys. Rev. E* **2005**, *71*, 026117. [CrossRef]
15. Paulau, P.V.; Feenders, C.; Blasius, B. Motif analysis in directed ordered networks and applications to foodwebs. *Sci. Rep.* **2015**, *5*, 11926. [CrossRef] [PubMed]
16. Móri, T.F.; Rokob, S. A random graph model driven by time-dependent branching dynamics. *Ann. Univ. Sci. Budapest. Sect. Comp.* **2017**, *46*, 191–213.
17. Fazekas, I.; Barta, A.; Noszály, C.; Porvázsnyik, B. A continuous-time network evolution model describing 3-interactions. *Commun. Stat. Theory Methods* **2021**. [CrossRef]
18. Fazekas, I.; Barta, A. Theoretical and simulation results for a 2-type network evolution model. In Proceedings of the 1st Conference on Information Technology and Data Science, Debrecen, Hungary, 6–8 November 2020; Fazekas, I., Hajdu, A., Tómács, T., Eds.; CEUR-WS.org: Aachen, Germany, 2021; pp. 104–114. Available online: http://ceur-ws.org/Vol-2874/ (accessed on 6 June 2021).
19. Mode, C.J. *Multitype Branching Processes; Theory and Applications*; American Elsevier: New York, NY, USA, 1971.
20. Fazekas, I.; Barta, A.; Noszály, C. Simulation results on a triangle-based network evolution model. *Ann. Math. Informaticae* **2020**, *51*, 7–15. [CrossRef]
21. Haccou, P.; Jagers, P.; Vatunin, V.A. *Branching Processes: Variation, Growth, and Extinction of Populations*; Cambridge University Press: Cambridge, UK, 2005.
22. Jagers, P. *Branching Processes with Biological Applications*; Wiley: London, UK, 1975.
23. Nerman, O. On the convergence of supercritical general (C-M-J) branching processes. *Z. Wahrsch. Verw. Geb.* **1981**, *57*, 365–395. [CrossRef]
24. Nerman, O. On the Convergence of Supercritical General Branching Processes. Ph.D. Thesis, University of Gothenburg, Gothenburg, Sweden, 1979.

 MDPI

Article

A Time-Inhomogeneous Prendiville Model with Failures and Repairs

Virginia Giorno *,† and Amelia G. Nobile †

Dipartimento di Informatica, Università degli Studi di Salerno, Via Giovanni Paolo II n. 132, 84084 Salerno, Italy; nobile@unisa.it
* Correspondence: giorno@unisa.it
† These authors contributed equally to this work.

Abstract: We consider a time-inhomogeneous Markov chain with a finite state-space which models a system in which failures and repairs can occur at random time instants. The system starts from any state j (operating, F, R). Due to a failure, a transition from an operating state to F occurs after which a repair is required, so that a transition leads to the state R. Subsequently, there is a restore phase, after which the system restarts from one of the operating states. In particular, we assume that the intensity functions of failures, repairs and restores are proportional and that the birth-death process that models the system is a time-inhomogeneous Prendiville process.

Keywords: continuous-time ehrenfest model; first-passage time densities; proportional intensity functions; asymptotic behaviors

MSC: 60J28; 60J35; 60K25; 60K20

Citation: Giorno, V.; Nobile, A.G. A Time-Inhomogeneous Prendiville Model with Failures and Repairs. *Mathematics* **2022**, *10*, 251. https://doi.org/10.3390/math10020251

Academic Editors: Alexander Zeifman, Victor Korolev and Alexander Sipin

Received: 20 December 2021
Accepted: 12 January 2022
Published: 14 January 2022

1. Introduction

Continuous-time Markov chains (CTMC) are usually used in various application fields related to queueing systems, mathematical biology, physics, and chemistry (cf., for instance, Anderson [1], Iosifescu and Tautu [2], Medhi [3], Bayley [4], van Kampen [5], Taylor and Karlin [6], Sericola [7]). In these cases, the stochastic process describes the evolution in continuous time of a Markov chain with a countable set of states that represent the number of customers in a queue, the number of molecules in a chemical reaction, the size of the population with births/deaths/immigrations/emigrations.

In the recent decades, particular attention has been paid to the study of these processes under the effect of random catastrophes that produce a sudden change of the state of a system. After such failure, one can think that the system is empty (total catastrophes) and then the dynamics immediately restart without delay (cf., for instance, Dharmaraja et al. [8], Giorno et al. [9–11], Di Crescenzo et al. [12], Economou and Fakinos [13,14], Chen et al. [15]). In more realistic cases, after a failure the system can be shipped for maintenance; in these cases, due to the extent of the failure, it is reasonable to assume random repair times. To introduce the effect of a catastrophe related to a failure of the system, one adds to the usual assumptions the existence of a non-zero probability of transition to an intermediate state from which the zero, or another operating state, can be reached at some randomly distributed instants (cf., for instance, Di Crescenzo et al. [16,17], Ye et al. [18], Mytalas and Zazanis [19], Krishna Kumar et al. [20]). In many cases, the times to failures and the times of repair are assumed to be exponential random variables. Some models consider the phase-type distributions for failure and repair times (see, for instance, Altiok [21–23], Dallery [24]).

Frequently, time-inhomogeneous Markov chains are used to model real dynamic systems. Research in this area are oriented to determine the transient and the limiting probability distribution, and to construct a continuous time diffusion approximation (cf., for instance,

Kendall [25], McNeil and Schach [26], Di Crescenzo et al. [27,28], Giorno et. al. [29,30]). Moreover, some studies on the ergodicity of time-inhomogeneous birth-death chains are considered in Ammar et al. [31], Zeifman et al. [32,33], Satin et al. [34]. For CTMC, the evaluation of first-passage time densities and their moments via analytical and numerical methods plays an important role (cf., for instance, Jouini [35], Giorno and Nobile [36] and references therein).

Various research have been devoted to stochastic "logistic models" that describe biological population growth in a limited environment or the number of customers in a queueing system with finite capacity. In particular, the logistic model proposed by Prendiville in 1949, and subsequently solved by Takashima in 1956, was applied in biology, in ecology and in queueing systems (cf. Prendiville [37], Takashima [38], Giorno et al. [39], Ricciardi [40]). The Prendiville process can be also viewed as the Ehrenfest model in continuous time (see, Karlin and McGregor [41], Flegg et al. [42]). Furthermore, Zheng [43] gives the extension of the Prendiville process to the inhomogeneous case. The Prendiville/Ehrenfest model has been also used to describe queueing systems in presence of catastrophes (cf. Dharmaraja [8], Giorno [44,45]). Moreover, Parthasarathy and Krishna Kumar [46] and Matis and Kiffe [47] consider stochastic compartment models with Prendiville growth mechanisms.

In the present paper, we consider a time-inhomogeneous birth-death process with a finite state-space and we assume that failures and repairs can occur at random time instants. Specifically, the state-space of the considered stochastic process, in addition to the operating states, includes two particular states, denoted by F and R. The dynamics system starts from any state j (operating, F, R). Due to a failure that occurs according to a non-stationary exponential distribution, a transition from an operating state to F occurs; after which a repair, that leads to the state R, starting from F, is required. Even the repair times are assumed to be random and they occur according to a non-stationary exponential distribution. After the system has been repaired, it restarts from one of the operating states.

The plan of the paper is as follows. In Section 2, we describe the stochastic model; we provide the Kolmogorov differential equations for the time-inhomogeneous CTMC with a finite state-space, assuming that the times of failures, repairs, and restores are exponentially distributed. In Section 3, we assume that the failures, repairs and restores intensity functions are proportional; we determine the transient probabilities that, starting from an arbitrary state j at time t_0, the system reaches the state F, or the state R or one of the operating states $0, 1, \ldots, \ell$ at time t. In Section 4, we analyze the time of first failure and determine its probability density function and related average. In Section 5, we obtain the probability generating function of the operating states of the system and the related conditional mean. In Section 6, the asymptotic behavior of the probabilities and of related average for the operating state is studied, under the assumption of proportional intensity functions.

2. The Model

Let $\{N(t), t \geq t_0\}$ be a time-inhomogeneous Markov chain with space-state $\mathcal{S} = \{-2, -1, 0, 1, \ldots, \ell\}$, where $n = -2$ corresponds to the failure state (F), $n = -1$ describes the repair state (R) from which the process can work again and $n = 0, 1, \ldots, \ell$ correspond to the operating states of the system (see, Figure 1). We assume that the arrival (upward jumps) and departures (downward jumps) at time t occur with intensity functions $\lambda_n(t)$ for $n = 0, 1, \ldots, \ell$ and $\mu_n(t)$ for $n = 1, 2, \ldots, \ell$, respectively. Moreover, the failures occur according to a non-homogeneous Poisson process, with intensity function $\xi_n(t)$, starting from the operating state n, with $n = 0, 1, \ldots, \ell$. If a failure occurs, then the system goes into the failure state F, and further, the completion of a repair occurs according to the intensity function $\varrho(t)$. After the repair, there is a restore phase after which the system restarts from an operating state n, with the intensity function $\gamma_n(t)$ for $n = 0, 1, \ldots, \ell$. Several cases can occur: *(a)* after the repair the system restarts from the state $n = 0$, so that we have $\gamma_0(t) = \gamma(t)$ and $\gamma_n(t) = 0$ for $n = 1, 2 \ldots, \ell$; *(b)* the state from which the system restarts is chosen randomly, by setting $\gamma_n(t) = \gamma(t)$ for $n = 0, 1, 2 \ldots, \ell$; *(c)* the intensity functions $\gamma_0(t), \gamma_1(t), \ldots, \gamma_\ell(t)$ are chosen by reflecting the priority of one state over the others.

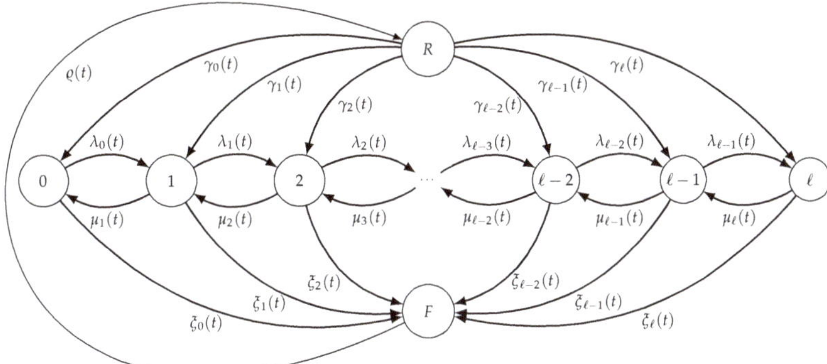

Figure 1. The state diagram of the Markov process $N(t)$ modeling failures and repairs.

Specifically, in any small interval $(t, t+\Delta t)$, $\Delta t > 0$, we assume that the transitions that regulate $N(t)$ occur according the following scheme:

- $n \to n+1$ with intensity function $\lambda_n(t)$ for $n = 0, 1, \ldots, \ell-1$,
- $n \to n-1$ with intensity function $\mu_n(t)$ for $n = 1, 2, \ldots, \ell$,
- $-1 \to n$ with intensity function $\gamma_n(t)$ for $n = 0, 1, \ldots, \ell$,
- $n \to -2$ with intensity function $\xi_n(t)$ for $n = 0, 1, \ldots, \ell$,
- $-2 \to -1$ with intensity function $\varrho(t)$,

where $\lambda_n(t), \mu_n(t), \gamma_n(t), \xi_n(t), \varrho(t)$ are positive, bounded and continuous functions for $t \geq 0$. In Buonocore et al. [48], a time-homogeneous similar model is considered in the biological context, assuming that $\lambda_n(t) = \lambda$, for $n = 0, 1, \ldots, \ell-1$, $\mu_n(t) = \mu$, for $n = 0, 1, \ldots, \ell-1$, $\gamma_n(t) = \gamma$ for $n = 0, 1, \ldots, \ell$, $\xi_n(t) = \xi$ for $n = 0, 1, \ldots, \ell$ and $\varrho(t) = \varrho$.

Let

$$p_{j,n}(t|t_0) = P\{N(t) = n | N(t_0) = j\}, \qquad j, n \in \mathcal{S} \tag{1}$$

be the transition probabilities of $N(t)$. Setting

$$\nu(t) = \sum_{n=0}^{\ell} \gamma_n(t), \tag{2}$$

one has:

$$\begin{aligned}
\frac{dp_{j,-2}(t|t_0)}{dt} &= \sum_{n=0}^{\ell} \xi_n(t)\, p_{j,n}(t|t_0) - \varrho(t)\, p_{j,-2}(t|t_0)\\
\frac{dp_{j,-1}(t|t_0)}{dt} &= -\nu(t)\, p_{j,-1}(t|t_0) + \varrho(t)\, p_{j,-2}(t|t_0),\\
\frac{dp_{j,0}(t|t_0)}{dt} &= \gamma_0(t)\, p_{j,-1}(t|t_0) - [\lambda_0(t) + \xi_0(t)]\, p_{j,0}(t|t_0) + \mu_1(t)\, p_{j,1}(t|t_0),\\
\frac{dp_{j,n}(t|t_0)}{dt} &= \gamma_n(t)\, p_{j,-1}(t|t_0) + \lambda_{n-1}(t)\, p_{j,n-1}(t|t_0)\\
&\quad -[\lambda_n(t) + \mu_n(t) + \xi_n(t)]\, p_{j,n}(t|t_0) + \mu_{n+1}(t)\, p_{j,n+1}(t|t_0), \qquad n = 1, 2, \ldots, \ell-1,\\
\frac{dp_{j,\ell}(t|t_0)}{dt} &= \gamma_\ell(t)\, p_{j,-1}(t|t_0) + \lambda_{\ell-1}(t)\, p_{j,\ell-1}(t|t_0) - [\mu_\ell(t) + \xi_\ell(t)]\, p_{j,\ell}(t|t_0),
\end{aligned} \tag{3}$$

to solve with the initial conditions

$$\lim_{t\downarrow t_0} p_{j,n}(t|t_0) = \delta_{j,n} \qquad j,n \in \mathcal{S}. \tag{4}$$

For $t \geq t_0$, denoting by

$$\mathcal{P}_j(t|t_0) = \sum_{n=0}^{\ell} p_{j,n}(t|t_0), \qquad j \in \mathcal{S}, \tag{5}$$

the probability that the system is in an operating state at time t, one has:

$$\mathcal{P}_j(t|t_0) + p_{j,-2}(t|t_0) + p_{j,-1}(t|t_0) = 1, \qquad j \in \mathcal{S}. \tag{6}$$

If $\xi_n(t) = \xi(t)$ for $n = 0, 1, \ldots, \ell$ and $t \geq t_0$, by virtue of (6), one obtains

$$\sum_{n=0}^{\ell} \xi_n(t)\, p_{j,n}(t|t_0) = \xi(t)\, [1 - p_{j,-1}(t|t_0) - p_{j,-2}(t|t_0)],$$

so that the first two equations of system (3) become:

$$\begin{aligned} \frac{dp_{j,-2}(t|t_0)}{dt} &= \xi(t)\, [1 - p_{j,-1}(t|t_0)] - [\xi(t) + \varrho(t)]\, p_{j,-2}(t|t_0), \\ \frac{dp_{j,-1}(t|t_0)}{dt} &= -\nu(t)\, p_{j,-1}(t|t_0) + \varrho(t)\, p_{j,-2}(t|t_0), \end{aligned} \tag{7}$$

to solve with the initial conditions

$$\lim_{t\downarrow t_0} p_{j,-2}(t|t_0) = \delta_{j,-2}, \qquad \lim_{t\downarrow t_0} p_{j,-1}(t|t_0) = \delta_{j,-1}. \tag{8}$$

Furthermore, if $\xi_n(t) = \xi(t)$ for $n = 0, 1, \ldots, \ell$ and $t \geq t_0$, by virtue of (3), one has that the probability $\mathcal{P}_j(t|t_0)$ satisfies the following differential equation

$$\frac{d\mathcal{P}_j(t|t_0)}{dt} = -\xi(t)\, \mathcal{P}_j(t|t_0) + \nu(t)\, p_{j,-1}(t|t_0) \tag{9}$$

to solve with the initial condition

$$\lim_{t\downarrow t_0} \mathcal{P}_j(t|t_0) = 1 - \delta_{j,-2} - \delta_{j,-1}. \tag{10}$$

Equation (9) shows that the probability that the system is in an operating state at time t does not depend on the intensity functions $\lambda_n(t)$ and $\mu_n(t)$ related to the birth-death process without failures and repairs.

3. Proportional Intensity Functions of Failures, Repairs and Restores

We assume that

$$\begin{aligned} &\varrho(t) = \varrho\, \varphi(t), \quad \xi_n(t) = \xi\, \varphi(t), \quad \gamma_n(t) = \gamma_n\, \varphi(t), \quad n = 0, 1, \ldots, \ell, \\ &\nu(t) = (\gamma_0 + \gamma_1 + \ldots + \gamma_\ell)\, \varphi(t), \end{aligned} \tag{11}$$

where $\varphi(t)$ is a positive, bounded and continuous function for $t \geq 0$. We denote by

$$\Phi(t|t_0) = \int_{t_0}^{t} \varphi(u)\, du, \qquad t \geq t_0 \tag{12}$$

and we assume that $\lim_{t\to+\infty} \Phi(t|t_0) = +\infty$.

3.1. Asymptotic Behavior of the System

Let

$$q_n = \lim_{t\to+\infty} p_{j,n}(t|t_0), \qquad j,n \in \mathcal{S}, \qquad Q = \sum_{n=0}^{\ell} q_n = 1 - q_{-2} - q_{-1}, \tag{13}$$

be the steady-state probabilities of the considered system.

Proposition 1. *Under the assumptions (11), one has:*

$$q_{-2} = \frac{\nu\,\xi}{\nu\,\varrho + \nu\,\xi + \varrho\,\xi}, \qquad q_{-1} = \frac{\varrho\,\xi}{\nu\,\varrho + \nu\,\xi + \varrho\,\xi}, \qquad Q = \frac{\varrho\,\nu}{\nu\,\varrho + \nu\,\xi + \varrho\,\xi}. \tag{14}$$

Proof. It follows from (7), by taking the limit as $t \to +\infty$. □

Note that the last identity in (14) is the probability that the system is in an operating state $n = 0, 1, \ldots, \ell$ in equilibrium regime.

3.2. Transient Behavior of the System

To determine the transient solution of system (7) with initial conditions (8), we denote by x_1 and x_2 the solutions of the following equation:

$$x^2 + (\nu + \varrho + \xi)\,x + \nu\,\varrho + \nu\,\xi + \varrho\,\xi = 0$$

and set

$$\Delta = (\nu - \varrho - \xi)^2 - 4\,\varrho\,\xi. \tag{15}$$

Since $x_1 + x_2 = -(\varrho + \xi + \nu) < 0$ and $x_1 x_2 = \nu(\varrho + \xi) + \xi\varrho > 0$, for $\Delta \geq 0$ one has that $x_1 < 0$ and $x_2 < 0$.

Proposition 2. *Under the assumptions (11), for $t \geq t_0$ the following results hold:*
(i) If $\Delta > 0$,

$$\begin{aligned}
p_{j,-2}(t|t_0) &= q_{-2} + [\delta_{j,-2} - q_{-2}]\, Z_1(t|t_0) + [\xi(1-\delta_{j,-1}) - (\xi+\varrho)\delta_{j,-2}]\, Z_2(t|t_0),\\
p_{j,-1}(t|t_0) &= q_{-1} + [\delta_{j,-1} - q_{-1}]\, Z_1(t|t_0) + [-\nu\,\delta_{j,-1} + \varrho\,\delta_{j,-2}]\, Z_2(t|t_0),\\
\mathcal{P}_j(t|t_0) &= Q + [1 - Q - \delta_{j,-2} - \delta_{j,-1}]\, Z_1(t|t_0) + [(\xi+\nu)\delta_{j,-1} - \xi(1-\delta_{j,-2})]\, Z_2(t|t_0),
\end{aligned}$$

with

$$Z_1(t|t_0) = \frac{x_1\, e^{x_2 \Phi(t|t_0)} - x_2\, e^{x_1 \Phi(t|t_0)}}{x_1 - x_2}, \qquad Z_2(t|t_0) = \frac{e^{x_1 \Phi(t|t_0)} - e^{x_2 \Phi(t|t_0)}}{x_1 - x_2}.$$

(ii) If $\Delta = 0$,

$$\begin{aligned}
p_{j,-2}(t|t_0) &= q_{-2} + e^{x_1 \Phi(t|t_0)} \Big\{ \delta_{j,-2} - q_{-2} + \Phi(t|t_0) \Big[\xi(1-\delta_{j,-1}) - (\xi+\varrho)\,\delta_{j,-2}\\
&\qquad - x_1\,(\delta_{j,-2} - q_{-2}) \Big] \Big\},\\
p_{j,-1}(t|t_0) &= q_{-1} + e^{x_1 \Phi(t|t_0)} \Big\{ \delta_{j,-1} - q_{-1} + \Phi(t|t_0) \Big[-\nu\,\delta_{j,-1} + \varrho\,\delta_{j,-2}\\
&\qquad - x_1\,(\delta_{j,-1} - q_{-1}) \Big] \Big\},\\
\mathcal{P}_j(t|t_0) &= Q + e^{x_1 \Phi(t|t_0)} \Big\{ 1 - Q - \delta_{j,-2} - \delta_{j,-1} + \Phi(t|t_0) \Big[(\xi+\nu)\,\delta_{j,-1} - \xi\,(1-\delta_{j,-2})\\
&\qquad - x_1\,(1 - Q - \delta_{j,-2} - \delta_{j,-1}) \Big] \Big\},
\end{aligned}$$

(iii) If $\Delta < 0$,

$$
\begin{aligned}
p_{j,-2}(t|t_0) &= q_{-2} + e^{a\,\Phi(t|t_0)}\Big\{(\delta_{j,-2} - q_{-2})\,\cos[b\,\Phi(t|t_0)] \\
&\quad + \frac{1}{b}\Big[-a\,(\delta_{j,-2} - q_{-2}) - (\xi + \varrho)\delta_{j,-2} + \xi\,(1 - \delta_{j,-1})\Big]\sin[b\,\Phi(t|t_0)]\Big\}, \\
p_{j,-1}(t|t_0) &= q_{-1} + e^{a\,\Phi(t|t_0)}\Big\{(\delta_{j,-1} - q_{-1})\,\cos[b\,\Phi(t|t_0)] \\
&\quad + \frac{1}{b}\Big[-a\,(\delta_{j,-1} - q_{-1}) - \nu\,\delta_{j,-1} + \varrho\,\delta_{j,-2}\Big]\sin[b\,\Phi(t|t_0)]\Big\}, \\
\mathcal{P}_j(t|t_0) &= Q + e^{a\,\Phi(t|t_0)}\Big\{(1 - Q - \delta_{j,-2} - \delta_{j,-1})\,\cos[b\,\Phi(t|t_0)] \\
&\quad + \frac{1}{b}\Big[a\,(1 - Q - \delta_{j,-2} - \delta_{j,-1}) + (\xi + \nu)\,\delta_{j,-1} - \xi\,(1 - \delta_{j,-2})\Big]\sin[b\,\Phi(t|t_0)]\Big\},
\end{aligned}
$$

where

$$
a = -\frac{\nu + \varrho + \xi}{2}, \qquad b = \frac{\sqrt{4\,\varrho\,\xi - (\nu - \varrho - \xi)^2}}{2}.
$$

Proof. From (7), with conditions (8), one has that $p_{j,-2}(t|t_0)$ is solution of the second order differential equation

$$
\begin{aligned}
&\frac{1}{\varphi(t)}\frac{d}{dt}\Big[\frac{1}{\varphi(t)}\frac{dp_{j,-2}(t|t_0)}{dt}\Big] + (\varrho + \xi + \nu)\,\frac{1}{\varphi(t)}\frac{dp_{j,-2}(t|t_0)}{dt} \\
&\quad + [\nu\,(\varrho + \xi) + \varrho\,\xi]\,p_{j,-2}(t|t_0) - \nu\,\xi = 0,
\end{aligned} \tag{16}
$$

to solve with the initial conditions:

$$
\lim_{t\downarrow t_0} p_{j,-2}(t|t_0) = \delta_{j,-2}, \qquad \lim_{t\downarrow t_0}\Big[\frac{1}{\varphi(t)}\frac{dp_{j,-2}(t|t_0)}{dt}\Big] = (1 - \delta_{j,-1})\,\xi - (\xi + \varrho)\,\delta_{j,-2}. \tag{17}
$$

Similarly, for $p_{j,-1}(t|t_0)$ one has

$$
\begin{aligned}
&\frac{1}{\varphi(t)}\frac{d}{dt}\Big[\frac{1}{\varphi(t)}\frac{dp_{j,-1}(t|t_0)}{dt}\Big] + (\varrho + \xi + \nu)\,\frac{1}{\varphi(t)}\frac{dp_{j,-1}(t|t_0)}{dt} \\
&\quad + [\nu\,(\varrho + \xi) + \varrho\,\xi]\,p_{j,-1}(t|t_0) - \varrho\,\xi = 0,
\end{aligned} \tag{18}
$$

to solve with the initial conditions:

$$
\lim_{t\downarrow t_0} p_{j,-1}(t|t_0) = \delta_{j,-1}, \qquad \lim_{t\downarrow t_0}\Big[\frac{1}{\varphi(t)}\frac{dp_{j,-1}(t|t_0)}{dt}\Big] = -\nu\,\delta_{j,-1} + \varrho\,\delta_{j,-2}. \tag{19}
$$

Results of theorem follow by using standard techniques to solve (16) and (18), with the initial conditions (17) and (19), respectively; then, recalling Equation (6), one determines $\mathcal{P}_j(t|t_0)$. □

In Figures 2–4 the probabilities $p_{j,-1}(t|0)$, $p_{j,-2}(t|0)$ and $\mathcal{P}_j(t|0)$ are plotted for $\varphi(t) = 1$, $\xi = 1$, $\nu = 4$ and some choices of the parameter ϱ. In particular, $\Delta = 3.36$ in Figure 2, $\Delta = 0$ in Figure 3 and $\Delta = -3.75$ in Figure 4.

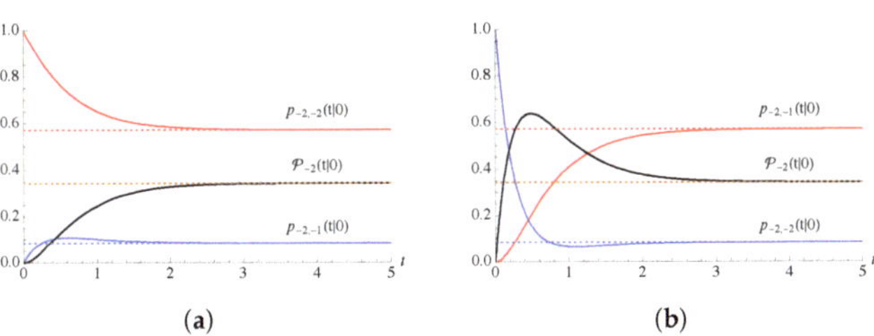

Figure 2. The probabilities $p_{j,-1}(t|0)$, $p_{j,-2}(t|0)$ and $\mathcal{P}_j(t|0)$ are plotted for $\varphi(t) = 1$ and for $\xi = 1.0$, $\varrho = 0.6$, $\nu = 4.0$. In (**a**) $j = -2$ (failure state) and in (**b**) $j = -1$ (repair state).

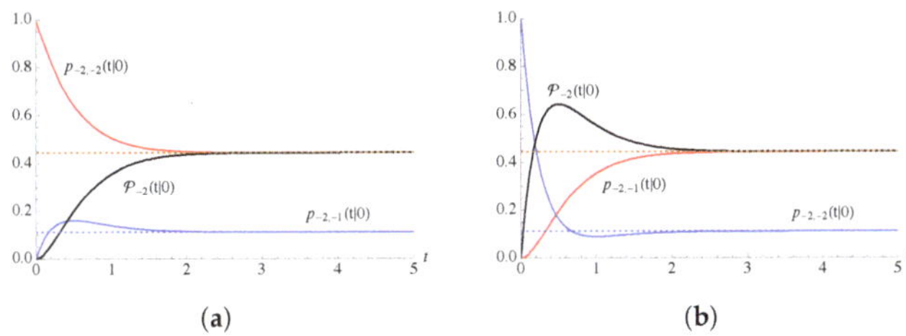

Figure 3. As in Figure 2, for $\varphi(t) = 1$ and for $\xi = 1.0$, $\varrho = 1.0$, $\nu = 4.0$. In (**a**) $j = -2$ (failure state) and in (**b**) $j = -1$ (repair state).

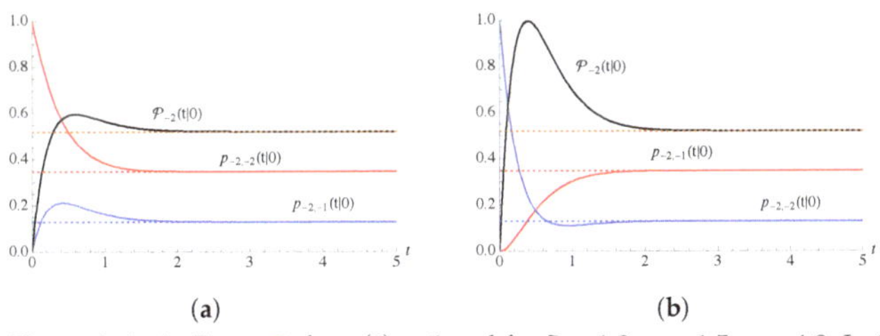

Figure 4. As in Figure 2, for $\varphi(t) = 1$ and for $\xi = 1.0$, $\varrho = 1.5$, $\nu = 4.0$. In (**a**) $j = -2$ (failure state) and in (**b**) $j = -1$ (repair state).

4. Time of First Failure

We denote by

$$\mathcal{T}_{j,-2}(t_0) = \inf\{t > t_0 : N(t) = -2\}, \qquad j \in \{-1, 0, 1, \dots, \ell\} \tag{20}$$

the random variable that describes the time of first failure of the system, i.e. the time in which the chain enters in the state F for the first time, starting from the state $j \in \{-1, 0, 1, \dots, \ell\}$ at time t_0. Let

$$g_{j,-2}(t|t_0) = \frac{d}{dt} P(\mathcal{T}_{j,-2}(t_0) \le t | N(t_0) = j), \qquad j \in \{-1, 0, 1, \dots, \ell\} \tag{21}$$

be the density of the time of first failure.

Proposition 3. *Under the assumptions (11), for $j \in \{-1, 0, 1, \dots, \ell\}$ one has*

$$g_{j,-2}(t|t_0) = \begin{cases} \xi\,\varphi(t)\,\dfrac{\nu\,\delta_{j,-1}\,e^{-\nu\,\Phi(t|t_0)} + \left[\xi\,(1-\delta_{j,-1}) - \nu\right] e^{-\xi\,\Phi(t|t_0)}}{\xi - \nu}, & \nu \neq \xi, \\[2ex] \xi\,\varphi(t)\,e^{-\xi\,\Phi(t|t_0)}\left[1 - \delta_{j,-1} + \xi\,\Phi(t|t_0)\,\delta_{j,-1}\right], & \nu = \xi. \end{cases} \tag{22}$$

Proof. We consider a time-inhomogeneous Markov process $\{\widehat{N}(t), t \geq t_0\}$ with state-space $\mathcal{S}$ obtained from $N(t)$ by setting an absorbing boundary into the state -2, that corresponds to the failure state F of the system and we denote by

$$\widehat{p}_{j,n}(t|t_0) = P\{\widehat{N}(t) = n | \widehat{N}(t_0) = j\}, \qquad j, n \in \mathcal{S}. \tag{23}$$

the probability that the system is in state n at time t and that no failure has yet occurred. Since,

$$P\{\mathcal{T}_{j,-2}(t_0) \leq t\} + \widehat{p}_{j,-1}(t|t_0) + \sum_{n=0}^{\ell} \widehat{p}_{j,n}(t|t_0) = 1, \qquad t \geq t_0,$$

one has $P\{\mathcal{T}_{j,-2}(t_0) \leq t\} = \widehat{p}_{j,-2}(t|t_0)$, so that for $t \geq t_0$ it results

$$g_{j,-2}(t|t_0) = \frac{d}{dt}\widehat{p}_{j,-2}(t|t_0), \qquad j \in \{-1, 0, 1, \ldots, \ell\}. \tag{24}$$

Hence, to determine the density of the time of first failure, it is necessary to consider the following differential equations

$$\begin{aligned}
\frac{d\widehat{p}_{j,-2}(t|t_0)}{dt} &= \xi\, \varphi(t)\, [1 - \widehat{p}_{j,-1}(t|t_0) - \widehat{p}_{j,-2}(t|t_0)],\\
\frac{d\widehat{p}_{j,-1}(t|t_0)}{dt} &= -\nu\, \varphi(t)\, \widehat{p}_{j,-1}(t|t_0),\\
\frac{d\widehat{p}_{j,0}(t|t_0)}{dt} &= \gamma_0\, \varphi(t)\, \widehat{p}_{j,-1}(t|t_0) - [\lambda_0(t) + \xi\, \varphi(t)]\, \widehat{p}_{j,0}(t|t_0) + \mu_1(t)\, \widehat{p}_{j,1}(t|t_0),\\
\frac{d\widehat{p}_{j,n}(t|t_0)}{dt} &= \gamma_n \varphi(t)\, \widehat{p}_{j,-1}(t|t_0) + \lambda_{n-1}(t)\, \widehat{p}_{j,n-1}(t|t_0)\\
&\quad -[\lambda_n(t) + \mu_n(t) + \xi\, \varphi(t)]\, \widehat{p}_{j,n}(t|t_0) + \mu_{n+1}(t)\, \widehat{p}_{j,n+1}(t|t_0), \qquad n = 1, 2, \ldots, \ell - 1,\\
\frac{d\widehat{p}_{j,\ell}(t|t_0)}{dt} &= \gamma_\ell\, \varphi(t)\, \widehat{p}_{j,-1}(t|t_0) + \lambda_{\ell-1}(t)\, \widehat{p}_{j,\ell-1}(t|t_0) - [\mu_\ell(t) + \xi\, \varphi(t)]\, \widehat{p}_{j,\ell}(t|t_0),
\end{aligned} \tag{25}$$

to solve with the initial conditions

$$\lim_{t \downarrow t_0} \widehat{p}_{j,n}(t|t_0) = \delta_{j,n}, \quad j, n \in \mathcal{S}, j \neq -2, \qquad \lim_{t \downarrow t_0} \widehat{p}_{-2,n}(t|t_0) = 0, \quad n \in \mathcal{S}. \tag{26}$$

Proceeding as in Proposition 2, one has:

$$\widehat{p}_{j,-2}(t|t_0) = \begin{cases} \dfrac{\xi\left[1-e^{-\nu\,\Phi(t|t_0)}\right] - \nu\left[1-e^{-\xi\,\Phi(t|t_0)}\right] + \xi\,(1-\delta_{j,-1})\left[e^{-\nu\,\Phi(t|t_0)} - e^{-\xi\,\Phi(t|t_0)}\right]}{\xi - \nu}, & \nu \neq \xi,\\[2ex] 1 - e^{-\xi\,\Phi(t|t_0)}\left[1 + \xi\,\Phi(t|t_0)\,\delta_{j,-1}\right], & \nu = \xi, \end{cases} \tag{27}$$

so that, by virtue of (24), Equation (22) holds. □

From (22) it follows that $P\{\mathcal{T}_{j,-2}(t_0) \leq +\infty\} = 1$, so that with certainty the system is destined to fail. By virtue of (24), for $j \in \{-1, 0, 1, \ldots, \ell\}$ the reliability of the system before the first repair is

$$\begin{aligned}
P\{\mathcal{T}_{j,-2}(t_0) > t\} &= \int_t^{+\infty} g_{j,-2}(\tau|t_0)\, d\tau = \int_t^{+\infty} \frac{d}{d\tau}\widehat{p}_{j,-2}(\tau|t_0)\, d\tau = 1 - \widehat{p}_{j,-2}(t|t_0)\\
&= \begin{cases} \dfrac{\xi\,\delta_{j,-1}\, e^{-\nu\,\Phi(t|t_0)} + [\xi\,(1-\delta_{j,-1}) - \nu]\, e^{-\xi\,\Phi(t|t_0)}}{\xi - \nu}, & \nu \neq \xi,\\[2ex] \left[1 + \xi\,\Phi(t|t_0)\,\delta_{j,-1}\right] e^{-\xi\,\Phi(t|t_0)}, & \nu = \xi. \end{cases}
\end{aligned} \tag{28}$$

Hence, for $j \in \{-1,0,1,\ldots,\ell\}$ the mean time to first failure is

$$E[\mathcal{T}_{j,-2}(t_0)] = \int_{t_0}^{+\infty} (t-t_0)\, g_{j,-2}(t|t_0)\, dt = \int_{t_0}^{+\infty} P\{\mathcal{T}_{j,-2}(t_0) > t\}\, dt$$
$$= \begin{cases} \frac{\xi}{\xi-\nu}\,\delta_{j,-1} \int_{t_0}^{+\infty} e^{-\nu\,\Phi(t|t_0)}\, dt + \frac{\xi\,(1-\delta_{j,-1})-\nu}{\xi-\nu} \int_{t_0}^{+\infty} e^{-\xi\,\Phi(t|t_0)}\, dt, & \nu \neq \xi, \\ \int_{t_0}^{+\infty} e^{-\nu\,\Phi(t|t_0)} \Big[1+\xi\,\delta_{j,-1}\,\Phi(t|t_0)\Big]\, dt, & \nu = \xi. \end{cases} \tag{29}$$

In particular, by setting $\varphi(t) = 1$, Equation (29) leads to

$$E[\mathcal{T}_{j,-2}] = \frac{1}{\nu}\,\delta_{j,-1} + \frac{1}{\xi}, \qquad j \in \{-1,0,1,\ldots,\ell\}.$$

In Figure 5 the density of the time of first failure is plotted for $\varphi(t) = 1$, $\xi = 1.0$, $\varrho = 0.6$, $\nu = 4.0$. If $j = -1$ one has $E[\mathcal{T}_{-1,-2}] = 1.25$, whereas $E[\mathcal{T}_{j,-2}] = 1$ if j is an operating state.

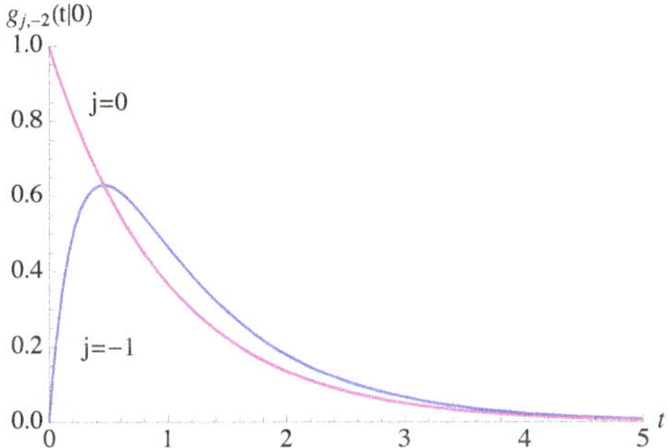

Figure 5. The density of the time of first failure is plotted for $\varphi(t) = 1$ and for $\xi = 1.0$, $\varrho = 0.6$, $\nu = 4.0$.

5. Operating States and Their Probabilities

For the birth-death chain $\{N(t), t \geq t_0\}$, in addition to the assumptions (11), we suppose that the birth and death intensity functions are

$$\lambda_n(t) = (\ell - n)\,\lambda(t), \quad n = 0,1,\ldots,\ell; \qquad \mu_n(t) = n\,\mu(t), \qquad n = 1,\ldots,\ell, \tag{30}$$

with $\lambda(t)$ and $\mu(t)$ positive, bounded and continuous functions for $t \geq 0$. Note that the birth-death intensity functions (30) define a time-inhomogeneous Prendiville process $\{\widetilde{N}(t), t \geq t_0\}$ with finite state-space $\{0,1,\ldots,\ell\}$. The process $\widetilde{N}(t)$ identifies with the process $N(t)$ in the absence of failures, repairs and restores.

Under the assumptions (11) and (30), the transition probabilities of $N(t)$ satisfy the following system:

$$\begin{aligned} \frac{dp_{j,0}(t|t_0)}{dt} &= \gamma_0\,\varphi(t)\,p_{j,-1}(t|t_0) - [\ell\,\lambda(t) + \xi\,\varphi(t)]\,p_{j,0}(t|t_0) + \mu(t)\,p_{j,1}(t|t_0), \\ \frac{dp_{j,n}(t|t_0)}{dt} &= \gamma_n\,\varphi(t)\,p_{j,-1}(t|t_0) + \lambda(t)\,(\ell-n+1)\,p_{j,n-1}(t|t_0) \\ &\quad -[\lambda(t)\,(\ell-n) + \mu(t)\,n + \xi\,\varphi(t)]\,p_{j,n}(t|t_0) + \mu(t)\,(n+1)\,p_{j,n+1}(t|t_0), \\ &\qquad\qquad n = 1,2,\ldots,\ell-1, \\ \frac{dp_{j,\ell}(t|t_0)}{dt} &= \gamma_\ell\,\varphi(t)\,p_{j,-1}(t|t_0) + \lambda(t)\,p_{j,\ell-1}(t|t_0) - [\ell\,\mu(t) + \xi\,\varphi(t)]\,p_{j,\ell}(t|t_0), \end{aligned} \tag{31}$$

to solve with the initial conditions

$$\lim_{t\downarrow t_0} p_{j,n}(t|t_0) = \delta_{j,n}, \qquad j \in \mathcal{S}, n \in \{0,1,\ldots,\ell\}. \tag{32}$$

Let

$$G_j(z,t) = \sum_{n=0}^{\ell} z^n\, p_{i,n}(t|t_0), \qquad j \in \mathcal{S} \tag{33}$$

be the probability generating function (PGF) of the operating states of $N(t)$. From (31) one has:

$$\begin{aligned}&\frac{\partial}{\partial t}G_j(z,t) + (z-1)\,[\lambda(t)\,z + \mu(t)]\,\frac{\partial}{\partial z}G_j(z,t)\\ &\qquad = [\ell\,(z-1)\lambda(t) - \xi\,\varphi(t)]\,G_j(z,t) + \varphi(t)p_{j,-1}(t|t_0)\sum_{i=0}^{\ell}\gamma_i\, z^i, \qquad j \in \mathcal{S},\end{aligned} \tag{34}$$

to solve with the conditions

$$\begin{aligned}&G_j(z,t_0) = \sum_{n=0}^{\ell}\delta_{j,n}\, z^n = \begin{cases} 0, & j=-1,-2\\ z^j, & j \in \{0,1,\ldots,\ell\},\end{cases}\\ &G_j(z,t_0) = \mathcal{P}(t|t_0) = 1 - p_{j,-2}(t|t_0) - p_{j,-1}(t|t_0).\end{aligned} \tag{35}$$

Proposition 4. *Under the assumption (11) and (30), the PGF of the operating states of $N(t)$ is*

$$\begin{aligned}G_j(z,t) &= e^{-\xi\,\Phi(t|t_0)}\sum_{i=0}^{\ell}\delta_{j,i}\,[1+(z-1)\,b_1(t|t_0)]^i\,[1+(z-1)\,b_2(t|t_0)]^{\ell-i}\\ &+\int_{t_0}^{t} du\;\varphi(u)\,p_{j,-1}(u|t_0)\,e^{-\xi\,\Phi(t|u)}\left[\frac{1+(z-1)\,b_2(t|t_0)}{1+(z-1)\,b_2(u|t_0)}\right]^{\ell}\\ &\quad\times\sum_{i=0}^{\ell}\gamma_i\left[\frac{1+(z-1)\,b_1(t|u)}{1+(z-1)\,b_2(t|u)}\right]^i, \qquad j\in\mathcal{S},\end{aligned} \tag{36}$$

where $\Phi(t|t_0)$ is given in (12) and where

$$b_1(t|t_0) = e^{-[\Lambda(t|t_0)+M(t|t_0)]}\,\big[1+B(t|t_0)\big], \qquad b_2(t|t_0) = e^{-[\Lambda(t|t_0)+M(t|t_0)]}\,B(t|t_0), \tag{37}$$

with

$$\Lambda(t|t_0) = \int_{t_0}^{t}\lambda(\tau)\,d\tau, \quad M(t|t_0) = \int_{t_0}^{t}\mu(\tau)\,d\tau, \quad B(t|t_0) = \int_{t_0}^{t}\lambda(\tau)\,e^{\Lambda(\tau|t_0)+M(\tau|t_0)}\,d\tau. \tag{38}$$

Proof. The proof is given in Appendix A. □

We remark that $0 \le b_1(t|t_0) \le 1$ and $0 \le b_2(t|t_0) \le 1$ for all $t \ge t_0$. Furthermore, we note that the function

$$\widetilde{G}_i(z,t) = \big[1+(z-1)\,b_1(t|t_0)\big]^i\,\big[1+(z-1)\,b_2(t|t_0)\big]^{\ell-i}, \qquad i \in \{0,1,\ldots,\ell\}, \tag{39}$$

which appears to the right-hand sides of (36), is the PGF of the time-inhomogeneous Prendiville process $\widetilde{N}(t)$, characterized by the birth-death intensity functions $\lambda_n(t)$ and $\mu_n(t)$, given in (30). The transition probabilities of $\widetilde{N}(t)$ are (cf. Zheng [43], Giorno and Nobile [49]):

$$
\begin{aligned}
\widetilde{p}_{0,n}(t|t_0) &= \binom{\ell}{n}[b_2(t|t_0)]^n\,[1-b_2(t|t_0)]^{\ell-n},\\
\widetilde{p}_{i,n}(t|t_0) &= [b_1(t|t_0)]^n\,[1-b_2(t|t_0)]^{\ell-i}\,[1-b_1(t|t_0)]^{i-n}\\
&\quad\times\sum_{r=\max(0,n-i)}^{\min(\ell-i,n)}\binom{\ell-i}{r}\binom{i}{n-r}\left\{\frac{b_2(t|t_0)\,[1-b_1(t|t_0)]}{b_1(t|t_0)\,[1-b_2(t|t_0)]}\right\}^r,\quad i=1,2,\dots,\ell-1,\\
\widetilde{p}_{\ell,n}(t|t_0) &= \binom{\ell}{n}[b_1(t|t_0)]^n\,[1-b_1(t|t_0)]^{\ell-n},
\end{aligned}
\tag{40}
$$

and the conditional mean and the conditional variance are:

$$
\begin{aligned}
&\mathrm{E}[\widetilde{N}(t)|\widetilde{N}(t_0)=i] = i\,b_1(t|t_0)+(\ell-i)\,b_2(t|t_0),\\
&\mathrm{Var}[\widetilde{N}(t)|\widetilde{N}(t_0)=i] = i\,b_1(t|t_0)\,[1-b_1(t|t_0)]+(\ell-i)\,b_2(t|t_0)\,[1-b_2(t|t_0)].
\end{aligned}
\tag{41}
$$

Under the assumptions (11) and (30), the probability that the system $N(t)$ is in the zero-state at time t can be determined from (33):

$$
\begin{aligned}
p_{j,0}(t|t_0) &= G_j(0,t) = e^{-\xi\,\Phi(t|t_0)}\sum_{i=0}^{\ell}\delta_{j,i}\,\widetilde{p}_{i,0}(t|t_0)\\
&+\sum_{i=0}^{\ell}\gamma_i\int_{t_0}^{t}du\;\varphi(u)\,p_{j,-1}(u|t_0)\,e^{-\xi\,\Phi(t|u)}\left[\frac{1-b_2(t|t_0)}{1-b_2(u|t_0)}\right]^{\ell}\left[\frac{1-b_1(t|u)}{1-b_2(t|u)}\right]^{i},\quad j\in\mathcal{S},
\end{aligned}
\tag{42}
$$

where

$$
\widetilde{p}_{i,0}(t|t_0) = [1-b_1(t|t_0)]^i\,[1-b_2(t|t_0)]^{\ell-i}
$$

is obtained from (40). Similarly, the probability that the system $N(t)$ is in the state $n=1$ at time t follows from (36):

$$
\begin{aligned}
p_{j,1}(t|t_0) &= \frac{dG_j(z,t)}{dz}\Big|_{z=0} = e^{-\xi\,\Phi(t|t_0)}\sum_{i=0}^{\ell}\delta_{j,i}\,\widetilde{p}_{i,1}(t|t_0)\\
&+\int_{t_0}^{t}du\;\varphi(u)\,p_{j,-1}(u|t_0)\,e^{-\xi\,\Phi(t|u)}\left[\frac{1-b_1(t|t_0)}{1-b_2(u|t_0)}\right]^{\ell-1}\left\{\frac{\ell\,[b_2(t|t_0)-b_2(u|t_0)]}{[1-b_2(u|t_0)]^2}\right.\\
&\times\sum_{i=0}^{\ell}\gamma_i\left[\frac{1-b_1(t|u)}{1-b_2(t|u)}\right]^{i}+e^{-[\Lambda(t|u)+M(t|u)]}\,\frac{1-b_1(t|t_0)}{1-b_2(u|t_0)}\sum_{i=0}^{\ell}i\,\gamma_i\left.\frac{[1-b_1(t|u)]^{i-1}}{[1-b_2(t|u)]^{i+1}}\right\},
\end{aligned}
\tag{43}
$$

where, by virtue of (40), one has:

$$
\begin{aligned}
\widetilde{p}_{i,1}(t|t_0) &= [1-b_1(t|t_0)]^{i-1}\,[1-b_2(t|t_0)]^{\ell-i-1}\\
&\quad\times\Big\{i\,b_1(t|t_0)[1-b_2(t|t_0)]+(\ell-i)\,b_2(t|t_0)[1-b_1(t|t_0)]\Big\}.
\end{aligned}
$$

For $r\in\mathbb{N}$, let us introduce the r-th conditional moment of $N(t)$:

$$
\mathrm{E}[N^r(t)|N(t)\geq 0, N(t_0)=j] = \frac{1}{\mathcal{P}_j(t|t_0)}\sum_{n=0}^{\ell}n^r\,p_{j,n}(t|t_0),\qquad j\in\mathcal{S}. \tag{44}
$$

From (36), we have

$$\mathrm{E}[N(t)|N(t) \geq 0, N(t_0) = j] = \frac{1}{\mathcal{P}_j(t|t_0)} \frac{dG_j(z,t)}{dz}\Big|_{z=1}$$

$$= \frac{1}{\mathcal{P}_j(t|t_0)} \left[e^{-\xi\,\Phi(t|t_0)} \sum_{i=0}^{\ell} \delta_{j,i}\, \mathrm{E}[\widetilde{N}(t)|\widetilde{N}(t_0) = i] + \int_{t_0}^{t} du\, \varphi(u)\, p_{j,-1}(u|t_0)\, e^{-\xi\,\Phi(t|u)} \right.$$

$$\left. \times \left\{ \ell\, \nu\, [b_2(t|t_0) - b_2(u|t_0)] + e^{-[\Lambda(t|u)+M(t|u)]} \sum_{i=0}^{\ell} i\, \gamma_i \right\} \right], \qquad j \in \mathcal{S}, \tag{45}$$

where $\mathrm{E}[\widetilde{N}(t)|\widetilde{N}(t_0) = i]$ is given in (41).

6. Asymptotic Distribution of Operating States

To study the asymptotic behavior of the probabilities for the operating states, we assume that the intensity functions of $N(t)$ are proportional. Specifically, in addition to the conditions (11), we suppose that

$$\lambda_n(t) = (\ell - n)\, \lambda\, \varphi(t), \quad n = 0, 1, \ldots, \ell; \qquad \mu_n(t) = n\, \mu\, \varphi(t), \quad n = 1, \ldots, \ell, \tag{46}$$

with $\varphi(t)$ positive, bounded and continuous function for $t \geq 0$.

Let

$$G(z) = \sum_{n=0}^{\ell} z^n q_n \tag{47}$$

be the asymptotic PGF of the operating states of $N(t)$. From (34) one has

$$(z-1)\, [\lambda\, z + \mu]\, \frac{dG(z)}{dz} = [\ell\, (z-1)\lambda - \xi]\, G(z) + q_{-1} \sum_{i=0}^{\ell} \gamma_i\, z^i, \qquad j \in \mathcal{S}, \tag{48}$$

to solve with the condition

$$G(1) = Q = 1 - q_{-2} - q_{-1}. \tag{49}$$

Proposition 5. *Under the assumptions (11) and (46), the asymptotic PGF of the operating states is:*

$$G(z) = (\lambda\, z + \mu)^{\xi/(\lambda+\mu)+\ell}\, (1-z)^{-\xi/(\lambda+\mu)}\, q_{-1}$$
$$\times \sum_{i=0}^{\ell} \gamma_i \int_z^1 x^i\, (\lambda\, x + \mu)^{-\xi/(\lambda+\mu)-\ell-1}\, (1-x)^{\xi/(\lambda+\mu)-1}\, dx. \tag{50}$$

Proof. The general solution of the differential Equation (48) is:

$$G(z) = (\lambda\, z + \mu)^{\xi/(\lambda+\mu)+\ell}\, (1-z)^{-\xi/(\lambda+\mu)}$$
$$\times \left[-q_{-1} \sum_{i=0}^{\ell} \gamma_i \int^z x^i\, (\lambda\, x + \mu)^{-\xi/(\lambda+\mu)-\ell-1}\, (1-x)^{\xi/(\lambda+\mu)-1}\, dx + c \right], \tag{51}$$

where c is an arbitrary constant. Making use of the condition (49), we note that the term in square brackets at the right-hand side of (51) must vanish when $z \to 1$, allowing to determine the constant c. Hence, from (51) we obtain (50). □

The knowledge of the asymptotic PGF (50) allows to calculate the asymptotic probabilities of the operating states, as

$$q_0 = G(0), \qquad q_n = \frac{1}{n!} \frac{d^n G(z)}{dz^n}\Big|_{z=0}, \qquad n = 1, 2, \ldots, \ell, \tag{52}$$

and the r-th asymptotic conditional moment of $N(t)$:

$$\mathrm{E}[N^r | N \geq 0] = \frac{1}{Q} \sum_{n=0}^{\ell} n^r \, q_n, \qquad r \in \mathbb{N}. \tag{53}$$

Proposition 6. *Under the assumptions (11) and (46), one has:*

$$\begin{aligned} q_0 &= \frac{1}{\lambda+\mu} \left(\frac{\mu}{\lambda+\mu}\right)^{\xi/(\lambda+\mu)+\ell} q_{-1} \sum_{i=0}^{\ell} \gamma_i \, B\Big(i+1, \frac{\xi}{\lambda+\mu}\Big) \\ &\quad \times F\Big(\frac{\xi}{\lambda+\mu}, \frac{\xi}{\lambda+\mu} + \ell + 1; \frac{\xi}{\lambda+\mu} + i + 1; \frac{\lambda}{\lambda+\mu}\Big), \\ q_1 &= \frac{1}{\mu} (\lambda \, \ell + \xi) \, q_0 - \frac{\gamma_0}{\mu} q_{-1}, \\ q_2 &= \frac{1}{2\mu^2} \{(\lambda \, \ell + \xi) \, [\lambda \, (\ell - 1) + \xi] + \xi \, \mu\} \, q_0 + \Big\{\frac{\gamma_0}{2\mu^2} [\lambda \, (\ell-1) + \xi + \mu] - \frac{\gamma_1}{2\mu}\Big\} q_{-1}, \end{aligned} \tag{54}$$

where

$$B(x,y) = \frac{\Gamma(x) \, \Gamma(y)}{\Gamma(x+y)} \tag{55}$$

denotes the beta function and

$$F(a,b;c;x) = \sum_{n=0}^{+\infty} \frac{(a)_n \, (b)_n}{(c)_n} \frac{x^n}{n!} \tag{56}$$

is the Gauss hypergeometric function.

Proof. Since $q_0 = G(0)$, by setting $z = 0$ in (50) one obtains:

$$q_0 = \mu^{\ell+\xi/(\lambda+\mu)} \, q_{-1} \sum_{i=0}^{\ell} \gamma_i \int_0^1 x^i \, (\lambda \, x + \mu)^{-\xi/(\lambda+\mu)-\ell-1} \, (1-x)^{\xi/(\lambda+\mu)-1} \, dx. \tag{57}$$

Recalling that (see, Gradshteyn and Ryzhik [50], p. 1005 and p. 1008, n. 9.131)

$$\begin{aligned} F(a,b;c;z) &= \frac{1}{B(b,c-b)} \int_0^1 x^{b-1} \, (1-x)^{c-b-1} (1-x\,z)^{-a} \, dx, \qquad \mathrm{Re}\, c > \mathrm{Re}\, b > 0, \\ F(a,b;c;z) &= (1-z)^{-a} \, F\Big(a, c-b; c; \frac{z}{z-1}\Big), \end{aligned}$$

by setting $a = \ell + 1 + \xi/(\lambda+\mu)$, $b = i+1$, $c = i + 1 + \xi/(\lambda+\mu)$ and $z = -\lambda/\mu$, for $i = 0, 1, \ldots, \ell$ one has

$$\begin{aligned} &\int_0^1 x^i \, (\lambda \, x + \mu)^{-\xi/(\lambda+\mu)-\ell-1} \, (1-x)^{\xi/(\lambda+\mu)-1} \, dx = \mu^{-\xi/(\lambda+\mu)-\ell-1} \\ &\qquad \times B\Big(i+1, \frac{\xi}{\lambda+\mu}\Big) \, F\Big(\frac{\xi}{\lambda+\mu} + \ell + 1, i+1; \frac{\xi}{\lambda+\mu} + i + 1 : -\frac{\lambda}{\mu}\Big) \\ &= (\lambda+\mu)^{-\xi/(\lambda+\mu)-\ell-1} \, B\Big(i+1, \frac{\xi}{\lambda+\mu}\Big) \, F\Big(\frac{\xi}{\lambda+\mu}, \frac{\xi}{\lambda+\mu} + \ell + 1; \frac{\xi}{\lambda+\mu} + i + 1; \frac{\lambda}{\lambda+\mu}\Big), \end{aligned}$$

where the symmetry property $F(a,b;c;z) = F(b,a;c;z)$ has been used in the last equality. Hence, the first equation in (54) follows from (57). Moreover, from (50) we have:

$$\frac{dG(z)}{dz} = \Big[\Big(\ell + \frac{\xi}{\lambda+\mu}\Big) \frac{\lambda}{\lambda \, z + \mu} + \frac{\xi}{\lambda+\mu} \frac{1}{1-z}\Big] \, G(z) - \frac{q_{-1}}{(\lambda \, z + \mu) \, (1-z)} \sum_{i=0}^{\ell} \gamma_i \, z^i, \tag{58}$$

so that the second equation in (54) follows from (52) for $n = 1$. Finally, from (58) one has:

$$\frac{d^2G(z)}{dz^2} = \Big[-\Big(\ell+\frac{\xi}{\lambda+\mu}\Big)\Big(\frac{\lambda}{\lambda z+\mu}\Big)^2 + \frac{\xi}{\lambda+\mu}\Big(\frac{1}{1-z}\Big)^2\Big]\,G(z) \\ +\Big[\Big(\ell+\frac{\xi}{\lambda+\mu}\Big)\frac{\lambda}{\lambda z+\mu} + \frac{\xi}{\lambda+\mu}\,\frac{1}{1-z}\Big]\frac{dG(z)}{dz} \\ +\frac{q_{-1}}{(\lambda z+\mu)\,(1-z)}\Big[\Big(\frac{\lambda}{\lambda z+\mu} - \frac{1}{1-z}\Big)\sum_{i=0}^{\ell}\gamma_i z^i - \sum_{i=0}^{\ell} i\,\gamma_i\,z^{i-1}\Big]. \tag{59}$$

Hence, by virtue of (52) for $n = 2$, from (59) the last equation in (54) follows. □

Proposition 7. *Under the assumptions (11) and (46), one obtain:*

$$\mathrm{E}[N|N \geq 0] = \frac{1}{\lambda+\mu+\xi}\Big\{\lambda\,\ell + \frac{\xi}{\nu}\sum_{i=0}^{\ell} i\,\gamma_i\Big\}, \tag{60}$$

with $\nu = \gamma_0 + \gamma_1 + \ldots + \gamma_\ell$.

Proof. By virtue of (53), from (58) one has

$$\mathrm{E}[N|N \geq 0] = \frac{1}{Q}\frac{dG(z)}{dz}\Big|_{z=1} \\ = \frac{1}{Q}\lim_{z\to 1}\frac{\Big[\Big(\ell+\frac{\xi}{\lambda+\mu}\Big)\frac{\lambda\,(1-z)}{\lambda z+\mu} + \frac{\xi}{\lambda+\mu}\Big]\,G(z) - \frac{q_{-1}}{\lambda z+\mu}\sum_{i=0}^{\ell}\gamma_i z^i}{1-z} \\ = \Big(\ell+\frac{\xi}{\lambda+\mu}\Big)\frac{\lambda}{\lambda+\mu} - \frac{\xi}{\lambda+\mu}\mathrm{E}[N|N \geq 0] - \frac{q_{-1}}{Q}\,\frac{\lambda\,\nu}{(\lambda+\mu)^2} + \frac{q_{-1}}{Q\,(\lambda+\mu)}\sum_{i=0}^{\ell} i\,\gamma_i$$

from which (60) follows. □

Example 1. *We assume that $\ell = 0$. Under the assumptions (11), the time-inhomogeneous Markov chain $N(t)$ is shown in Figure 6.*

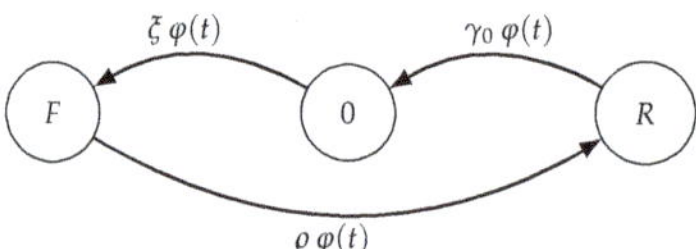

Figure 6. The state diagram of the Markov process $N(t)$ with $\ell = 0$.

In this case, there is only one operating state in zero, the intensity functions of failure $\xi(t) = \xi\,\varphi(t)$, of repair $\varrho(t) = \varrho\,\varphi(t)$ and of restore $\gamma_0(t) = \gamma_0\,\varphi(t)$ are proportional and $p_{j,0}(t|t_0) + p_{j,-2}(t|t_0) + p_{j,-1}(t|t_0) = 1$. From (42), one has:

$$p_{j,0}(t|t_0) = e^{-\xi\,\Phi(t|t_0)}\,\delta_{j,0} + \gamma_0\int_{t_0}^{t}\varphi(u)\,p_{j,-1}(u|t_0)\,e^{-\xi\,\Phi(t|u)}\,du, \qquad j = -2,-1,0. \tag{61}$$

Of course, the conditional mean (45) is equal to zero for all $t \geq t_0$.
From Proposition 6, one obtains:

$$q_0 = \frac{1}{\xi}\Big(\frac{\mu}{\lambda+\mu}\Big)^{\xi/(\lambda+\mu)}\,q_{-1}\,F\Big(\frac{\xi}{\lambda+\mu}, \frac{\xi}{\lambda+\mu}+1; \frac{\xi}{\lambda+\mu}+1; \frac{\lambda}{\lambda+\mu}\Big). \tag{62}$$

Since,

$$F(a,b;b;z) = (1-z)^{-a}, \tag{63}$$

from (62) one clearly has

$$q_0 = \frac{q_{-1}}{\xi}\,\gamma_0 = \frac{\varrho\,\gamma_0}{\gamma_0\,\varrho + \gamma_0\,\xi + \varrho\,\xi},$$

that identifies with the probability Q, being $\nu = \gamma_0$.

Example 2. *We assume that $\ell = 1$. Under the assumption (11) and (46), the time-inhomogeneous Markov chain $N(t)$ is shown in Figure 7.*

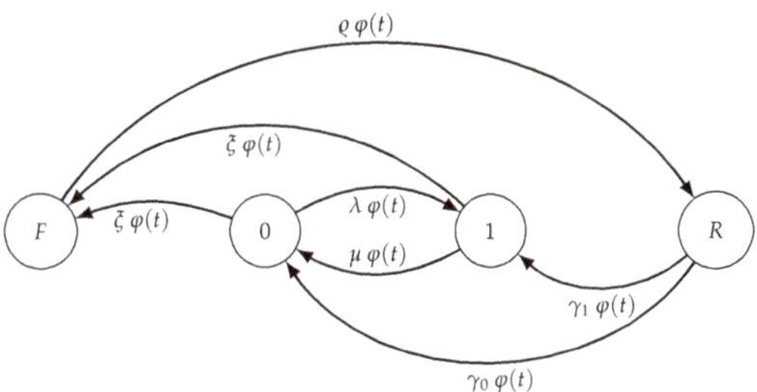

Figure 7. The state diagram of the Markov chain $N(t)$ with $\ell = 1$.

In this case, there are two operating states 0 and 1, with intensity functions of failure $\xi(t) = \xi\,\varphi(t)$, of repair $\varrho(t) = \varrho\,\varphi(t)$ and of restores $\gamma_i(t) = \gamma_i\,\varphi(t)$ for $i = 0, 1$; the birth-death intensity functions are $\lambda_0(t) = \lambda\,\varphi(t)$ and $\mu_1(t) = \mu\,\varphi(t)$. By setting $\ell = 1$ in the first equation in of (54) one has

$$\begin{aligned} q_0 = {} & \frac{1}{\xi}\Big(\frac{\mu}{\lambda+\mu}\Big)^{\xi/(\lambda+\mu)+1} q_{-1}\,\Big[\gamma_0\,F\Big(\frac{\xi}{\lambda+\mu}, \frac{\xi}{\lambda+\mu}+2; \frac{\xi}{\lambda+\mu}+1; \frac{\lambda}{\lambda+\mu}\Big) \\ & +\gamma_1\,\frac{\lambda+\mu}{\lambda+\mu+\xi}\,F\Big(\frac{\xi}{\lambda+\mu}, \frac{\xi}{\lambda+\mu}+2; \frac{\xi}{\lambda+\mu}+2; \frac{\lambda}{\lambda+\mu}\Big)\Big]. \end{aligned} \tag{64}$$

Recalling the Gauss' recursion function (see, Gradshteyn and Ryzhik [50], p. 1010, n. 9.137.17)

$$c\,F(a,b;c;z) - (c-b)\,F(a,b;c+1;z) - b\,F(a,b+1;c+1;z) = 0 \tag{65}$$

and the relation (63), one obtains:

$$F\Big(\frac{\xi}{\lambda+\mu}, \frac{\xi}{\lambda+\mu}+2; \frac{\xi}{\lambda+\mu}+1; \frac{\lambda}{\lambda+\mu}\Big) = \frac{\lambda+\mu}{\lambda+\mu+\xi}\,\frac{\mu+\xi}{\mu}\,\Big(\frac{\mu}{\lambda+\beta}\Big)^{-\xi/(\lambda+\mu)} \tag{66}$$

Making use of (66) and of the relation (63) in Equation (64), for $\ell = 1$ it follows

$$\begin{aligned} q_0 &= \frac{\mu}{\xi}\,\frac{1}{\lambda+\mu+\xi}\,\Big[\Big(1+\frac{\xi}{\mu}\Big)\,\gamma_0 + \gamma_1\Big]\,q_{-1}, \\ q_1 &= \frac{\lambda+\xi}{\mu}\,q_0 - \frac{\gamma_0}{\mu}\,q_{-1}. \end{aligned} \tag{67}$$

Of course, $q_0 + q_1 = Q = \varrho\,\nu/(\varrho\,\nu + \nu\,\xi + \varrho\,\xi)$, with $\nu = \gamma_0 + \gamma_1$. From (53) we have

$$E(N|N \geq 0) = \frac{q_1}{Q} = \frac{\lambda+\xi}{\mu}\,\frac{q_0}{Q} - \frac{\gamma_0}{\mu}\,\frac{\xi}{\gamma_0+\gamma_1} = \frac{1}{\lambda+\mu+\xi}\,\Big(\lambda + \xi\,\frac{\gamma_1}{\gamma_0+\gamma_1}\Big).$$

that identifies with (60) for $\ell = 1$.

Example 3. *We assume that $\ell = 2$. Under the assumption (11) and (46), the time-inhomogeneous Markov chain $N(t)$ is shown in Figure 8.*

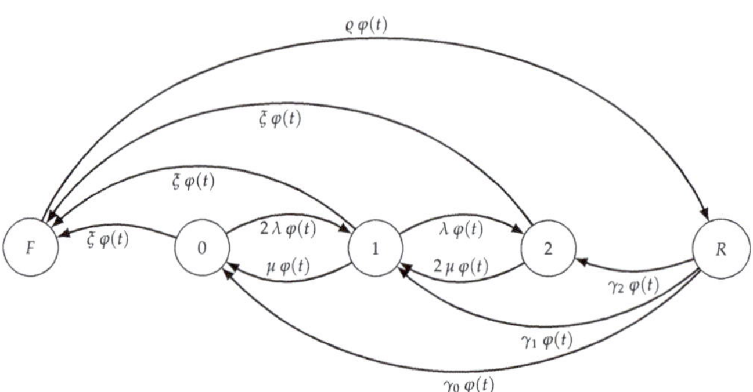

Figure 8. The state diagram of the Markov chain $N(t)$ with $\ell = 2$.

In this case, there are three operating states 0, 1 and 2, with the intensity functions of failure $\xi(t) = \xi\,\varphi(t)$*, of repair* $\varrho(t) = \varrho\,\varphi(t)$ *and of restores* $\gamma_i(t) = \gamma_i\,\varphi(t)$ *for* $i = 0,1,2$*; the birth-death intensity functions are* $\lambda_n(t) = (2-n)\,\lambda\,\varphi(t)$ *for* $n = 0,1$ *and* $\mu_n(t) = n\,\mu\,\varphi(t)$ *for* $n = 1,2$*. By setting* $\ell = 2$ *in the first equation in of (54) one obtains*

$$
\begin{aligned}
q_0 = \frac{1}{\xi}\Big(\frac{\mu}{\lambda+\mu}\Big)^{\xi/(\lambda+\mu)+2} q_{-1}\,\Big[&\gamma_0\,F\Big(\frac{\xi}{\lambda+\mu},\frac{\xi}{\lambda+\mu}+3;\frac{\xi}{\lambda+\mu}+1;\frac{\lambda}{\lambda+\mu}\Big)\\
&+\gamma_1\,\frac{\lambda+\mu}{\lambda+\mu+\xi}\,F\Big(\frac{\xi}{\lambda+\mu},\frac{\xi}{\lambda+\mu}+3;\frac{\xi}{\lambda+\mu}+2;\frac{\lambda}{\lambda+\mu}\\
&+2\,\gamma_2\,\frac{(\lambda+\mu)^2}{(\lambda+\mu+\xi)\,[2(\lambda+\mu)+\xi]}\,F\Big(\frac{\xi}{\lambda+\mu},\frac{\xi}{\lambda+\mu}+3;\frac{\xi}{\lambda+\mu}+3;\frac{\lambda}{\lambda+\mu}\Big)\Big].
\end{aligned}
\tag{68}
$$

By virtue of (65), one has:

$$
\begin{aligned}
F\Big(\frac{\xi}{\lambda+\mu},\frac{\xi}{\lambda+\mu}+3;\frac{\xi}{\lambda+\mu}+1;\frac{\lambda}{\lambda+\mu}\Big) &= \frac{(\lambda+\mu)^2}{2(\lambda+\mu)+\xi}\,\Big(\frac{\mu}{\lambda+\beta}\Big)^{-\xi/(\lambda+\mu)}\\
&\quad\times\Big[\frac{\xi}{\mu^2}+\frac{2}{\lambda+\mu+\xi}\,\Big(1+\frac{\xi}{\mu}\Big)\Big],\\
F\Big(\frac{\xi}{\lambda+\mu},\frac{\xi}{\lambda+\mu}+3;\frac{\xi}{\lambda+\mu}+2;\frac{\lambda}{\lambda+\mu}\Big) &= \frac{\lambda+\mu}{2(\lambda+\mu)+\xi}\,\Big(\frac{\mu}{\lambda+\beta}\Big)^{-\xi/(\lambda+\mu)}\,\Big(2+\frac{\xi}{\mu}\Big),
\end{aligned}
\tag{69}
$$

Making use of (69) and of the relation (63) in Equation (68), for $\ell = 2$ *it follows*

$$
\begin{aligned}
q_0 &= \frac{\mu^2}{\xi}\,\frac{1}{(\lambda+\mu+\xi)\,[2(\lambda+\mu)+\xi]}\,q_{-1}\\
&\qquad\times\Big\{\gamma_0\,\Big[\frac{\xi(\lambda+\mu+\xi)}{\mu^2}+2\Big(1+\frac{\xi}{\mu}\Big)\Big]+\gamma_1\,\Big(2+\frac{\xi}{\mu}\Big)+2\,\gamma_2\Big\},\\
q_1 &= \frac{2\lambda+\xi}{\mu}\,q_0-\frac{\gamma_0}{\mu}\,q_{-1},\\
q_2 &= \frac{(\xi+\lambda)\,(\xi+2\lambda)+\xi\,\mu}{2\,\mu^2}\,q_0-\Big[\gamma_0\frac{\xi+\lambda+\mu}{2\,\mu^2}+\frac{\gamma_1}{2\,\mu}\Big]\,q_{-1}.
\end{aligned}
\tag{70}
$$

Clearly, $q_0 + q_1 + q_2 = Q = \varrho\nu/(\varrho\nu + \nu\xi + \varrho\xi)$*, with* $\nu = \gamma_0 + \gamma_1 + \gamma_2$*. Finally, from* (53) *one obtains*

$$\mathrm{E}(N|N \geq 0) = \frac{q_1 + 2q_2}{Q} = \Big[\frac{2\lambda + \xi}{\mu} + \frac{(\xi+\lambda)(\xi+2\lambda) + \xi\mu}{\mu^2}\Big]\frac{q_0}{Q}$$
$$-\Big[\gamma_0\frac{\xi+\lambda+2\mu}{\mu^2} + \frac{\gamma_1}{\mu}\Big]\frac{\xi}{\gamma_0+\gamma_1+\gamma_2} = \frac{1}{\lambda+\mu+\xi}\Big\{2\lambda + \xi\,\frac{\gamma_1+2\gamma_2}{\gamma_0+\gamma_1+\gamma_2}\Big\},$$

that identifies with (60) *for* $\ell = 2$.

7. Conclusions

In the present paper, we have considered a time-inhomogeneous CTMC with a finite space-state in which failures and repairs can occur at random times. In addition to the operating states, the space of the states includes two particular ones, denoted by F and R, representing the failure state and the repair one, respectively. The failures occur according to a non-stationary exponential distribution and they produce a transition from an operating state to F. Subsequently, a repair is required that involves a transition from F to R. Even the repair times are assumed to be random and occurring according to a non-stationary exponential distribution. After the reparation, the system restarts from one of the operating states.

Assuming that the failures, repairs and restores are characterized by proportional intensity functions, we determine the transition probabilities that, starting from an arbitrary state j at time t_0, the system reaches the state F, or the state R, or one of the operating states at time t. The obtained results show that that the probability that the system is in an operating state at time t does not depend on the intensity functions related to the birth-death process without failures and repairs. In other words, the transition probabilities related to the states F, R, as well as the transition probability that the system occupies an operating state, are independent of the dynamics existing between the operating states. We determine the density of the time of first failure and the related average. Moreover, we focus on the transition probabilities of operating states by determining the PGF and the conditioned mean. Finally, under the assumption of proportional intensity functions, we analyze the asymptotic behavior for the probabilities of the operating states by calculating the asymptotic PGF and the asymptotic conditional mean.

Author Contributions: Conceptualization, V.G. and A.G.N.; methodology, V.G. and A.G.N.; software, V.G. and A.G.N.; validation, V.G. and A.G.N.; formal analysis, V.G. and A.G.N.; investigation, V.G. and A.G.N.; resources, V.G. and A.G.N.; data curation, V.G. and A.G.N.; visualization, V.G. and A.G.N.; supervision, V.G. and A.G.N. All authors have read and agreed to the published version of the manuscript.

Funding: This research is partially supported by MIUR—PRIN 2017, Project "Stochastic Models for Complex Systems".

Institutional Review Board Statement: Not applicable.

Informed Consent Statement: Not applicable.

Acknowledgments: The authors are members of the research group GNCS of INdAM.

Conflicts of Interest: The authors declare that they have no conflict of interest.

Appendix A. Proof of Proposition 4

Equation (34) with the conditions (35) can be solved by using the method of characteristics (cf., for instance, Williams [51]). We consider the following differential equations:

$$\frac{dt}{d\psi}=1, \qquad \frac{dz}{d\psi}=(z-1)\big[\lambda(t)\,z+\mu(t)\big],$$
$$\frac{dG_j}{d\psi}=[\ell\,(z-1)\,\lambda(t)-\xi\,\varphi(t)]\,G_j+\varphi(t)\,p_{j,-1}(t|t_0)\sum_{i=0}^{\ell}\gamma_i\,z^i, \tag{A1}$$

with the initial conditions:

$$t(s,\psi=t_0)=t_0, \qquad z(s,\psi=t_0)=s, \qquad G_j(s,\psi=t_0)=\sum_{i=0}^{\ell}\delta_{j,i}\,s^i. \tag{A2}$$

The first equation of (A1), with the related initial condition in (A2), leads to $t=\psi$. By setting $t=\psi$ in the second equation of (A1) and by using the second of (A2) one obtains:

$$z-1=\frac{(s-1)\,e^{\Lambda(\psi|t_0)+M(\psi|t_0)}}{1-(s-1)\,B(\psi|t_0)}, \tag{A3}$$

with $\Lambda(t|t_0)$, $M(t|t_0)$ and $B(t|t_0)$ defined in (38). Moreover, solving the third equation in (A1) with $t=\psi$ and z obtained from (A3) we have

$$G_j(s,\psi)=e^{-\xi\,\Phi(\psi|t_0)}\exp\left\{\ell\,(s-1)\int_{t_0}^{\psi}\frac{\lambda(u)\,e^{\Lambda(u|t_0)+M(u|t_0)}}{1-(s-1)\,B(u|t_0)}\,du\right\}\sum_{i=0}^{\ell}\delta_{j,i}\,s^i$$
$$+\int_{t_0}^{\psi}du\;\varphi(u)\,p_{j,-1}(u|t_0)\,e^{-\xi\,\Phi(\psi|u)}\exp\left\{\ell\,(s-1)\int_{u}^{\psi}\frac{\lambda(\vartheta)\,e^{\Lambda(\vartheta|t_0)+M(\vartheta|t_0)}}{1-(s-1)\,B(\vartheta|t_0)}\,d\vartheta\right\}$$
$$\times\sum_{i=0}^{\ell}\gamma_i\left[1+\frac{(s-1)\,e^{\Lambda(u|t_0)+M(u|t_0)}}{1-(s-1)\,B(u|t_0)}\right]^i, \tag{A4}$$

where the use of the third of (A2) has been made. From (A3) with $\psi=t$, we also obtain

$$s=\frac{1+(z-1)\,b_1(t|t_0)}{1+(z-1)\,b_2(t|t_0)}, \tag{A5}$$

with $b_1(t|t_0)$ and $b_2(t|t_0)$ defined in (37). By virtue of (A5), one has:

$$(s-1)\int_{t_0}^{t}\frac{\lambda(u)\,e^{\Lambda(u|t_0)+M(u|t_0)}}{1-(s-1)\,B(u|t_0)}\,du=\ln\big[1+(z-1)\,b_2(t|t_0)\big],$$
$$1+\frac{(s-1)\,e^{\Lambda(u|t_0)+M(u|t_0)}}{1-(s-1)\,B(u|t_0)}=\frac{1+(z-1)\,b_1(t|u)}{1+(z-1)\,b_2(t|u)}. \tag{A6}$$

Finally, recalling that $\psi=t$ and making use of (A5) and (A6), from (A4) one derives (36).

References

1. Anderson, W.J. *Continuous-Time Markov Chains: An Applications-Oriented Approach*; Springer Series in Statistics; Springer: New York, NY, USA, 1991.
2. Iosifescu, M.; Tautu, P. *Stochastic Processes and Applications in Biology and Medicine II. Models*; Springer: Berlin/Heidelberg, Germany, 1973.
3. Medhi, J. *Stochastic Models in Queueing Theory*; Academic Press: Amsterdam, The Netherlands, 2003.
4. Bailey, N.T.J. *The Elements of Stochastic Processes with Applications to the Natural Sciences*; John Wiley & Sons, Inc.: New York, NY, USA, 1964.
5. van Kampen, N.G. *Stochastic Processes in Physics and Chemistry*; Elsevier Science: Amsterdam, The Netherlands, 1992.
6. Taylor, H.M.; Karlin, S. *An Introduction to Stochastic Modeling*; Academic Press: Boston, MA, USA, 1994.
7. Sericola, B. *Markov Chains: Theory, Algorithms and Applications*; John Wiley & Sons, Inc.: Hoboken, NJ, USA, 2013.
8. Dharmaraja, S.; Di Crescenzo, A.; Giorno, V.; Nobile, A.G. A continuous-time Ehrenfest model with catastrophes and its jump-diffusion approximation. *J. Stat. Phys.* **2015**, *161*, 326–345. [CrossRef]
9. Giorno, V.; Nobile, A.G. On a bilateral linear birth and death process in the presence of catastrophe. In *Computer Aided Systems Theory-EUROCAST 2013, Part I*; Moreno-Diaz, R., Pichler, F., Quesada-Arencibia, A., Eds.; LNCS 8111; Springer: Berlin/Heidelberg, Germany, 2013; pp. 28–35.

10. Giorno, V.; Nobile, A.G.; Spina, S. On some time non-homogeneous queueing systems with catastrophes. *Appl. Math. Comp.* **2014**, *245*, 220–234. [CrossRef]
11. Giorno, V.; Nobile, A.G.; Pirozzi, E. A state-dependent queueing system with asymptotic logarithmic distribution. *J. Math. Anal. Appl.* **2018**, *458*, 949–966. [CrossRef]
12. Di Crescenzo, A.; Giorno, V.; Nobile, A.G. Constructing transient birth-death processes by means of suitable transformations. *Appl. Math. Comp.* **2016**, *281*, 152–171. [CrossRef]
13. Economou, A.; Fakinos, D. A continuous-time Markov chain under the influence of a regulating point process and applications in stochastic models with catastrophes. *Eur. J. Oper. Res.* **2003**, 149, 625–640. [CrossRef]
14. Economou, A.; Fakinos, D. Alternative approaches for the transient analysis of Markov chains with catastrophes. *J. Stat. Theory Pract.* **2008**, *2*, 183–197. [CrossRef]
15. Chen, A.; Zhang, H.; Liu, K.; Rennolls, K. Birth-death processes with disasters and instantaneous resurrection. *Adv. Appl. Probab.* **2004**, *36*, 267–292. [CrossRef]
16. Di Crescenzo, A.; Giorno, V.; Krishna Kumar, B.; Nobile, A.G. A double-ended queue with catastrophes and repairs, and a jump-diffusion approximation. *Method. Comput. Appl. Probab.* **2012**, *14*, 937–954. [CrossRef]
17. Di Crescenzo, A.; Giorno, V.; Krishna Kumar, B.; Nobile, A.G. A time-non-homogeneous double-ended queue with failures and repairs and its continuous approximation. *Mathematics* **2018**, *6*, 81. [CrossRef]
18. Ye, J.; Liu, L.; Jiang, T. Analysis of a Single-Sever Queue with Disasters and Repairs Under Bernoulli Vacation Schedule. *J. Syst. Sci. Inf.* **2016**, *4*, 547–559. [CrossRef]
19. Mytalas, G.C.; Zazanis, M.A. An $M^X/G/1$ queueing system with disasters and repairs under a multiple adapted vacation policy. *Nav. Res. Logist.* **2015**, *62*, 171–189. [CrossRef]
20. Krishna Kumar, B.K.; Krihnamoorthy, A.; Madheswari, S.P.; Basha, S.S. Transient analysis of a single server queue with catastrophes, failures, and repairs. *Queueing Syst.* **2007**, *56*, 133–141. [CrossRef]
21. Altiok, T. On the phase-type approximations of general distributions. *IIE Trans.* **1985**, *17*, 110–116. [CrossRef]
22. Altiok, T. Queueing modeling of a single processor with failures. *Perform. Eval.* **1989**, *9*, 93–102. [CrossRef]
23. Altiok, T. *Performance Analysis of Manufacturing Systems*; Springer Series in Operations Research; Springer: New York, NY, USA, 1997.
24. Dallery, Y. On modeling failure and repair times in stochastic models of manufacturing systems using generalized exponential distributions. *Queueing Syst.* **1994**, *15*, 199–209. [CrossRef]
25. Kendall, D.G. On the generalized "birth-and-death"process. *Ann. Math. Stat.* **1948**, *19*, 1–15. [CrossRef]
26. McNeil, D.R.; Schach, S. Central limit analogues for Markov population processes. *J. R. Stat. Soc. Ser. B (Methodol.)* **1973**, *35*, 1–23. [CrossRef]
27. Di Crescenzo, A.; Nobile, A.G. Diffusion approximation to a queueing system with time dependent arrival and service rates. *Queueing Syst.* **1995**, *19*, 41–62. [CrossRef]
28. Di Crescenzo, A.; Giorno, V.; Krishna Kumar, B.; Nobile, A.G. $M/M/1$ queue in two alternating environments and its heavy traffic approximation. *J. Math. Anal. Appl.* **2018**, *458*, 973–1001. [CrossRef]
29. Giorno, V.; Nobile, A.G.; Ricciardi, L.M. On some time-nonhomogeneous diffusion approximations to queueing systems. *Adv. Appl. Prob.* **1987**, *19*, 974–994. [CrossRef]
30. Giorno, V.; Nobile, A.G. On a class of birth-death processes with time-varying intensity functions. *Appl. Math. Comput.* **2020**, *379*, 125255. [CrossRef]
31. Ammar, S.I.; Zeifman, A.; Satin, Y.; Kiseleva, K.; Korolev, V. On limiting characteristics for a non stationary two processor heterogeneous system with catastrophes, server failures and repairs. *J. Ind. Manag. Optim.* **2021**, *17*, 1057–1068. [CrossRef]
32. Zeifman, A.I.; Isaacson, D.L. On strong ergodicity for nonhomogeneous continuous-time Markov chains. *Stoch. Process. Appl.* **1994**, *50*, 263–273. [CrossRef]
33. Zeifman, A.; Satin, Y.; Korolev, V.; Shorgin, S. On truncations for weakly ergodic inhomogeneous birth and death processes. *Int. J. Appl. Math. Comput. Sci.* **2014**, *24*, 503–518. [CrossRef]
34. Satin, Y.A.; Zeifman, A.I.; Shilova, G.N. On approaches to constructing limiting regimes for some queuing models. *Inform. Primen.* **2020**, *14*, 3–9.
35. Jouini, O.; Dallery, Y. Moments of first passage times in general birth-death processes. *Math. Meth. Oper. Res.* **2008**, *68*, 49–76. [CrossRef]
36. Giorno, V.; Nobile, A.G. First-passage times and related moments for continuous-time birth-death chains. *Ric. Mat.* **2019**, *68*, 629–659. [CrossRef]
37. Prendiville, B.J. Discussion: Symposium on stochastic processes. *J. Roy. Statist. Soc. B* **1949**, *11*, 273.
38. Takashima, M. Note on evolutionary processes. *Bull. Math. Stat.* **1956**, *7*, 18–24. [CrossRef]
39. Giorno, V.; Negri, C.; Nobile, A.G. A solvable model for a finite capacity queueing system. *J. Appl. Prob.* **1985**, *22*, 903–911. [CrossRef]
40. Ricciardi, L.M. Stochastic Population Theory: Birth and Death Processes. In *Mathematical Ecology. Biomathematics*; Hallam, T.G., Levin, S.A., Eds.; Springer: Berlin/Heidelberg, Germany, 1986; Volume 17, pp. 155–190.
41. Karlin, S.; McGregor, J. Ehrenfest Urn Model. *J. Appl. Prob.* **1965**, *2*, 352–376. [CrossRef]

42. Flegg, M.B.; Pollett, P.K.; Gramotnev, D.K. Ehrenfest model for condensation and evaporation processes in degrading aggregates with multiple bonds. *Phys. Rev. E* **2008**, *78*, 031117. [CrossRef]
43. Zheng, Q. Note on the non-homogeneous Prendiville process. *Math. Biosci.* **1998**, *148*, 1–5. [CrossRef]
44. Giorno, V.; Nobile, A.G.; Saura, A. Prendiville Stochastic Growth Model in the Presence of Catastrophes. In *Cybernetics and Systems 2004, Proceedings of the 17th European Meeting on Cybernetics and Systems Research, Vienna, Austria, 13–16 April 2004*; Trappl, R., Ed.; Austrian Society for Cybernetic Studies: Vienna, Austria, 2004; pp. 151–156.
45. Giorno, V.; Nobile, A.G.; Spina, S. Some Remarks on the Prendiville Model in the Presence of Jumps. In *Computer Aided Systems Theory–EUROCAST 2019*; Moreno-Díaz, R., Pichler, F., Quesada-Arencibia, A., Eds.; LNCS 12013; Springer: Berlin/Heidelberg, Germany, 2020; pp. 150–157.
46. Parthasarathy, P.R.; Krishna Kumar, B. Stochastic Compartmental models with Prendiville growth mechanisms. *Math. Biosci.* **1995**, *125*, 51–60. [CrossRef]
47. Matis, J.H.; Kiffe, T.R. Stochastic Compartment models with Prendiville growth rates. *Math. Biosci.* **1996**, *138*, 31–43. [CrossRef]
48. Buonocore, A.; Di Crescenzo, A.; Giorno, V.; Nobile, A.G.; Ricciardi, L.M. A Markov chain-based model for actomyosin dynamics. *Sci. Math. Jpn.* **2009**, *70*, 159–174.
49. Giorno, V.; Nobile, A.G. Bell polynomial approach for time-inhomogeneous linear birth-death process with immigration. *Mathematics* **2020**, *8*, 1123. [CrossRef]
50. Gradshteyn, I.S.; Ryzhik, I.M. *Table of Integrals, Series and Products*; Academic Press Inc.: Cambridge, MA, USA, 2014.
51. Williams, W.E. *Partial Differential Equations*; Clarendon Press: Oxford, UK, 1980.

MDPI

Article

Conditions for Existence of Second-Order and Third-Order Filters for Discrete Systems with Additive Noises

Mikhail Kamenshchikov [1,2,3]

[1] Department of Nonlinear Dynamical Systems and Control Processes, Faculty of Computational Mathematics and Cybernetics, Lomonosov Moscow State University, 119991 Moscow, Russia; mkamenshchikov@cs.msu.ru or kamenshchikov.ma@misis.ru
[2] Laboratory of Terminal Control Systems, V. A. Trapeznikov Institute of Control Sciences, Russian Academy of Sciences, 117997 Moscow, Russia
[3] Department of Engineering Cybernetics, National University of Science and Technology «MISiS», 119049 Moscow, Russia

Abstract: The problem of constructing functional optimal observers (filters) for stochastic control systems with additive noises in discrete time are studied in this work. Under the assumption that there is no filter of the first order, necessary and sufficient conditions for the existence of filters of the second and third order are obtained in the canonical basis. Analytical expressions of the transfer function matrix from the input noise to the estimation error are presented. A numerical example is given to compare the performance of filters by the quadratic criterion in the steady state.

Keywords: discrete time functional filter; optimal unbiased estimation; steady state

Citation: Kamenshchikov, M. Conditions for Existence of Second-Order and Third-Order Filters for Discrete Systems with Additive Noises. *Mathematics* **2022**, *10*, 370. https://doi.org/10.3390/math10030370

Academic Editor: Leonid Piterbarg

Received: 12 December 2021
Accepted: 22 January 2022
Published: 25 January 2022

1. Introduction

The reduced-order filtering problem occupies an important place in the theory of optimal state estimation. Instead of the traditionally used Kalman filter, which forms an estimate of the total system state vector and has an order that coincides with the order of the system, it is proposed to construct its analogue, a functional filter with a reduced dimension. In this case, the computational effort to implement a functional filter is reduced. In addition, the reduced order of the filter simplifies the analysis of the dynamic system.

The problem under study is at the intersection of two classical problems of state estimation theory: the full order filtering problem for stochastic systems and the design functional observer problem for deterministic systems. The first problem relates to the filtration theory and was solved for non-linear case (even for the non-stationary case) in 1959–1960 by Stratonovich [1,2], and for linear case in 1960–1961 by Kalman and Bucy [3,4], for both continuous and discrete time. The solution to the second problem of constructing functional observers for linear stationary fully defined systems was proposed in 1966 by Luenberger [5]. The further development of the theory of functional observers is reflected in detail in the books by O'Reilly [6] and Korovin and Fomichev [7]. In particular, in [7], the conditions of existence and algorithms for the synthesis of functional observers for linear stationary fully deterministic systems are given for various cases, namely scalar and vector output; and scalar and vector functional. Two methods of solving the design functional observer problem are also proposed: the pseudo-input method and the scalar observer method. Both methods allow one to obtain necessary and sufficient conditions for the existence of functional observers of order k ($k < \nu - 1$, where ν is the observability index of the system), which were first proposed in [8,9].

Much attention has been paid to the construction of the reduced-order filters for linear systems. Minimizing the quadratic error criterion over the interval and using the solution of the two-point boundary value problem, the reduced-order filter is designed in [10,11]. Based on the quasi-diagonal matrix decomposition and the solution of the Riccati and

Lyapunov matrix equations, a method is proposed in [12] for determining the parameters of continuous and digital linear filters of reduced order which ensures their asymptotic stability provided that the estimated system is stabilizable and detectable. In [13–15], the proof of the uniqueness of the optimal unbiased reduced-order filter and the properties of the reduced-order innovation process in continuous and discrete time are proposed. Developing the results obtained in [13], the necessary and sufficient conditions for existence, stability, and convergence of the designed filter are obtained for both continuous and discrete stochastic systems in [14], and for discrete stochastic systems with unknown inputs in [15]. In [16], a method for the synthesis of functional optimal observers in the frequency domain using spectral factorization in continuous and discrete time is proposed, and the transfer function of the filter and the properties of its associated innovation sequence are obtained. In [17], using a model reduction of the original system and solving the Lyapunov equations involved in each iteration of the optimum search algorithm, a simple method for reduced-order $\mathcal{H}_2$ filter design is proposed. An approach to design a sliding mode control-based functional observer for discrete-time stochastic systems, existence conditions, and stability analysis of the proposed observer are given in [18]. In [19], a generalization of the classical unbiasedness condition in the joint problem of stabilization and optimal filtering is presented, and an alternative method for constructing reduced-order filters is proposed, based on reduction to a non-linear optimization problem. Conditions for existence of second-order and third-order filters for systems in continuous time with additive noises are proposed in [20]. In [21,22], the frequency-weighted $\mathcal{H}_2$-optimal model order reduction problem is investigated, and algorithms are proposed, that constructs a reduced-order model, which nearly satisfies the first-order optimality conditions.

In practice, reduced-order optimal filters are used in signal processing of inertial navigation systems, in health parameter estimation for an aircraft turbofan engine, in induction motor state estimation, in dynamic image analysis, in restoration of progressive and interlaced video, in separation of heart and respiratory sounds, in meteorology and oceanography applications (see [23] and references therein).

This article proposes an approach to constructing reduced-order filters, which differs from the methods in [13–16] in that the filter order does not necessarily coincide with the dimension of the estimated functional. Unlike the methods in [12,17,21,22], where Lyapunov equations are used to calculate the quality criterion, this article uses the method of integral quadratic performance measures, which makes it possible to determine the dependence on parameters in an explicit form.

The problem formulation is presented in Section 2, where the scalar linear functional of the state vector is estimated from the measured scalar output. Perturbations are white random processes with a priori known probabilistic characteristics, uncorrelated with each other at different times and with the initial state of the system. The root-mean-square error in the steady state is chosen as a criterion of optimality. In Section 3, necessary and sufficient conditions for the existence of filters of the second and third order are obtained using canonical forms. Analytical expressions for the transfer function matrix are given in Section 4. The dependence of the parameters number of second-order and third-order filters on the order of original system is presented. Section 5 contains an illustrative example of comparing second and third order filters by quadratic criterion in steady state. Section 6 summarizes the article.

The mathematical notations used in this text are listed in Table 1.

Table 1. Mathematical notations.

Notation	Meaning
0	zero; zero vector; zero matrix
$\in$	belongs to
$\mathbb{R}$	the set of real numbers
$\cdot^{\mathsf{T}}$	transposition of a vector or matrix
i	integer
j	$\sqrt{-1}$; integer
$\mathbb{E}[\cdot]$	the mathematical expectation operator
δ_{ij}	Kronecker delta
$\triangleq$	equal by definition
$\det(\cdot)$	determinant of the square matrix
z	z-transform variable
I_k	identity matrix of dimension $k \times k$
θ	normalized angular frequency
ω	angular frequency
$\cdot^{-1}$	inverse of the square matrix
$\lvert \cdot \rvert$	magnitude or absolute value of a complex scalar

2. Problem Statement

Consider an n–dimensional linear discrete system with stochastic perturbations and with a scalar output:

$$\begin{aligned} x_{i+1} &= Ax_i + Bu_i + w_i, \\ y_i &= Cx_i + v_i, \end{aligned} \qquad i \geq 0, \tag{1}$$

where $x_i \in \mathbb{R}^n$ is the unknown phase vector, $u_i \in \mathbb{R}^m$ is the known input of the system, and $y_i \in \mathbb{R}$ is the measured output of the system; A, B, C are constant matrices of appropriate sizes, w_i, v_i are discrete uncorrelated, and zero mean white noise processes of dimensions n and 1, respectively, with given covariance matrices $\mathbb{E}[w_i w_j^{\mathsf{T}}] = Q\delta_{ij}$, $\mathbb{E}[v_i v_j] = R\delta_{ij}$; the initial state x_0 is a random variable uncorrelated with noises w_i, v_i, and has $\mathbb{E}[x_0] = \bar{x}_0$, $\mathbb{E}[(x_0 - \bar{x}_0)(x_0 - \bar{x}_0)^{\mathsf{T}}] = P_0$. Here, Q, P_0 are positive semidefinite matrices; and $R > 0$. These assumptions can be represented as

$$\mathbb{E}\left[\begin{pmatrix} w_i \\ v_i \\ x_0 \end{pmatrix} \begin{pmatrix} w_j^{\mathsf{T}} & v_j & x_0^{\mathsf{T}} & 1 \end{pmatrix}\right] = \begin{bmatrix} Q\delta_{ij} & 0 & 0 & 0 \\ 0 & R\delta_{ij} & 0 & 0 \\ 0 & 0 & P_0 + \bar{x}_0\bar{x}_0^{\mathsf{T}} & \bar{x}_0 \end{bmatrix}, \quad i \geq 0,\ j \geq 0.$$

It is also assumed that the matrices Q, R, P_0 are known a priori. The first equation in the system (1) can be understood in the sense of the stochastic difference equation [24].

Required based on the observation of the output y_i and the known input u_i, define an unbiased estimate $\tilde{\sigma}_i$ for scalar functional

$$\sigma_i = Fx_i, \quad i \geq 0, \tag{2}$$

with the known matrix $F \in \mathbb{R}^{1\times n}$, providing the minimum of the steady state mean value of the squared observation error $e_i \triangleq \sigma_i - \tilde{\sigma}_i$:

$$J = \lim_{i\to\infty} \mathbb{E}[e_i^2] \to \min. \tag{3}$$

3. Filter Design

Let matrix F have the standard decomposition [6,7]

$$F = PT + VC,$$

where $P \in \mathbb{R}^{1\times k}$, $T \in \mathbb{R}^{k\times n}$, and $V \in \mathbb{R}$. Then, $\sigma_i = Pq_i + Vy_i - Vv_i$, where $q_i = Tx_i \in \mathbb{R}^k$ is an unknown vector to be estimated. To reconstruct it, we use an observer of order k

$$\begin{aligned} \widetilde{q}_{i+1} &= N\widetilde{q}_i + TBu_i + My_i, \quad \widetilde{q}_0 = T\bar{x}_0, \\ \widetilde{\sigma}_i &= P\widetilde{q}_i + Vy_i, \end{aligned} \qquad i \geq 0, \tag{4}$$

where $\widetilde{q}_i \in \mathbb{R}^k$ is the phase vector of the observer; N, M are constant matrices of appropriate sizes. In the second equation of the observer, the output y_i of the original system (1) appears, which makes it possible to obtain an advantage in terms of the quadratic criterion over the filter without it.

Without loss of generality, we make the following standard [6] assumptions regarding the original system (1) and the desired filter (4).

Assumption 1. *The pair $\{C, A\}$ is observable and is given in the second canonical form of observability [7]*

$$A = \begin{pmatrix} 0 & 0 & \dots & 0 & -\alpha_1 \\ 1 & 0 & \dots & 0 & -\alpha_2 \\ \dots & \dots & \dots & \dots & \dots \\ 0 & 0 & \dots & 1 & -\alpha_n \end{pmatrix}, \quad C = \begin{pmatrix} 0 & \dots & 0 & 1 \end{pmatrix}, \tag{5}$$

where α_i is the coefficients of the characteristic polynomial of the matrix A, i.e.,

$$\alpha(z) = \det(zI_n - A) = z^n + \alpha_n z^{n-1} + \ldots + \alpha_1.$$

The matrix F in the canonical basis has the form:

$$F = \begin{pmatrix} f_1 & f_2 & \dots & f_n \end{pmatrix}.$$

Assumption 2. *The pair $\{P, N\}$ is observable and is given in the first canonical form of observability [7]*

$$N = \begin{pmatrix} 0 & 1 & 0 & \dots & 0 \\ 0 & 0 & 1 & \dots & 0 \\ \dots & \dots & \dots & \dots & \dots \\ 0 & 0 & 0 & \dots & 1 \\ -l_1 & -l_2 & -l_3 & \dots & -l_k \end{pmatrix}, \quad P = \begin{pmatrix} 1 & 0 & \dots & 0 \end{pmatrix}, \tag{6}$$

where l_i is the coefficients of the characteristic polynomial of the matrix N, i.e.,

$$\beta(z) = \det(zI_k - N) = z^k + l_k z^{k-1} + \ldots + l_1.$$

Let us investigate the question of when linear filters of the second ($k = 2$) and third ($k = 3$) order can estimate the functional (2) from the state vector. In addition, it is assumed that there is no first-order ($k = 1$) filter giving an unbiased estimate for the functional (2). Discrete-time filters of various orders starting from the first order were considered in [25].

Theorem 1. *For system (1) of order higher than the third ($n > 3$) with stochastic perturbations and filters (4) of the second and third order, giving an unbiased estimate of the functional (2) from the state vector, it is true that:*

(1) the necessary and sufficient conditions for the existence of a second-order filter have the form

$$T = \begin{pmatrix} f_1 & f_2 & \dots & f_{n-1} & f_n - V \\ f_2 & f_3 & \dots & f_n - V & t_{2n} \end{pmatrix},$$

$$M = \begin{pmatrix} -\sum_{i=1}^{n} \alpha_i f_i - t_{2n} + \alpha_n V \\ -\sum_{i=1}^{n-1} \alpha_i f_{i+1} - (\alpha_n - l_2)t_{2n} + l_1(f_n - V) + \alpha_{n-1} V \end{pmatrix},$$

$$V = f_n + l_1 f_{n-2} + l_2 f_{n-1},\ t_{2n} = -l_1 f_{n-1} - l_2(f_n - V),\ f_1 f_3 - f_2^2 \neq 0,$$

$$l_1 = \frac{f_2 a - f_3^2}{f_1 f_3 - f_2^2},\ l_2 = \frac{f_2 f_3 - f_1 a}{f_1 f_3 - f_2^2},\ 1 - l_1 > 0,\ 1 - l_2 + l_1 > 0,\ 1 + l_1 + l_2 > 0, \tag{7}$$

$$a = \begin{cases} f_4, & \text{if } n > 4, \\ f_4 - V, & \text{if } n = 4; \end{cases} \quad f_i = -l_1 f_{i-2} - l_2 f_{i-1},\ i = 5, \dots, n-1, \text{ for } n > 5,$$

where the condition $f_1 f_3 - f_2^2 \neq 0$ means that the observer (4) of the first order cannot reconstruct the unbiased estimate of the functional (2);
(2) the necessary and sufficient conditions for the existence of a third-order filter have the form

$$T = \begin{pmatrix} f_1 & f_2 & \dots & f_{n-2} & f_{n-1} & f_n - V \\ f_2 & f_3 & \dots & f_{n-1} & f_n - V & t_{2n} \\ f_3 & f_4 & \dots & f_n - V & t_{2n} & t_{3n} \end{pmatrix},$$

$$M = \begin{pmatrix} -\sum_{i=1}^{n} \alpha_i f_i - t_{2n} + \alpha_n V \\ -\sum_{i=1}^{n-1} \alpha_i f_{i+1} - \alpha_n t_{2n} - t_{3n} + \alpha_{n-1} V \\ -\sum_{i=1}^{n-2} \alpha_i f_{i+2} - (\alpha_{n-1} - l_2)t_{2n} - (\alpha_n - l_3)t_{3n} + l_1(f_n - V) + \alpha_{n-2} V \end{pmatrix},$$

$$V = f_n + l_1 f_{n-3} + l_2 f_{n-2} + l_3 f_{n-1},$$

$$t_{2n} = -l_1 f_{n-2} - l_2 f_{n-1} - l_3(f_n - V),\ t_{3n} = -l_1 f_{n-1} - l_2(f_n - V) - l_3 t_{2n},$$

$$b \neq \frac{f_3(f_3^2 - f_2 a) + a(f_1 a - f_2 f_3)}{f_1 f_3 - f_2^2}, \tag{8}$$

$$l_1 = \frac{a(a^2 - f_3 b) + b(f_2 b - f_3 a) + c(f_3^2 - f_2 a)}{b(f_1 f_3 - f_2^2) - f_3(f_3^2 - f_2 a) - a(f_1 a - f_2 f_3)},$$

$$l_2 = \frac{a(f_2 b - f_3 a) + b(f_3^2 - f_1 b) + c(f_1 a - f_2 f_3)}{b(f_1 f_3 - f_2^2) - f_3(f_3^2 - f_2 a) - a(f_1 a - f_2 f_3)},$$

$$l_3 = \frac{a(f_3^2 - f_2 a) + b(f_1 a - f_2 f_3) + c(f_2^2 - f_1 f_3)}{b(f_1 f_3 - f_2^2) - f_3(f_3^2 - f_2 a) - a(f_1 a - f_2 f_3)},$$

$$1 - l_1^2 > 0,\ l_1^2 - 1 < l_1 l_3 - l_2,\ 1 + l_3 + l_2 + l_1 > 0,\ -1 + l_3 - l_2 + l_1 < 0, \tag{9}$$

$$a = \begin{cases} f_4, & \text{if } n > 4, \\ f_4 - V, & \text{if } n = 4; \end{cases} \quad b = \begin{cases} f_5, & \text{if } n > 5, \\ f_5 - V, & \text{if } n = 5, \\ t_{24}, & \text{if } n = 4; \end{cases} \quad c = \begin{cases} f_6, & \text{if } n > 6, \\ f_6 - V, & \text{if } n = 6, \\ t_{25}, & \text{if } n = 5, \\ t_{34}, & \text{if } n = 4; \end{cases}$$

$$f_i = -l_1 f_{i-3} - l_2 f_{i-2} - l_3 f_{i-1},\ i = 7, \dots, n-1, \text{ for } n > 7,$$

where the condition (8) for the case $n > 5$ means that the observer (4) of the second order cannot reconstruct the unbiased estimate of the functional (2).

Proof. Using the stochastic difference equations of the original system (1) and the observer (4), it is not difficult to obtain that the estimation error $e_i^q \triangleq q_i - \widetilde{q}_i$ is described by the equation

$$e_{i+1}^q = q_{i+1} - \widetilde{q}_{i+1} = Tx_{i+1} - N\widetilde{q}_i - TBu_i - My_i$$
$$= TAx_i - N(q_i - e_i^q) - MCx_i + Tw_i - Mv_i$$
$$= Ne_i^q + (TA - MC - NT)x_i + Tw_i - Mv_i. \quad (10)$$

The equation for the error $e_i = \sigma_i - \widetilde{\sigma}_i$ has the form

$$e_i = \sigma_i - \widetilde{\sigma}_i = Fx_i - P\widetilde{q}_i - Vy_i = PTx_i + VCx_i - P\widetilde{q}_i - VCx_i - Vv_i = Pe_i^q - Vv_i. \quad (11)$$

Based on the known results [6], we can conclude that the estimates $\widetilde{q}_i$ and $\widetilde{\sigma}_i$ are unbiased for q_i and σ_i, respectively, if and only if the following conditions are satisfied:

$$F = PT + VC, \quad TA - MC - NT = 0, \quad N \text{ is a Schur matrix.} \quad (12)$$

Moreover, if the matrix N is a Schur matrix, then [26] the observation error e_i^q in the steady state is a stationary in the wide sense random process, in which the mathematical expectation is constant, and the correlation function depends on one variable.

Both statements of the theorem are obtained in a similar way from the conditions (12), Assumption 1 about the canonical representations of the original system (5) and Assumption 2 about the canonical representations of the desired filter (6). □

Remark 1. *Inequalities in Formulas (7) and (9) are discrete stability constraints for the filter (4) obtained using the simplified stability criterion [27] for linear discrete systems.*

Remark 2. *If the condition (8) is violated, then there is a set of the degenerate third-order observers whose coefficients of the characteristic polynomial according to Vieta's formulas have the form*

$$l_1 = z_1 z_2 (C_1 + z_1 + z_2),\ l_2 = z_1 z_2 - (z_1 + z_2)(C_1 + z_1 + z_2),\ l_3 = C_1$$

and are located at the intersection of the domain of discrete stability of the matrix N and the solution set for the system of equations

$$z_1^3 + l_3 z_1^2 + l_2 z_1 + l_1 = 0, \quad z_2^3 + l_3 z_2^2 + l_2 z_2 + l_1 = 0, \quad (13)$$

where z_1, z_2 are the roots of the characteristic polynomial $z^2 + l_2 z + l_1$ with coefficients l_1, l_2 satisfying (7); and they are determined by the quadratic formula

$$z_{1,2} = \frac{f_1 a - f_2 f_3 \pm \sqrt{(f_2 f_3 - f_1 a)^2 - 4(f_2 a - f_3^2)(f_1 f_3 - f_2^2)}}{2(f_1 f_3 - f_2^2)}.$$

C_1 is a free unknown, which is chosen so that the stability conditions (9) are satisfied:

$$-1 - z_1 - z_2 < C_1 < 1 - z_1 - z_2,$$

and the variable a is determined according to the second statement of Theorem 1.

4. Transfer Function Matrix of the Estimation Error System

This section discusses a method for calculating the optimality criterion (3) by interpreting [28] the steady state root-mean-square error as $\mathcal{H}_2$ norm of the weighted transfer matrix of the estimation error system (10) and (11)

$$J = \lim_{i\to\infty} \mathbb{E}[e_i^2] = \frac{1}{2\pi}\int_{-\pi}^{\pi} W_{e\bar{u}}(e^{j\theta})\begin{pmatrix} Q & 0 \\ 0 & R \end{pmatrix} W_{e\bar{u}}^{\mathsf{T}}(e^{-j\theta})d\theta,$$

where the transfer function matrix $W_{e\bar{u}}(z)$ from the vector noise $\bar{u}_i \triangleq (w_i^{\mathsf{T}} \quad v_i)^{\mathsf{T}}$ to the estimation error e_i must be stable and can be found using the following theorem.

Theorem 2. *If the conditions of Theorem 1 are satisfied, then the transfer function matrix $W_{e\bar{u}}(z)$ of the estimation error system has the form*

$$W_{e\bar{u}}(z) = \frac{1}{\beta(z)}(W_{ew}^1(z) \quad \dots \quad W_{ew}^n(z) \quad W_{ev}(z)),$$

(1) in which, for the case of a second-order filter:

$$\begin{gathered} W_{ew}^i(z) = f_i(z+l_2) + f_{i+1},\ i = 1,\dots,n-2, \\ W_{ew}^{n-1}(z) = f_{n-1}(z+l_2) + f_n - V,\ W_{ew}^n(z) = (f_n - V)(z+l_2) + t_{2n}, \\ W_{ev}(z) = -V\beta(z) + (\sum_{i=1}^{n}\alpha_i f_i + t_{2n} - \alpha_n V)(z+l_2) \\ + \sum_{i=1}^{n-1}\alpha_i f_{i+1} + (\alpha_n - l_2)t_{2n} - l_1(f_n - V) - \alpha_{n-1}V, \\ \beta(z) = z^2 + l_2 z + l_1; \end{gathered}$$

(2) in which, for the case of a third-order filter:

$$\begin{gathered} W_{ew}^i(z) = f_i(z^2 + l_3 z + l_2) + f_{i+1}(z+l_3) + f_{i+2},\ i = 1,\dots,n-3, \\ W_{ew}^{n-2}(z) = f_{n-2}(z^2 + l_3 z + l_2) + f_{n-1}(z+l_3) + f_n - V, \\ W_{ew}^{n-1}(z) = f_{n-1}(z^2 + l_3 z + l_2) + (f_n - V)(z+l_3) + t_{2n}, \\ W_{ew}^n(z) = (f_n - V)(z^2 + l_3 z + l_2) + t_{2n}(z+l_3) + t_{3n}, \\ W_{ev}(z) = -V\beta(z) + (\sum_{i=1}^{n}\alpha_i f_i + t_{2n} - \alpha_n V)(z^2 + l_3 z + l_2) \\ +(\sum_{i=1}^{n-1}\alpha_i f_{i+1} + \alpha_n t_{2n} + t_{3n} - \alpha_{n-1}V)(z+l_3) \\ + \sum_{i=1}^{n-2}\alpha_i f_{i+2} + (\alpha_{n-1} - l_2)t_{2n} + (\alpha_n - l_3)t_{3n} - l_1(f_n - V) - \alpha_{n-2}V, \\ \beta(z) = z^3 + l_3 z^2 + l_2 z + l_1. \end{gathered}$$

Proof. The estimation error system (10) and (11) can be written as follows:

$$\begin{gathered} e_{i+1}^q = Ne_i^q + \bar{B}\bar{u}_i, \quad e_i = Pe_i^q + \bar{D}\bar{u}_i, \quad i \geq 0, \\ \bar{B} = (T \quad -M), \quad \bar{D} = (0 \quad -V). \end{gathered}$$

For this system, the transfer function matrix from the input $\bar{u}_i$ to the output e_i is equal to

$$W_{e\bar{u}}(z) = P(zI_k - N)^{-1}\bar{B} + \bar{D}. \tag{14}$$

Using Formula (14), the necessary and sufficient existence conditions of a filter of the appropriate order from Theorem 1, and Assumption 2 on the canonical representation of the filter, we obtain both statements of Theorem 2. Moreover, the pair $\{P, N\}$ is observable by Assumption 2, and the pair $\{N, \bar{B}\}$ is controllable by the condition $f_1 f_3 - f_2^2 \neq 0$ for the second-order filter and by the condition (8) for the third-order filter. Consequently, using the properties [29,30] of the concept of controllability and observability, we obtain that the specified transfer function matrix is irreducible. □

Remark 3. *Depending on the order n of the original system (1) and the order k of the desired filter (4), the transfer function matrix* $W_{e\bar{u}}(z)$ *has unknown parameters indicated in Table 2.*

Table 2. Transfer function matrix parameters.

The Order of the Original System	The Order of the Desired Filter k = 2	k = 3
$n = 4$	V	V, t_{24}, t_{34}
$n = 5$	no parameters	V, t_{25}
$n = 6$		V
$n > 6$		no parameters

Remark 4. *If the condition (8) of Theorem 1 is violated for a third-order filter, then the transfer matrix of the error system can be calculated according to the first statement of Theorem 2.*

There are various ways to find the optimality criterion without calculating the poles of the transfer function. Firstly, calculation J can be reduced to the calculation of integrals

$$\frac{1}{2\pi}\int_{-\pi}^{\pi}\left|\frac{b_0 e^{j\theta k} + b_1 e^{j\theta(k-1)} + \ldots + b_k}{a_0 e^{j\theta k} + a_1 e^{j\theta(k-1)} + \ldots + a_k}\right|^2 d\theta, \tag{15}$$

where the coefficients a_i, b_i depend on unknown parameters of the filter (4) according to Theorem 2 and Remark 3. There are special formulas and tables [31–35] for calculating integrals (15).

Secondly, by bilinear transformation [36], calculation J can be reduced to the calculation of integrals

$$\frac{1}{2\pi}\int_{-\infty}^{\infty}\left|\frac{\widetilde{b}_0(j\omega)^k + \widetilde{b}_1(j\omega)^{k-1} + \ldots + \widetilde{b}_k}{\widetilde{a}_0(j\omega)^{k+1} + \widetilde{a}_1(j\omega)^k + \ldots + \widetilde{a}_{k+1}}\right|^2 d\omega. \tag{16}$$

There are also special formulas and tables [35,37–39] for calculating integrals (16).

Thirdly, a discrete Lyapunov matrix equation can be used [40] to calculate J.

5. Numerical Example

This section presents a numerical example of comparing second and third order filters in terms of the asymptotic quadratic mean observation error.

We consider the system (1) and (2) of the fourth order, in which the matrices A, C are given in the canonical form (5) with $\alpha_1 = 1/16$, $\alpha_2 = 1/2$, $\alpha_3 = 3/2$, $\alpha_4 = 2$; the matrix B is zero matrix; the elements of the matrix F are equal to $f_1 = 1$, $f_2 = 1/2$, $f_3 = 1/3$, $f_4 = 1/4$; the probabilistic characteristics are $Q = P_0 = I_4$, $R = 1$, $\bar{x}_0 = \begin{bmatrix} 1 & 0 & 0 & 0 \end{bmatrix}^{\mathrm{T}}$.

There is no first-order filter reconstructing the unbiased estimate of the scalar functional (2). To find the unknown parameter (V) of the second-order filter (4), we solve the problem of minimizing the optimality criterion (3), which, according to Section 4, is

$$J(V) = \frac{492{,}687{,}360V^4 + 143{,}928{,}576V^3 + 55{,}244{,}160V^2 - 6{,}303{,}956V - 396{,}985}{768(36V+1)(36V+5)(108V-13)}, \tag{17}$$

where the parameter V must be such that the characteristic polynomial of the observer is stable, i.e., $V \in (-1/36, 13/108)$. The function (17) defined over the open interval $(-1/36, 13/108)$ has a global minimum at $V \approx 0.1148$. Figure 1 shows the graph of the function $J(V)$.

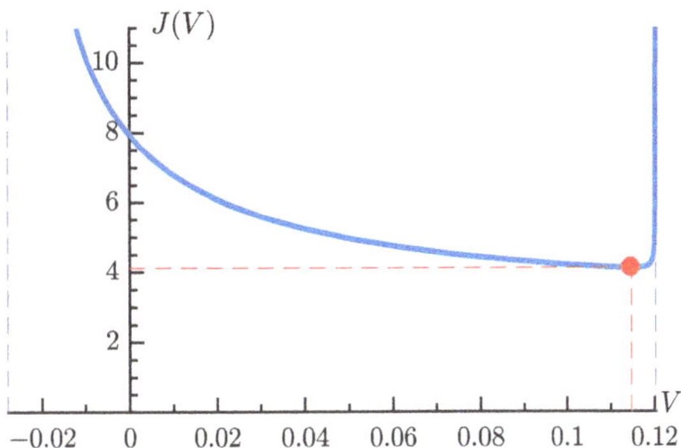

Figure 1. Graph of the function (17). The global minimum of the function $J(V)$ over the open interval $(-1/36, 13/108)$ is the red point ($V \approx 0.1148, J(V) \approx 4.1223$). The two blue dashed lines are the asymptotes $V = -1/36$ and $V = 13/108$.

The numerical values of the second-order filter matrices are

$$P = \begin{pmatrix} 1 & 0 \end{pmatrix}, \; N \approx \begin{pmatrix} 0 & 1 \\ 0.5224 & -0.378 \end{pmatrix},$$

$$T \approx \begin{pmatrix} 1 & 0.5 & 0.3333 & 0.1352 \\ 0.5 & 0.3333 & 0.1352 & 0.123 \end{pmatrix}, \; M \approx \begin{pmatrix} -1.2058 \\ -0.6708 \end{pmatrix}, \; V \approx 0.1148.$$

The steady state mean value of the squared observation error in this case is

$$J \approx 4.1223.$$

If the condition (8) is violated ($t_{24} = 12V^2 - 2V + 7/36$), then, by Remark 2, degenerate third-order observers have the form (4), in which

$$P = \begin{pmatrix} 1 & 0 & 0 \end{pmatrix}, \; N \approx \begin{pmatrix} 0 & 1 & 0 \\ 0 & 0 & 1 \\ -0.1975 + 0.5224C_1 & 0.6653 - 0.378C_1 & -C_1 \end{pmatrix},$$

$$T \approx \begin{pmatrix} 1 & 0.5 & 0.3333 & 0.1352 \\ 0.5 & 0.3333 & 0.1352 & 0.123 \\ 0.3333 & 0.1352 & 0.123 & 0.0241 \end{pmatrix}, \; M \approx \begin{pmatrix} -1.2058 \\ -0.6708 \\ -0.3763 \end{pmatrix}, \; V \approx 0.1148,$$

where the unknown C_1 is chosen so that the stability conditions (9) are satisfied, i.e., $C_1 \in (\underline{C_1}, \overline{C_1})$, $\underline{C_1} \approx -0.622$, $\overline{C_1} \approx 1.378$. According to Remark 4, the transfer function matrix and the optimality criterion in this case are found in the same way as for the second-order filter.

Therefore, if the condition (8) is violated, then there is a set of degenerate observers whose coefficients of the characteristic polynomial $\beta(z)$ are at the intersection of the linear manifold of solutions of the system (13) in which

$$z_1 = \frac{3 - 36V + \sqrt{3}\sqrt{1 + 432V^2}}{6} \approx 0.558,$$
$$z_2 = \frac{3 - 36V - \sqrt{3}\sqrt{1 + 432V^2}}{6} \approx -0.9361;$$

and the domain of discrete stability of the matrix N.

If the condition (8) is satisfied $(t_{24} \neq 12V^2 - 2V + 7/36)$ that there exists a functional optimal observer (4) of the third order solving the optimal filtering problem. In this case, to find unknown variables (V, t_{24}, t_{34}), the problem of minimizing the optimality criterion is solved with a restriction on the parameters that must be such that the characteristic polynomial of the observer is stable. Figure 2 illustrates solution of this problem in discrete stability regions given by the inequalities (9) in coordinates (l_1, l_2, l_3) on Figure 2a and in coordinates (V, t_{24}, t_{34}) on Figure 2b. As one can see, solution paths of the sequential quadratic programming method [41] from different starting points converge to the common minimum of the optimality criterion (4), which has the following coordinates

$$l_1 \approx -0.0119, \quad l_2 \approx 0.2303, \quad l_3 \approx 0.0591;$$

$$V \approx 0.373, \quad t_{24} \approx -0.0636, \quad t_{34} \approx 0.0361.$$

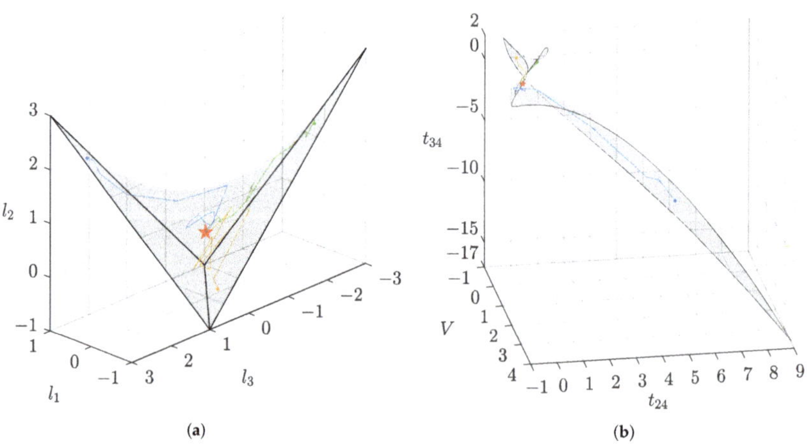

Figure 2. Three convergence paths (blue, green and orange arrows) to the common minimum (red star) of the optimality criterion J from various starting points in the discrete stability regions in coordinates (**a**) l_1, l_2, l_3; (**b**) V, t_{24}, t_{34}.

The numerical values of the third-order filter matrices are

$$P = \begin{pmatrix} 1 & 0 & 0 \end{pmatrix},\ N \approx \begin{pmatrix} 0 & 1 & 0 \\ 0 & 0 & 1 \\ 0.0119 & -0.2303 & -0.0591 \end{pmatrix},$$

$$T \approx \begin{pmatrix} 1 & 0.5 & 0.3333 & -0.123 \\ 0.5 & 0.3333 & -0.123 & -0.0636 \\ 0.3333 & -0.123 & -0.0636 & 0.0361 \end{pmatrix},\ M \approx \begin{pmatrix} -0.503 \\ 0.0776 \\ 0.0528 \end{pmatrix},\ V \approx 0.373.$$

The optimality criterion in this case is

$$J \approx 2.3179.$$

Thus, the optimality criterion (3) for the third-order filter turned out to be less than for the second-order filter. Previously, second and third order filters were compared from both practical and theoretical points of view. In the context of satellite signal processing, it has been shown [42] that increasing the order of the filters led to an improvement in dynamic stress performance. In [17], a smaller value of the $\mathcal{H}_2$ norm for a third-order filter over a second-order filter was obtained on a numerical experiments. Moreover, it has recently been theoretically explained [22] that, as the order of the reduced model was increased, the deviation in the satisfaction of the optimality conditions further reduced.

6. Conclusions

Necessary and sufficient conditions for the existence of discrete unbiased filters of the second and third order are proposed. In the canonical basis, analytical expressions are obtained both for the transfer function matrix of the estimation error system and for the coefficients of the characteristic polynomial of functional filters. In a numerical experiment, filters of the second and third order are constructed for a linear stochastic discrete system of the fourth order. The comparison was carried out according to the root-mean-square optimality criterion. It is shown that, in comparison with the second-order filter, the third-order filter is more optimal in terms of the quadratic criterion in the steady state.

Funding: This research was funded by Russian Foundation for Basic Research grant number 20-37-90065 and grant number 20-08-00073.

Institutional Review Board Statement: Not applicable.

Informed Consent Statement: Not applicable.

Data Availability Statement: Not applicable.

Conflicts of Interest: The author declares no conflict of interest.

References

1. Stratonovich, R.L. Optimum nonlinear systems which bring about a separation of a signal with constant parameters from noise. *Radiofizika* **1959**, *2*, 892–901.
2. Stratonovich, R.L. Application of the Markov processes theory to optimal filtering. *Radio Eng. Electron. Phys.* **1960**, *5*, 1–19.
3. Kalman, R.E. A New Approach to Linear Filtering and Prediction Problems. *J. Basic Eng.* **1960**, *82*, 35–45. [CrossRef]
4. Kalman, R.E.; Bucy, R.S. New Results in Linear Filtering and Prediction Theory. *J. Basic Eng.* **1961**, *83*, 95–108. [CrossRef]
5. Luenberger, D. Observers for multivariable systems. *IEEE Trans. Autom. Control* **1966**, *11*, 190–197. [CrossRef]
6. O'Reilly, J. *Observers for Linear Systems*; Academic Press: London, UK, 1983.
7. Korovin, S.K.; Fomichev, V.V. *State Observers for Linear Systems with Uncertainty*; de Gruyter: Berlin, Germany, 2009.
8. Korovin, S.K.; Il'in, A.V.; Fomichev, V.V.; Hlawenka, A. Asymptotic observers for uncertain vector linear systems. *Dokl. Math.* **2004**, *69*, 483–487.
9. Korovin, S.K.; Medvedev, I.S.; Fomichev, V.V. Functional observers for linear uncertain stationary dynamical systems. *Dokl. Math.* **2006**, *74*, 934–938. [CrossRef]
10. Dombrovskij, V.V. Method of synthesizing suboptimal filters of reduced order for digital linear dynamic systems. *Autom. Remote Control* **1981**, *42*, 1483–1489.
11. Sims, C.S. Reduced-Order Modeling and Filtering. In *Advances in Theory and Applications*; Control and Dynamic Systems; Leondes, C., Ed.; Academic Press: New York, NY, USA, 1982; Volume 18, pp. 55–103. [CrossRef]
12. Dombrovskij, V.V. Synthesis and stability of continuous and discrete low-order linear filters. *Autom. Remote Control* **1989**, *50*, 1533–1539.
13. Nagpal, K.M.; Helmick, R.E.; Sims, C.S. Reduced-order estimation Part 1. Filtering. *Int. J. Control* **1987**, *45*, 1867–1888. [CrossRef]
14. Darouach, M. On the optimal unbiased functional filtering. *IEEE Trans. Autom. Control* **2000**, *45*, 1374–1379. [CrossRef]
15. Darouach, M.; Boutayeb, M.; Zasadzinski, M. Reduced order filtering for stochastic discrete-time systems with unknown inputs. *IFAC Proc. Vol.* **2002**, *35*, 133–138. [CrossRef]
16. Darouach, M.; Ali, H.S. Optimal unbiased functional filtering in the frequency domain. *Syst. Sci. Control Eng.* **2014**, *2*, 308–315. [CrossRef]

17. Aalto, H. Simplified Design of $\mathcal{H}_2$-suboptimal Reduced-order Filters. *IFAC Proc. Vol.* **2013**, *46*, 230–235. [CrossRef]
18. Singh, S.; Janardhanan, S. Sliding mode control-based linear functional observers for discrete-time stochastic systems. *Int. J. Syst. Sci.* **2017**, *48*, 3246–3253. [CrossRef]
19. Kamenshchikov, M.A.; Kapalin, I.V. A procedure for constructing optimum functional filters for linear stationary stochastic systems. *Mosc. Univ. Comput. Math. Cybern.* **2018**, *42*, 163–170. [CrossRef]
20. Fomichev, V.V.; Kamenshchikov, M.A. Comparative analysis of optimal filters of the second and third order for continuous-time systems. *Differ. Equ.* **2021**, *57*, 1527–1535. [CrossRef]
21. Zulfiqar, U.; Sreeram, V.; Ahmad, M.I.; Du, X. Frequency-weighted $\mathcal{H}_2$-pseudo-optimal model order reduction. *IMA J. Math. Control Inf.* **2021**, *38*, 622–653. [CrossRef]
22. Zulfiqar, U.; Sreeram, V.; Ahmad, M.I.; Du, X. Frequency-weighted $\mathcal{H}_2$-optimal model order reduction via oblique projection. *Int. J. Syst. Sci.* **2022**, *53*, 182–198. [CrossRef]
23. Simon, D. Reduced order kalman filtering without model reduction. *Control Intell. Syst.* **2007**, *35*. [CrossRef]
24. Åström, K.J. *Introduction to Stochastic Control Theory*; Academic Press: New York, NY, USA, 1970.
25. Kamenshchikov, M.A. Transfer functions of optimum filters of different dynamic orders for discrete systems. *Mosc. Univ. Comput. Math. Cybern.* **2021**, *45*, 60–70. [CrossRef]
26. Kwakernaak, H.; Sivan, R. *Linear Optimal Control Systems*; Wiley-Interscience: New York, NY, USA, 1972.
27. Jury, E. A Simplified Stability Criterion for Linear Discrete Systems. *Proc. IRE* **1962**, *50*, 1493–1500. [CrossRef]
28. Saberi, A.; Stoorvogel, A.A.; Sannuti, P. *Filtering Theory*; Birkhäuser: Boston, MA, USA, 2007.
29. Kailath, T. *Linear Systems*; Prentice-Hall: Hoboken, NJ, USA, 1980.
30. Chen, C.T. *Linear System Theory and Design*; Oxford University Press: New York, NY, USA, 1999.
31. Tsypkin, Y.Z. *Sampling Systems Theory and Its Application*; Pergamon: Oxford, UK, 1964.
32. Jury, E.I. *Theory and Application of the z-Transform Method*; Wiley: New York, NY, USA, 1964.
33. Jury, E. A note on the evaluation of the total square integral. *IEEE Trans. Autom. Control* **1965**, *10*, 110–111. [CrossRef]
34. Astrom, K.; Jury, E.; Agniel, R. A numerical method for the evaluation of complex integrals. *IEEE Trans. Autom. Control* **1970**, *15*, 468–471. [CrossRef]
35. Jury, E.; Gutman, S. The inner formulation for the total square integral (sum). *Proc. IEEE* **1973**, *61*, 395–397. [CrossRef]
36. Greaves, C.; Gagne, G.; Bordner, G. Evaluation of integrals appearing in minimization problems of discrete-data systems. *IEEE Trans. Autom. Control* **1966**, *11*, 145–148. [CrossRef]
37. James, H.M.; Nichols, N.B.; Phillips, R.S. *Theory of Servomechanisms*; McGraw-Hill: New York, NY, USA, 1947.
38. Popov, E.P. *The Dynamics of Automatic Control Systems*; Pergamon: Oxford, UK, 1962.
39. Jury, E.; Dewey, A. A general formulation of the total square integrals for continuous systems. *IEEE Trans. Autom. Control* **1965**, *10*, 119–120. [CrossRef]
40. Man, F. Evaluation of time-weighted quadratic performance indices for linear discrete systems. *IEEE Trans. Autom. Control* **1970**, *15*, 496–497. [CrossRef]
41. Nocedal, J.; Wright, S. *Numerical Optimization*; Springer Science & Business Media: New York, NY, USA, 2006.
42. Kaplan, E.; Hegarty, C. *Understanding GPS: Principles and Applications*; Artech House: Boston, MA, USA, 2005.

mathematics

MDPI

Article

Optimal Prefetching in Random Trees

Kausthub Keshava [1,†], Alain Jean-Marie [2,*] and Sara Alouf [3]

1 Deloitte India (Offices of the US), Hyderabad 500032, Telangana, India; kkeshava@deloitte.com
2 Inria, University of Montpellier, 34095 Montpellier, France
3 Inria, Université Côte d'Azur, 06902 Sophia Antipolis, France; Sara.Alouf@inria.fr
* Correspondence: Alain.Jean-Marie@inria.fr
† The work of this author was performed partly as a Master's student at IISER Mohali, India, and as an intern at Inria.

Abstract: We propose and analyze a model for optimizing the prefetching of documents, in the situation where the connection between documents is discovered progressively. A random surfer moves along the edges of a random tree representing possible sequences of documents, which is known to a controller only up to depth d. A quantity k of documents can be prefetched between two movements. The question is to determine which nodes of the known tree should be prefetched so as to minimize the probability of the surfer moving to a node not prefetched. We analyzed the model with the tools of Markov decision process theory. We formally identified the optimal policy in several situations, and we identified it numerically in others.

Keywords: prefetching; optimization; Markov decision processes; random trees; Galton–Watson

Citation: Keshava, K.; Jean-Marie, A.; Alouf, S. Optimal Prefetching in Random Trees. *Mathematics* **2021**, *9*, 2437. https://doi.org/10.3390/math9192437

Academic Editors: Alexander Zeifman, Victor Korolev and Alexander Sipin

Received: 31 August 2021
Accepted: 22 September 2021
Published: 1 October 2021

1. Introduction

Prefetching is a basic technique underlying many computer science applications. Its main purpose is to reduce the time needed to access some information by loading it in advance and concurrently with the process that needs this information. From prefetching of data and code in CPUs and memory architectures, to prefetching of web pages and video segments in Internet-based applications, this technique is ubiquitous. Yet, the technique fundamentally involves a tradeoff between access latency and the consumption of resources (memory, network), and the optimization of this tradeoff is not completely understood.

Clearly, the issue here is randomness: the entity in charge of prefetching, let us call it the "controller", does not know in advance what is the precise data access sequence of the process needing the data. It must therefore make decisions based on the current state of said process and its knowledge of the possible evolution. The adequate formalism for modeling optimal decisions in such a context is that of Markov Decision Processes (MDPs). The principle of using Markov decision processes to optimize prefetching in the context of video applications was first demonstrated in [1,2]. The model was extended in [3,4] and further extended in [5].

The basic principle of these models is that the set of (video) documents to be viewed is represented by a directed graph. The nodes represent the documents, and the edges represent the possible transitions: which documents can be viewed after the viewing of the current document is completed. The edges can be labeled with probabilities or frequencies. A random "surfer" alternates viewing periods and moves to another node/document according to the probabilities of the edges. The controller knows where the surfer stands and knows all about the graph, but does not know which way the surfer will go: only the odds. Its decision is to choose which nodes to download during the time the surfer views the current document. The amount of nodes that can be downloaded is constrained by network resources and is called the "prefetching budget". The amount of storage memory available to the controller is assumed to be sufficient: no memory management is involved in the decision. The criterion to be minimized is typically the average number of times the surfer moves to a document that has not been prefetched: this is a measure of the user's dissatisfaction. The criterion might also involve some measure of the waste of network and memory resources. An optimal policy can, in principle, be computed using dynamic programming.

In practical situations, the probabilities that the surfer moves to some new document after viewing the current document are not known a priori. However, these probabilities can be learned from data using Markov models, as in [6–11]. Moreover, the optimal control of a prefetching agent can be approximated using machine learning techniques such as reinforcement learning, as in [2]. A way to evaluate the efficiency of a machine learning algorithm is to test it on a problem for which the exact solution is known. The purpose of this paper is to provide such a benchmark situation, by determining the optimal policy and the minimal possible cost it induces, to which heuristics and learning algorithms can be compared.

While these previous modeling attempts demonstrated that the MDP formalism is flexible enough to take into account many features of a real system, they also illustrate that finding an optimal policy is a complex problem. Indeed, computing an optimal prefetching policy is very hard in general, that is when the graph of documents does not have a particular property. In [12], the authors studied the *feasibility* variant of the problem. There, it was assumed that the controller has a prefetching budget k, representing the number of documents that can be prefetched while some document is viewed. The question is to decide whether k is large enough so that there exists a policy that prefetches all nodes of a graph before the random surfer tries to access them. This is a subproblem of the Markov optimization model: if such a policy exists, the MDP model should find it as a policy that realizes a zero cost. The results of [12,13] concluded that finding the minimum possible k is difficult when the graph is general. Computing the optimal policy in the corresponding MDP must be even more difficult.

However, if the underlying graph is a tree, it was proven in [12] that the minimal budget k that ensures the existence of a costless policy can be computed in polynomial time. The corresponding prefetching strategy is also easy to compute and has the feature of being "connected". This property, which we also call "greedy", means that the controller can choose the documents to download in the set of neighbors (in the document graph) of the documents already downloaded.

The models and results reviewed thus far assumed that the complete space of documents is known to the controller. This ideal situation may be either unrealistic or undesirable. For instance, if the documents are pages on the global web, storing all the knowledge about this graph is probably impossible and also useless since the web surfer will not visit all the graph during a surfing session. Furthermore, since the complexity of the decision grows exponentially with the size of the graph, it may help the controller to limit, on purpose, the size of the known graph to a neighborhood of the current document.

The current literature lacks a model for the optimal prefetching problem, which features a dynamic graph of documents. It also lacks situations where an optimal control can be formally (and not just numerically) identified, even when the underlying graph of documents is static. We fill these gaps in two ways. First, we propose a new optimal prefetching model in which the graph of documents is dynamic. We focused on trees, since those are the simplest graphs, with the potential for having computable solutions as the literature review suggests. Second, we compute exactly the optimal control for some instances of this model. We proceed with an informal description of the model, then we highlight our contribution.

1.1. The Model

We propose to use the modeling described above, but replace the graph of known documents with a tree of depth d. The root of this tree is the current position of the surfer. The tree represents all the possible sequences of d moves of the surfer. After the surfer has moved to one of the neighbors of its current position, a discovery phase adds a new generation of documents at depth d. The rest of the tree is then ignored: the possibility that the surfer moves back to a node already viewed is neglected, as well as the possibility that several paths exist from one node to another. If any of these possibilities happens in practice, the task will become easier to the controller.

In the discovery phase, we assume that a random number of new documents is attached to every leaf of the current tree, with a uniform distribution between 1 and some integer p, which we refer to as the "fanout". In practical graphs of documents, this assumption is not very realistic. The advantages of making such an assumption are that the space of possible configurations will remain finite and that probabilities relative to objects in this space will be easier to write.

As in previous models, the controller is assumed to have a fixed prefetching budget: some integer number k. Given a tree of documents with some nodes already downloaded, the problem is to decide which k nodes to download so as to minimize the cost. We chose as criterion the stationary probability that the surfer moves to a node that is not prefetched. All these elements are converted in the specification of a Markov decision process, with criterion the infinite-horizon average cost. The model has only three parameters: the depth d and fanout p of trees and the prefetching budget k. The question is whether there is a simple rule based on these three parameters that leads to an optimal decision.

1.2. Contribution

The first contribution of this paper is the precise specification of this MDP, in Section 3. This specification is based on sets of trees, presented in Section 2, together with their basic properties.

We then turn to the identification of optimal prefetching policies, in Sections 5–7. The results we obtained include: (a) A bound on the optimal cost in general trees; (b) The characterization of optimal policies in trees with depth 1, arbitrary fanout, and arbitrary budget; (c) The characterization of optimal policies in trees with depth 2, arbitrary fanout, and budget 1; (d) An exploration of the optimal policy in trees with depth 2, budget 2 and fanout less than 5. In the process of obtaining these results, we show, in Section 4, the properties of underlying Markov chains on the "shape of trees", which do not depend on the specific policy used and are of independent interest. We discuss the results and the modeling assumptions in Section 8 and conclude in Section 9.

The main notation that is used throughout the paper is summarized in Table A1 in Appendix A.

2. Preliminaries: Sets of Trees

This section is devoted to the presentation of mathematical objects that are used in the definition of our problem and in its solution.

The state space of the MDP we are about to construct is a set of marked trees. We introduce it now, together with other sets of trees that will be useful in the analysis. We shall use "with fanout p" as a shorthand for "with nodes having between 1 and p sons".

Definition 1 (Trees and marked trees). *Define:*

(a) $\mathcal{T}_{p,d}$ *the set of rooted trees of depth d with fanout p;*

(b) $\mathcal{M}_{p,d}$ *the set of rooted trees of depth d with fanout p and a mark in the set* $\{0,1\}$*;*

(c) $\mathcal{M}^{+}_{p,d}$ *the set of rooted trees of depth d with fanout p and a mark in the set* $\{0,1\}$ *except for leaves that have the mark 0.*

These sets are represented mathematically with the following recursive formulas:

$$\mathcal{T}_{p,0} = \{0\} \tag{1}$$
$$\mathcal{T}_{p,d} = \{0\} \times \text{SEQ}_{1..p}(\mathcal{T}_{p,d-1}) \qquad d \geq 1 \tag{2}$$
$$\mathcal{M}_{p,0} = \{0,1\} \tag{3}$$
$$\mathcal{M}_{p,d} = \{0,1\} \times \text{SEQ}_{1..p}(\mathcal{M}_{p,d-1}) \qquad d \geq 1 \tag{4}$$
$$\mathcal{M}^{+}_{p,1} = \mathcal{M}_{p,0} \times \text{SEQ}_{1..p}(\mathcal{T}_{p,0}) \tag{5}$$
$$\mathcal{M}^{+}_{p,d} = \mathcal{M}_{p,0} \times \text{SEQ}_{1..p}(\mathcal{M}^{+}_{p,d-1}) \qquad d \geq 2. \tag{6}$$

In these expressions, $\text{SEQ}_{1..p}(A)$ denotes, in the notation of [14], a sequence of objects in the set A, with the length between 1 and p. In (1), we associate the constant mark "0" with nodes in unmarked trees. At the risk of being confusing, we shall say that a node in marked trees of $\mathcal{M}_{p,d}$ or $\mathcal{M}^{+}_{p,d}$ is "unmarked" if it has the mark 0. With this convention, we can say that $\mathcal{T}_{p,d} \subset \mathcal{M}^{+}_{p,d} \subset \mathcal{M}_{p,d}$.

A tree t in $\mathcal{M}_{p,d}$ is represented as follows. If the depth is $d = 0$, $t = (\mu)$ where μ is the mark. If $d > 0$, $t = (\mu, s)$ where $\mu = 0$ or $\mu \in \{0,1\}$ depending on the set, and $s = (s_1, \ldots, s_m)$ is a list of length $m \in [1..p]$. The elements of s are called "subtrees". The root nodes of these subtrees are called "sons" of t. The following notation will be useful to designate the components of a tree. Figure 1 illustrates this terminology.

Definition 2 (Mark, subtrees, internal nodes, leaves). *For a tree represented as $t = (\mu, s)$, let $\mu(t) : \mathcal{M}_{p,d} \to \{0,1\}$ denote the mark of the root and $s(t) : \mathcal{M}_{p,d} \to \text{SEQ}_{1..k}(\mathcal{M}_{p,d-1})$ denote the list of subtrees of t. Let also* inode(t) *denote the number of internal nodes in t and* leaves(t) *denote the number of leaves in t.*

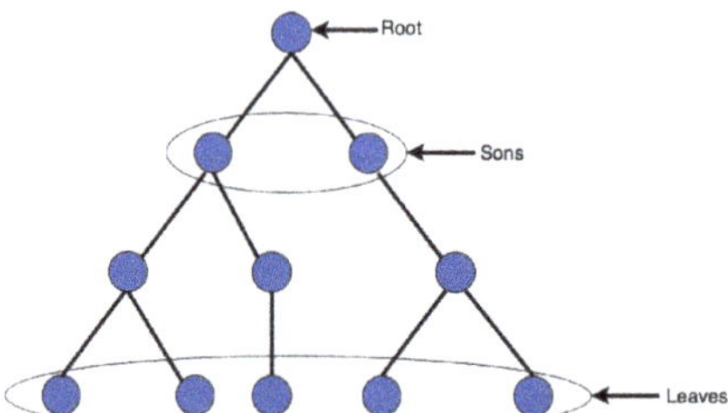

Figure 1. A tree t of depth $d = 3$ with fanout $p = 2$. There are 2 subtrees ($|s(t)| = 2$), 6 internal nodes (inode$(t) = 6$), and 5 leaves (leaves$(t) = 5$).

The cardinal of the sets defined in Definition 1 is important to know, in case we want to turn to numerical experiments. From the recursive definition of the different sets of trees, the following result is easily established.

Lemma 1. *Let $T_{p,d}$, $M_{p,d}$, and $M^{+}_{p,d}$ denote respectively the cardinals of sets $\mathcal{T}_{p,d}$, $\mathcal{M}_{p,d}$ and $\mathcal{M}^{+}_{p,d}$. Then:*

- $T_{p,0} = 1$, $T_{p,1} = p$ *and for* $d \geq 2$,

$$T_{p,d} = \sum_{m=1}^{p} (T_{p,d-1})^m = \frac{(T_{p,d-1})^{p+1} - T_{p,d-1}}{T_{p,d-1} - 1};$$

- $M_{p,0} = 2$ *and for* $d \geq 1$,

$$M_{p,d} = 2\sum_{m=1}^{p}(M_{p,d-1})^m = 2\frac{(M_{p,d-1})^{p+1} - M_{p,d-1}}{M_{p,d-1} - 1};$$

- $M^+_{p,1} = 2p$ *and for* $d \geq 2$,

$$M^+_{p,d} = 2\sum_{m=1}^{p}(M^+_{p,d-1})^m = 2\frac{(M^+_{p,d-1})^{p+1} - M^+_{p,d-1}}{M^+_{p,d-1} - 1}.$$

Similar formulas can be established for generating functions of the size of trees in each set. We shall not develop this analysis further. Table 1 shows the values of $T_{p,d}$, $M_{p,d}$, and $M^+_{p,d}$ for small values of p and d. Clearly, these numbers grow extremely fast with d. The sets remain manageable for small values of p and d. For instance, Figure 2 lists the 6 trees of $\mathcal{T}_{2,2}$.

Table 1. Instances of the cardinals of the sets of rooted trees of depth d with fanout p, when marks are ignored ($T_{p,d}$), when marks are in $\{0,1\}$ ($M_{p,d}$), and when leaves have mark 0 and other nodes have marks in $\{0,1\}$ ($M^+_{p,d}$), for different d and p.

Fanout	Cardinals of Sets $d = 1$			Cardinals of Sets $d = 2$			Cardinals of Sets $d = 3$			Cardinals of Sets $d = 4$		
p	$T_{p,d}$	$M_{p,d}$	$M^+_{p,d}$	$T_{p,d}$	$M_{p,d}$	$M^+_{p,d}$	$T_{p,d}$	$M_{p,d}$	$M^+_{p,d}$	$T_{p,d}$	$M_{p,d}$	$M^+_{p,d}$
1	1	4	2	1	8	4	1	16	8	1	32	16
2	2	12	4	6	312	40	42	195,312	3280	1806	$> 10^{10}$	$> 10^7$
3	3	28	6	39	45,528	516	60,879	$> 10^{14}$	$> 10^8$	$> 10^{14}$	$> 10^{43}$	$> 10^{25}$
4	4	60	8	340	$> 10^7$	9360	$> 10^{10}$	$> 10^{29}$	$> 10^{16}$	$> 10^{40}$	$> 10^{120}$	$> 10^{65}$

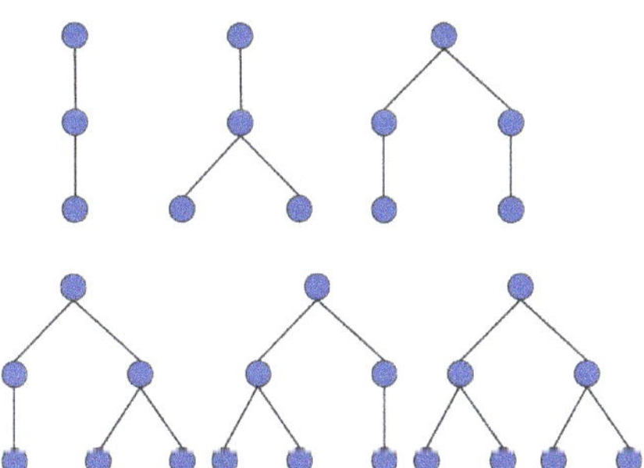

Figure 2. There are 6 trees in $\mathcal{T}_{2,2}$, the set of rooted trees with depth 2 and fanout 2.

3. The MDP Model

In this section, we formally describe the five elements of the prefetching MDP. An MDP model is formally defined by a state space, an action space, transitions, costs, and an evaluation criterion. The state is usually made of numerical variables or discrete structures that summarize the information needed for the following definitions. Actions and transitions specify what controls are allowed to the controller, how they modify the states, and with what probabilities. The cost function, which depends on the states and actions, quantifies the impact of actions. The costs incurred at different time steps are aggregated into a numerical criterion: the objective is to minimize this criterion.

3.1. *Prefetching Process Flow*

The prefetching process flow is summarized as follows. The current state of the prefetching program is the currently known graph of depth d, together with the knowledge of nodes that have been already prefetched. This is represented as a marked tree of depth d. The surfer is assumed to stand at the root. The controller then prefetches up to k

documents, which is represented by marking the corresponding nodes in the tree. Then, the surfer moves randomly to one of the sons, with uniform probabilities among them. If the document corresponding to this node is not already prefetched, then some cost is incurred. Finally, the controller discovers a new generation of nodes. We assume that every possible exploration/discovery of the current subtree is equally likely. After discovery, the controller is back at the beginning of this decision loop.

3.2. State Space and Action Space

According to the prefetching process flow described previously, the state space S of the MDP is $\mathcal{M}^+_{p,d}$ as defined in Definition 1, since the leaves of a state were just discovered and are not prefetched: their mark is 0. For a given tree $t \in S$, the action space A_t is the set of all subsets of the vertices of t with cardinal at most k. An action $a \in A_t$ will have the effect of marking the nodes in a. Some actions will not make much sense: those that mark already marked nodes. We choose to include them nevertheless as possible actions, which greatly simplifies the description of the set A_t. The parameter $k \geq 1$ is called the prefetching budget.

3.3. Transitions

We first formalize the transition between the different trees using random variables and set mappings. Then, we quantify the probabilities of these transitions and the costs.

Definition 3 (Prefetching process random variables). *Define:*

(a) *t the random variable denoting the state (tree) of the prefetching MDP; t takes values in $\mathcal{M}^+_{p,d}$;*
(b) *t_a the random variable denoting the tree after some marking action has been performed; t_a takes values in $\mathcal{M}_{p,d}$;*
(c) *t_b the random variable denoting the tree after the movement of the surfer and before the exploration of the tree; t_b takes values in $\mathcal{M}_{p,d-1}$.*

In the evolution of the controlled process, these three random variables will depend on the time step n. When necessary, we use the notation $t(0), t_a(0), t_b(0), t(1), t_a(1), t_b(1), \ldots, t(n), t_a(n), t_b(n), \ldots$ to denote this (random) succession of trees.

Definition 4 (Discovery). *Let $\mathcal{D} : \mathcal{M}_{p,d-1} \to \mathcal{P}(\mathcal{M}^+_{p,d})$ denote the mapping such that $\mathcal{D}(t)$ is the set of trees that can be discovered from tree $t \in \mathcal{M}_{p,d-1}$. Elements of $\mathcal{D}(t)$ are in $\mathcal{M}^+_{p,d}$ and are obtained from the tree t by updating its leaves according to the following rule: for a leaf $l = (\mu)$ (i.e, depth-0 tree) in t, update it as:*

$$(\mu) \to (\mu, l_{\text{new}}) \text{ where } l_{\text{new}} \in \{(0), (0,0), \cdots, (0)^p\}. \tag{7}$$

Definition 5 (Successors after discovery). *Let $\mathcal{SD} : \mathcal{M}^+_{p,d} \to \mathcal{P}(\mathcal{M}^+_{p,d})$ be defined by:*

$$\mathcal{SD}((\mu, s)) = \bigsqcup_{t_s \in s} \mathcal{D}(t_s) \tag{8}$$

where $\sqcup$ refers to the disjoint union of sets.

The set $\mathcal{SD}(t)$ contains all trees that are the possible results from a combination of surfer movement and the discovery of new leaves.

For a tree $t = (\mu, s) \in \mathcal{M}^+_{p,d}$, let $a(t) \in \mathcal{M}_{p,d}$ denote the tree after marking according to action $a \in A_t$. Then, $s(a(t))$ is the set of subtrees of t after marking action a. The transition probability of moving from a tree t to a tree t' under action a is:

$$P(t,a,t') = \begin{cases} \dfrac{1}{|s(t)||\mathcal{D}(t_b)|} & \text{if } t' \in \mathcal{D}(t_b) \text{ and } t_b \in s(a(t)) \\ 0 & \text{otherwise.} \end{cases} \tag{9}$$

3.4. Immediate Cost

The immediate cost of moving to tree t' by choosing action a while in tree t is:

$$c(t,a,t') = \begin{cases} 0 & \text{if } \mu(t') = 1 \\ 1 & \text{if } \mu(t') = 0. \end{cases}$$

Accordingly, the expected cost incurred when applying action a to tree t is:

$$c(t,a) = \sum_{t' \in \mathcal{SD}(t)} P(t,a,t')c(t,a,t')$$

but a simpler expression is available by substituting the explicit values for the probability term and the cost term, as stated in the next lemma.

Lemma 2. *The expected cost can be written as:*

$$c(t,a) = 1 - \frac{1}{|s(t)|} \sum_{t' \in s(a(t))} \mu(t'). \tag{10}$$

Given a budget or a specific family of policies, there are states in S that will never feature in the MDP. Thus, we shall focus only on the "usable states".

Definition 6 (Usable states). *The states in S that are attained through transitions given a specific value of budget k or a family of policies are called usable states. Denote this set of states as $\mathcal{U}$.*

For example, budget dependent states for $k = 1$, $d = 2$, and $p \geq 2$ will not include states where both nodes at depth 1 (if two exist) are marked. We will come back to this in Section 6.

3.5. Policies and Criterion

We choose as evaluation criterion the expected average cost over an infinite horizon. The class of policies in which we optimize is, in the terminology of [15] (Section 2.1.4), that of History-dependent, Randomized strategies (HR). However, the classical theory allows focusing on stationary strategies only. Some of our definitions are valid in the general class HR. In this context, $\gamma_n(t)$ denotes the action prescribed by the policy at time step n for tree t.

Definition 7 (Sensible policies, greedy policies). *A policy $\gamma \in HR$ is called:*

Sensible *if, for every state t and every time instant n, $\gamma_n(t)$ marks k unmarked nodes, or the number of unmarked nodes in t;*

Greedy *if it is sensible and if for every t, $\gamma_n(t)$ marks: either k of the unmarked sons of t if these are more than k or else all unmarked sons of t plus possibly nodes at a depth of at least 2.*

Sensible policies do not waste the marking budget on marked nodes, unless they have no other possibility. Among them, greedy policies mark sons as priority.

The following observation will be useful. Its proof is immediate by unrolling the marking/surfing/discovering cycle.

Lemma 3. *Consider a stationary policy γ such that for any t, $\gamma(t)$ marks nodes only up to some depth d_m. Let $\mathcal{M}_\gamma$ be the Markov chain generated by this policy. Then, the usable states $\mathcal{U}$, and in particular the recurrent classes of $\mathcal{M}_\gamma$, contain only trees with nodes marked up to depth $d_m - 1$.*

Remark 1. *Given the rules of surfing (uniform choice) and of discovery (independence of different subtrees), it seems possible to further reduce the size of the state space by exploiting symmetries. For instance, in Figure 2, the fourth and fifth trees will have exactly the same properties. We chose not to exploit these symmetries, because this would lead to an extra complexity in the formalism. Furthermore, it would render the enumeration of state spaces more complex and, as a consequence, complicate the description of the process on tree shapes; see the following section.*

4. The Markov Chain of Tree Shapes

In this section, we temporarily forget the control part of the MDP and focus on the process of trees generated by the surfing/discovering mechanism. It turns out to contain two Markov chains, which we identify and analyze.

4.1. Definition and Basic Properties

An important feature of the MDP constructed in Section 3 is that the *shape* of the successive trees does not depend on the marking strategy. In order to formalize this, we first define the shape of trees.

Definition 8 (Shape of trees). *Consider the mapping $\sigma_{p,d} : \mathcal{M}_{p,d} \to \mathcal{T}_{p,d}$, defined for all p and d recursively with:*

$$\sigma_{p,0}(t) = 0, \qquad \sigma_{p,d}(\mu, (s_1, \ldots, s_m)) = (0, (\sigma_{p,d-1}(s_1), \ldots, \sigma_{p,d-1}(s_m))), \text{if } d \geq 1.$$

The tree $\sigma_{p,d}(t)$ is called the shape of tree t.

We now state the aforementioned property of the shape of trees in the MDP. Observe that, by the definition of the succession of trees t, t_a, t_b, we have $\sigma_{p,d}(t(n)) = \sigma_{p,d}(t_a(n))$ for all n.

Proposition 1. *In the MDP defined in Section 3:*

(i) The distribution of the sequence $\{\sigma_{p,d}(t(0)), \sigma_{p,d-1}(t_b(0)), \sigma_{p,d}(t(1)), \ldots\}$ does not depend on the strategy $\gamma \in HR$;

(ii) The processes $\{\sigma_{p,d}(t(n)); n \in \mathbb{N}\}$, and $\{\sigma_{p,d-1}(t_b(n)); n \in \mathbb{N}\}$ are homogeneous Markov chains on the state spaces $\mathcal{T}_{p,d}$ and $\mathcal{T}_{p,d-1}$, respectively.

Proof. The proof of (i) follows from the fact that transition probabilities in (9) do not depend on the action a. Then, the Markov nature of embedded sequences $\sigma_{p,d}(t(n)) = \sigma_{p,d}(t_a(n))$ and $\sigma_{p,d-1}(t_b(n))$ is clear since random moves of the surfer and discoveries depend only on the shape of the current tree. □

We proceed with the identification of the stationary distributions of the Markov chains featured in Proposition 1. We first introduce the family of candidate distributions and state their basic properties. We then prove the result about Markov chains.

Definition 9. *Let $\pi_{p,d} : \mathcal{T}_{p,d} \to [0,1]$ be the sequence of functions defined recursively by:*

$$\pi_{p,0}(t) = 1 \tag{11}$$

$$\pi_{p,d}(0, (s_1, \ldots, s_m)) = \frac{1}{p} \prod_{k=1}^{m} \pi_{p,d-1}(s_k) \qquad d \geq 1. \tag{12}$$

Lemma 4. *The functions $\pi_{p,d}$ introduced in Definition 9 have the following properties:*

(*i*) *For each fixed d, $\pi_{p,d}$ is a probability distribution on $\mathcal{T}_{p,d}$;*

(*ii*) *The probabilities $\pi_{p,d}$ can be expressed as:*

$$\pi_{p,d}(t) = \frac{1}{p^{\mathrm{inode}(t)}}.$$

The interpretation of Definition 9 and Lemma 4 (ii) is that the probability of a tree t is the probability that this tree is generated by a Galton–Watson process with branching probabilities uniform in $\{1, \ldots, p\}$, stopped at generation d.

Proof. The proof of (i) proceeds by recurrence. For $d = 0$, the property is trivial. Assume then the property proven up to d. Then, selecting the number of sons of a tree and using (12), we have:

$$\sum_{t \in \mathcal{T}_{p,d+1}} \pi_{p,d+1}(t) = \sum_{m=1}^{p} \sum_{s_1,\ldots,s_m \in \mathcal{T}_{p,d}} \pi_{p,d+1}(0, (s_1, \ldots, s_m)) = \sum_{m=1}^{p} \sum_{s_1,\ldots,s_m \in \mathcal{T}_{p,d}} \frac{1}{p} \prod_{j=1}^{m} \pi_{p,d}(s_j)$$

$$= \sum_{m=1}^{p} \frac{1}{p} \left(\sum_{s \in \mathcal{T}_{p,d}} \pi_{p,d}(s) \right)^{m} = \sum_{m=1}^{p} \frac{1}{p} = 1.$$

We continue with the proof of (ii). Recursively, $\mathrm{inode}(t) = 1 + \sum_{t' \in s(t)} \mathrm{inode}(t')$. The result then follows from the definition (12). □

The following properties of the distribution $\pi_{p,d}$ will be useful. The proofs of the first two of them are straightforward and omitted.

Lemma 5. *Let t be a random tree in $\mathcal{T}_{p,d}$ distributed according to $\pi_{p,d}$. Then, $|s(t)|$ is uniformly distributed in $\{1, \ldots, p\}$.*

Lemma 6. *Let t be a random tree in $\mathcal{T}_{p,d}$ distributed according to $\pi_{p,d}$. Then, conditioned on the fact that $|s(t)| = m$, the subtrees $s_1, \ldots, s_m$ are independent and uniformly distributed according to $\pi_{p,d-1}$.*

Proposition 2. *Consider the Markov chains $\mathcal{M} = \{\sigma_{p,d}(t(n)); n \in \mathbb{N}\}$ and $\mathcal{M}_b = \{\sigma_{p,d-1}(t_b(n)); n \in \mathbb{N}\}$:*

(*i*) *Both chains are ergodic;*

(*ii*) *The stationary distribution of $\mathcal{M}$ is $\pi_{p,d}$;*

(*iii*) *The stationary distribution of $\mathcal{M}_b$ is $\pi_{p,d-1}$.*

Proof. The property (i) is proven if we can show that both chains are irreducible and aperiodic. Irreducibility follows from the fact that there is a sequence of transitions with a nonzero probability leading to, say, the tree that is a chain (all its internal nodes have only one son, let us name it $c_{p,d}$): if the discovery phase adds just one leaf to every leaf of t_b (this happens with positive probability), after d steps, the tree is $c_{p,d}$, whatever the random surfing moves. The tree t_b itself is $c_{p,d-1}$. Aperiodicity also follows from this construction since the transition from $c_{p,d}$ to $c_{p,d}$ has a nonzero probability.

In order to prove (ii), we check that the distribution $\pi_{p,d}$ satisfies the equation $\pi = \pi P$. Since $\mathcal{M}$ is ergodic, this will be the unique solution. We first identify the set of trees that have a positive probability of transition to a given tree $t \in \mathcal{T}_{p,d}$. To that end, we have to reverse the process of the transformation of one tree into another. Reversing the discovery phase, we are led to define $\mathrm{top}(t) \in \mathcal{T}_{p,d-1}$ as the tree deduced from t by removing the leaves. Then, reversing the surfer movement, we conclude that t' can be transformed into

t if and only if t' has $\mathrm{top}(t)$ as one of its subtrees. Let $\mathrm{Prec}(t) \subset \mathcal{T}_{p,d}$ be this set. For any $t' \in \mathrm{Prec}(t)$, we have:

$$P(t',t) = \frac{\mathrm{card}\{\tau \in s(t') | \tau = \mathrm{top}(t)\}}{|s(t')|} \frac{1}{p^{\mathrm{leaves}(\mathrm{top}(t))}} .$$

Accordingly, it is convenient to partition the set $\mathrm{Prec}(t)$ into "blocks" of states as follows:

$$\begin{aligned} &\mathrm{Prec}(t) = \cup_{m=1}^{p} \cup_{n=1}^{m} \mathcal{P}(m,n) \\ &\mathcal{P}(m,n) := \{t' \in \mathcal{T}_{p,d}, |s(t')| = m, s(t') \text{ contains } \mathrm{top}(t) \text{ exactly } n \text{ times}\}. \end{aligned}$$

Trees that have the same number of sons and the same number of occurrences of $\mathrm{top}(t)$ among their sons are grouped together. By construction, we have:

$$P(t',t) = \frac{n}{m} \frac{1}{p^{\mathrm{leaves}(\mathrm{top}(t))}} \qquad \forall t' \in \mathcal{P}(m,n).$$

This transition probability is therefore constant in the block $\mathcal{P}(m,n)$. Then, we can write:

$$\begin{aligned} \sum_{t' \in \mathcal{T}_{p,d}} \pi_{p,d}(t') P(t',t) &= \sum_{t' \in \mathrm{Prec}(t)} \pi_{p,d}(t') P(t',t) \\ &= \sum_{m=1}^{p} \sum_{n=1}^{m} \sum_{t' \in \mathcal{P}(m,n)} \pi_{p,d}(t') \frac{n}{m} \frac{1}{p^{\mathrm{leaves}(\mathrm{top}(t))}} \\ &= \frac{1}{p^{\mathrm{leaves}(\mathrm{top}(t))}} \sum_{m=1}^{p} \sum_{n=1}^{m} \frac{n}{m} \sum_{t' \in \mathcal{P}(m,n)} \pi_{p,d}(t') . \end{aligned} \tag{13}$$

We evaluate the inner sum, which is the total probability of the block $\mathcal{P}(m,n)$ under the distribution $\pi_{p,d}$. According to Lemma 6, the distribution of subtrees of $t' \in \mathcal{P}(m,n)$ is that of m independent trees in $\mathcal{T}_{p,d-1}$. Therefore, the probability, conditioned on m, that exactly n subtrees of t' are $\mathrm{top}(t)$ is the following binomial distribution (resulting from picking the n locations for trees $\mathrm{top}(t)$ among the m possibilities):

$$\binom{m}{n} \pi_{p,d-1}(\mathrm{top}(t))^n (1 - \pi_{p,d-1}(\mathrm{top}(t)))^{m-n} . \tag{14}$$

We conclude that:

$$\sum_{t' \in \mathcal{P}(m,n)} \pi_{p,d}(t') = \frac{1}{p} \binom{m}{n} \pi_{p,d-1}(\mathrm{top}(t))^n (1 - \pi_{p,d-1}(\mathrm{top}(t)))^{m-n} . \tag{15}$$

Using this result, we can evaluate the product of the distribution $\pi_{p,d}$ and the matrix P through the following computation:

$$\sum_{t' \in \mathcal{T}_{p,d}} \pi_{p,d}(t') P(t',t)$$

$$= \frac{1}{p^{\mathrm{leaves}(\mathrm{top}(t))}} \sum_{m=1}^{p} \sum_{n=1}^{m} \frac{n}{m} \frac{1}{p} \binom{m}{n} \pi_{p,d-1}(\mathrm{top}(t))^n (1 - \pi_{p,d-1}(\mathrm{top}(t)))^{m-n} \tag{16a}$$

$$= \frac{1}{p^{1+\mathrm{leaves}(\mathrm{top}(t))}} \sum_{m=1}^{p} \sum_{n=1}^{m} \binom{m-1}{n-1} (\pi_{p,d-1}(\mathrm{top}(t)))^n \left(1 - \pi_{p,d-1}(\mathrm{top}(t))\right)^{m-n} \tag{16b}$$

$$
\begin{aligned}
&= \frac{\pi_{p,d-1}(\text{top}(t))}{p^{1+\text{leaves}(\text{top}(t))}} \sum_{m=1}^{p} \sum_{n=0}^{m-1} \binom{m-1}{n} (\pi_{p,d-1}(\text{top}(t)))^n \left(1 - \pi_{p,d-1}(\text{top}(t))\right)^{m-1-n} \\
&= \frac{\pi_{p,d-1}(\text{top}(t))}{p^{1+\text{leaves}(\text{top}(t))}} \sum_{m=1}^{p} 1 && (17a) \\
&= \frac{\pi_{p,d-1}(\text{top}(t))}{p^{1+\text{leaves}(\text{top}(t))}} \cdot p = \frac{1}{p^{\text{inode}(\text{top}(t))+\text{leaves}(\text{top}(t))}} = \frac{1}{p^{\text{inode}(t)}} && (17b) \\
&= \pi_{p,d}(t).
\end{aligned}
$$

We used (13) and (14) to obtain (16a). The binomial expansion theorem was used in (17a). Finally, in (17b), we note that $\text{inode}(t)$ is the sum of $\text{inode}(\text{top}(t))$ and $\text{leaves}(\text{top}(t))$, which gives us the desired result, $\pi_{p,d}(t)$.

Finally, we prove (iii). We know that $\sigma(t_b)$ results from $\sigma(t)$ through the random choice of a son of t. Invoking again Lemma 6, we have: conditioned on the event $\{|s(t)| = m\}$, each subtree is τ with probability $\pi_{p,d-1}(\tau)$, and the probability that a uniform random choice picks τ has probability $\pi_{p,d-1}$ as well. Since this does not depend on m, the result is true also when removing the conditioning. □

4.2. Application to Greedy Policies

As an application of Lemma 5, we obtain an upper bound on the optimal cost that can be realized by any policy. This was based on the following result.

Lemma 7. *Consider a random variable $t \in \mathcal{T}_{p,d}$ distributed as $\pi_{p,d}$. Define the random variable:*

$$C = \frac{[\,|s(t)| - k\,]^+}{|s(t)|}.$$

Then, $C = 0$ with probability 1 (in particular, $\mathbb{E}C = 0$) if $k \le p$ and:

$$\mathbb{E}C \;=\; \frac{1}{p}\,\mathbb{H}_{pk}, \qquad k \ge p, \tag{18}$$

where:

$$\mathbb{H}_{pk} \;:=\; \sum_{m=k+1}^{p} \frac{m-k}{m} \;=\; p - k - k(\mathbb{H}_p - \mathbb{H}_k). \tag{19}$$

Proof. According to Lemma 5, $|s(t)|$ is uniformly distributed. Then.

$$
\begin{aligned}
\mathbb{E}C &= \sum_{m=1}^{p} \frac{1}{p} \frac{[m-k]^+}{m} \;=\; \frac{1}{p} \sum_{m=k+1}^{p} \frac{m-k}{m} \;=\; \frac{1}{p}\,\mathbb{H}_{pk} \\
&= \frac{1}{p} \sum_{m=k+1}^{p} \left(1 - \frac{k}{m}\right) \;=\; \frac{1}{p}\left(p - k - k \sum_{m=k+1}^{p} \frac{1}{m}\right) \\
&= \frac{1}{p}(p - k - k(\mathbb{H}_p - \mathbb{H}_k)).
\end{aligned}
$$

□

We now state the bound announced.

Proposition 3. *The optimal expected cost g^* of the MDP described in Section 3 satisfies:*

$$g^* \;\le\; 1 - \frac{k}{p}(1 + \mathbb{H}_p - \mathbb{H}_k). \tag{20}$$

The cost of any greedy policy satisfies the same bound.

Proof. Consider the policy that marks k sons of the current tree or all the sons if k is larger than this. This is a greedy policy, in the sense of Definition 7. The average cost of this policy is precisely given by $\mathbb{E}C$ as in Lemma 7, which is equal to the right-hand side in (20). Any greedy policy will mark at least the same nodes: it will necessarily have a smaller average cost. The optimal cost of the MDP is necessarily smaller than the cost of this specific policy. □

4.3. Metrics of Tree Shapes

We applied the results of Section 4.1 to evaluate several simple metrics in tree shapes generated by the MDP. This may be useful in particular for estimating the quantity of memory needed in a simulation of this process.

4.3.1. Average Number of Nodes

Let $N_{p,d} = \mathbb{E}(|t|)$ be the average size of a tree t distributed according to $\pi_{p,d}$. According to Lemma 6, we can write, conditional on the event $\{|s(t)| = m\}$: $|t| = 1 + |s_1| + \ldots + |s_m|$ so that:

$$\mathbb{E}(|t| \mid |s(t)| = m) = 1 + \sum_{j=1}^{m} \mathbb{E}(|s_j|) \; = \; 1 + m\, N_{p,d-1}$$

$$N_{p,d} \; = \; \mathbb{E}(|t|) = 1 + \mathbb{E}(|s(t)|) N_{p,d-1} \; = \; 1 + \frac{p+1}{2} N_{p,d-1}\,. \tag{21}$$

Since the initial condition is $N_{p,0} = 1$, the recurrence in (21) has the solution:

$$N_{p,d} \; = \; \frac{2}{p-1} \left(\left(\frac{p+1}{2} \right)^{d+1} - 1 \right), \quad p \geq 2, \tag{22}$$

with $N_{1,d} = d + 1$. We have proven the following result.

Lemma 8. *The average number of nodes in a tree $t \in \mathcal{T}_{p,d}$ distributed according to $\pi_{p,d}$ is given by* (22).

4.3.2. Average Number of Leaves

Let $L_{p,d}$ be the average number of leaves, in trees t distributed according to $\pi_{p,d}$. The reasoning of Section 4.3.1 can be reproduced: according to Lemma 6, we can write, conditional on the event $\{|s(t)| = m\}$: $|t| = |s_1| + \ldots + |s_m|$. If follows that:

$$L_{p,d} \; = \; \mathbb{E}(|s(t)|)\, L_{p,d-1} \; = \; \frac{p+1}{2}\, L_{p,d-1}.$$

Since $L_{p,0} = 1$, we have the following result.

Lemma 9. *The average number of leaves in a tree $t \in \mathcal{T}_{p,d}$ distributed according to $\pi_{p,d}$ is:*

$$L_{p,d} \; = \; \left(\frac{p+1}{2} \right)^{d}.$$

Note that the maximal number of leaves of a tree in $\mathcal{T}_{p,d}$ is p^2.

4.3.3. Average Number of Nodes Created

Let $C_{p,d}$ be the average number of nodes created at some time step in the Markov chain $\{\sigma_{p,d}(t(n)); n \in \mathbb{N}\}$ in its stationary regime. It is equal to the expected number of nodes created from some tree $t \in \mathcal{T}_{p,d}$ distributed according to $\pi_{p,d}$. This number itself is equal to the expected number of leaves in t, since t results from the addition of leaves to some t', also distributed according to $\pi_{p,d}$. We therefore have, with Lemma 9:

$$C_{p,d} = \left(\frac{p+1}{2}\right)^d.$$

4.3.4. Average Number of Nodes Deleted

Let $D_{p,d}$ be the average number of nodes deleted at some time step in the Markov chain $\{\sigma_{p,d}(t(n)); n \in \mathbb{N}\}$ in its stationary regime. By stationarity, it is expected that $D_{p,d} = C_{p,d}$. We verify this with a direct computation.

Consider a tree $t \in \mathcal{T}_{p,d}$. Conditioned on the events $\{|s(t)| = m\}$ and the surfer moving to the jth son, the number of nodes deleted is: $1 + \sum_{\ell \neq j} |s_\ell|$ (the root and the other subtrees). Using Lemma 5 and the fact that each s_ℓ is distributed according to $\pi_{p,d-1}$ (Lemma 6), we have then:

$$\begin{aligned} D_{p,d} &= \sum_{m=1}^{p} \frac{1}{p} \sum_{j=1}^{m} \frac{1}{m}\Big(1 + (m-1)N_{p,d-1}\Big) = \frac{1}{p}\sum_{m=1}^{p}\Big(1 + (m-1)N_{p,d-1}\Big) \\ &= 1 + \frac{p(p-1)}{2p} N_{p,d-1} = 1 + \frac{p-1}{2} N_{p,d-1} = \left(\frac{p+1}{2}\right)^d \end{aligned}$$

where the last equality results from (22). This is equal to the average number of nodes created $C_{p,d}$, as expected.

5. Trees of Depth $d = 1$ with an Arbitrary Marking Budget

In this section, we consider trees of depth 1, and we prove the following result.

Theorem 1. *When $d = 1$, any greedy policy is optimal.*

It is quite clear intuitively that, indeed, no reasonable alternative exists. The exercise here is to check that the theory does provide a way to prove the result formally. In the process, we identify arguments that will be useful for the proofs of stronger results.

For the purpose of the forthcoming proof, we rename the elements of $\mathcal{M}^+_{p,1}$ as: $\mathcal{M}^+_{p,1} = \{t_{\mu,j} : \mu \in \{0,1\}, 1 \le j \le p\}$. In the notation of Section 2, we have $t_{\mu,j} = (\mu; \underbrace{(0,\ldots,0)}_{j\text{ times}})$. Furthermore, observe that trees t_b belong to $\mathcal{M}_{p,0} = \{0,1\}$: these are trees reduced to a root with a mark.

Proof. We shall prove the result using Theorem A1. Define the constant g and the function $f : \mathcal{T}_{p,1} \to \mathbb{R}$ as:

$$g = \frac{\mathbb{H}_{pk}}{p} \tag{23}$$

$$f(t_{\mu,j}) = \frac{(j-k)^+}{j}, \qquad \mu \in \{0,1\}. \tag{24}$$

The symbol $\mathbb{H}_{pk}$ was defined in (19). We shall check that this g, f satisfies the optimality Equation (A1). For every state $s = t_{\mu,j}$ and every action a, we write the quantity to be minimized in the right-hand side of this equation:

$$\begin{aligned} Q(s,a) &:= c(s,a) + \sum_{s' \in \mathcal{M}^+_{p,1}} P(s,a,s') f(s') \\ &= c(t_{\mu,j}, a) + \sum_{s' \in \mathcal{M}^+_{p,1}} P(t_{\mu,j}, a, s') f(s') \\ &= c(t_{\mu,j}, a) + \sum_{\mu' \in \{0,1\}} \overline{P}(t_{\mu,j}, a, \mu') \sum_{j'=1}^{p} \frac{1}{p} f(t_{\mu',j'}). \end{aligned} \tag{25}$$

We obtained (25) by conditioning the transition $s \to s'$ on the value of the tree t_b. The new notation $\overline{P}(t, a, t_b)$ stands for the probability of moving from t to t_b when action a is applied. Given the definition of $f(\cdot)$ in (24), we have further:

$$\begin{aligned} Q(s,a) &= c(t_{\mu,j}, a) + \sum_{\mu' \in \{0,1\}} \overline{P}(t_{\mu,j}, a, \mu') \left(\sum_{j'=1}^{p} \frac{1}{p} \frac{(j'-k)^+}{j'} \right) \\ &= c(t_{\mu,j}, a) + \sum_{\mu' \in \{0,1\}} \overline{P}(t_{\mu,j}, a, \mu') \frac{\mathbb{H}_{pk}}{p} \\ &= c(t_{\mu,j}, a) + g. \end{aligned} \tag{26}$$

The actions a can be grouped according to the number of sons they mark in the tree t_a: this number ranges from 0 to $\min\{j,k\}$. When t_a has ℓ sons marked, this determines the cost as:

$$c(t_{\mu,j}, a) = \frac{j-\ell}{j}.$$

Finally, the minimization with respect to a amounts to the following minimization with respect to ℓ:

$$\begin{aligned} \min_a \left\{ c(s,a) + \sum_{s' \in \mathcal{M}^+_{p,1}} P(s,a,s') f(s') \right\} &= \min_{0 \le \ell \le j \wedge k} \left\{ \frac{j-\ell}{j} + g \right\} \\ &= \frac{(j-k)^+}{j} + g = f(s) + g. \end{aligned} \tag{27}$$

The constant g and the function f therefore solve Equation (A1). This function is bounded since the state space is finite. Therefore, there exists an optimal policy γ^* with cost g. Clearly, this policy consists of marking up to k sons of any tree: a greedy policy in the sense of Definition 7. □

From the proof of Theorem 1 and also from Lemma 7, we have the corollary:

Corollary 1. *The average value of any tree of depth 1 is* $\mathbb{H}_{pk}/p$.

Remark 2. *The fact that, in the present case,* $f(s) = \min_a c(s,a)$ *is a consequence of the fact that the cost of the future tree* s' *resulting from the transition is actually independent of the action* a.

Remark 3. *It was proven in [16] that the finite-horizon, total-cost optimal-value function is given by:*

$$W_N^*(t_{\mu,j}) = \frac{(j-k)^+}{j} + \frac{(N-1)}{p}\mathbb{H}_{pk} \,,$$

and is realized by any greedy policy. We obtained from this result the value of $g = \lim_{N\to+\infty} W_N^*(t_{\mu,j})/N$ *and the form of function* f *that must satisfy* $f(s) - f(s') = \lim_{N\to+\infty}(W_N^*(s) - W_N^*(s'))$.

6. Trees of Depth $d = 2$ with Marking Budget $k = 1$

In this section, we consider trees of depth 2, and we prove the following result.

Proposition 4. *When* $k = 1$ *and* $d = 2$*, any greedy policy is optimal.*

We begin with some notation and preliminary results. Then, we provide the proof.

6.1. Preliminaries

As a preliminary, observe that as a consequence of the marking/moving/discovery cycle, the subtrees of $\mathcal{T}_{1,p}$ that appear have at most one leaf marked. Indeed, at the beginning of the cycle, trees $t \in \mathcal{M}_{2,p}^+$ have all leaves unmarked. The marking with budget $k = 1$ marks at most one of these leaves. Then, the surfer moves to one subtree $t_b \in \mathcal{M}_{1,p}$, which inherits this property. The discovery phase merely adds unmarked leaves at depth 2. With this observation, we can restrict our attention to the usable set $\mathcal{U}$ of trees with at most one leaf marked, since only those can appear recurrently when some stationary policy is applied.

A second preparation is to calculate the average cost under some greedy policy γ, which therefore marks one node at depth one in any tree t. The choice of this node does not matter. According to Lemma 3, the Markov chain $\mathcal{M}_\gamma$ generated by this policy has recurrent states with marks only at depth 0, that is at the root. Therefore, the cost (10) is always given by $c(t, \gamma(t)) = 1 - 1/|s(t)|$. It is then of the form assumed in Lemma 7, and the application of (18) yields the expected cost for policy γ:

$$J_\gamma = \frac{1}{p}(p - 1 - (\mathbb{H}_p - \mathbb{H}_1)) = 1 - \frac{\mathbb{H}_p}{p} \,. \tag{28}$$

6.2. Notation and Terminology

When $d = 2$, the trees of interest are simpler than in the general case, and it is convenient to devise an appropriate notation. For trees in $\mathcal{M}_{p,2}^+$, all subtrees have depth one and unmarked leaves. We shall adopt the simplified notation for such trees: $(\mu; m)$ denotes a depth one tree with root marked with μ and m unmarked leaves. A typical tree of $\mathcal{M}_{p,2}^+$ is then denoted by $t = (\mu; (\mu_1, j_1), \ldots, (\mu_m, j_m))$ for some $m \in [1..p]$.

After marking, the subtrees of depth one will have at most one leaf marked: then $(\mu; m^+)$ will denote this tree with one marked leaf. Which leaf exactly is marked does not make a difference in the following reasoning.

In the analysis, the number of unmarked sons of a tree is a key criterion. Accordingly, we introduce the following typology for $t = (\mu; (\mu_1, j_1), \ldots, (\mu_m, j_m)) \in \mathcal{M}_{p,2}^+$:

$$\text{Type 1: } \sum_{r=1}^{m} \mu_r \le m - 1, \qquad \text{Type 2: } \sum_{r=1}^{m} \mu_r = m.$$

6.3. Proof

The proof uses the optimality equations and Theorem A1, as in Section 5. The "g" value needed for this was computed as J_γ in (28). The next step is to evaluate the "f" function in the optimality Equation (A1) in Theorem A1. It is sufficient to provide a value to states that are usable in the sense of Definition 6. For other states, the value of f is defined by (A1), since the right-hand side only contains values of reachable states.

The function f that is proposed is the following:

$$f(\mu;(\mu_1,j_1),\ldots,(\mu_m,j_m)) = \begin{cases} -\dfrac{1}{m}\left(1+\sum_{r=1}^{m}\mu_r+\sum_{r=1}^{m}\dfrac{1}{j_r}\right) & \sum_{r=1}^{m}\mu_r \le m-1 \quad \text{(type 1)} \\ -1-\dfrac{1}{m}\left(2\sum_{r=1}^{m}\dfrac{1}{j_r}+SL_1(t)\dfrac{\mathbb{H}_p-p}{p-1}\right) & \sum_{r=1}^{m}\mu_r = m \quad \text{(type 2).} \end{cases} \tag{29}$$

In this last line, $SL_1(t) = |\{r|j_r = 1\}|$ is the number of subtrees of t that have exactly one leaf.

Proof of Proposition 4. We apply Theorem A1 by checking that the function f, in (29), and the constant $g = 1 - \mathbb{H}_p/p$ satisfy the optimality equations. To that end, we first evaluate the expected value of trees t_b in $\mathcal{M}_{p,1}$. Denote by $P_{\text{dis}} : \mathcal{M}_{p,1} \to \mathcal{M}^+_{p,2}$ the transition probability from a tree t_b to a "discovered" tree t'. We can write:

$$\tilde{f}(t_b) = \sum_{t'\in\mathcal{M}^+_{p,2}} P_{\text{dis}}(t_b,t')f(t').$$

We have, for any $\mu \in \{0,1\}$:

$$\begin{aligned} \tilde{f}(\mu;j) &= \sum_{\ell_1=1}^{p}\cdots\sum_{\ell_j=1}^{p}\frac{1}{p^j}f(\mu;(0,\ell_1),\ldots,(0,\ell_j))) \\ &= -\frac{1}{j}\sum_{\ell_1=1}^{p}\cdots\sum_{\ell_j=1}^{p}\frac{1}{p^j}\left(1+\sum_{r=1}^{j}\frac{1}{\ell_r}\right) \\ &= -\frac{1}{j}\left(1+\sum_{r=1}^{j}\sum_{\ell_1=1}^{p}\cdots\sum_{\ell_j=1}^{p}\frac{1}{p^j}\frac{1}{\ell_r}\right) \\ &= -\frac{1}{j}\left(1+\sum_{r=1}^{j}\frac{p^{j-1}}{p^j}\sum_{\ell_r=1}^{p}\frac{1}{\ell_r}\right) = -\frac{1}{j}\left(1+\sum_{r=1}^{j}\frac{\mathbb{H}_p}{p}\right) \\ &= -\left(\frac{1}{j}+\frac{\mathbb{H}_p}{p}\right) \end{aligned} \tag{30}$$

$$\begin{aligned} [j\ge 2] \quad \tilde{f}(\mu;j^+) &= \sum_{\ell_1=1}^{p}\cdots\sum_{\ell_j=1}^{p}\frac{1}{p^j}f(\mu;(0,\ell_1),\ldots,(1,\ell_q),\ldots,(0,\ell_j))) \\ &= -\frac{1}{j}\sum_{\ell_1=1}^{p}\cdots\sum_{\ell_j=1}^{p}\frac{1}{p^j}\left(2+\sum_{r=1}^{j}\frac{1}{\ell_r}\right) \\ &= -\left(\frac{2}{j}+\frac{\mathbb{H}_p}{p}\right) \end{aligned} \tag{31}$$

$$\begin{aligned} \tilde{f}(\mu;1^+) &= \sum_{\ell=1}^{p}\frac{1}{p}f(\mu;(1,\ell)) = \frac{1}{p}\sum_{\ell=1}^{p}\left(-1-\frac{2}{\ell}-\mathbf{1}_{\ell=1}\frac{\mathbb{H}_p-p}{p-1}\right) \\ &= -\frac{1}{p}\left(p+2\mathbb{H}_p+\frac{\mathbb{H}_p-p}{p-1}\right) \\ &= -\left(1+2\frac{\mathbb{H}_p}{p}+\frac{\mathbb{H}_p-p}{p(p-1)}\right). \end{aligned} \tag{32}$$

From these formulas, the following identities are obtained: for any μ and μ',

$$\tilde{f}(\mu;j) - \tilde{f}(\mu';j^+) = \frac{1}{j} \qquad j \geq 2$$
$$\tilde{f}(\mu;1) - \tilde{f}(\mu';1^+) = \frac{\mathbb{H}_p}{p} + \frac{\mathbb{H}_p - p}{p(p-1)} = \frac{\mathbb{H}_p - 1}{p-1}.$$

The following result then immediately follows:

Lemma 10. *For all $j \in \{1,\ldots,p\}$, $\mu,\mu' \in \{0,1\}$ and all $p \geq 2$, $0 \leq \tilde{f}(\mu;j) - \tilde{f}(\mu';j^+) \leq 1/2$.*

We now proceed with checking that f and g solve the optimality equations. We start with Type 1 trees. Let $t = (\mu;(\mu_1,j_1),\ldots,(\mu_m,j_m))$ with $\sum_{r=1}^m \mu_m < m$. The alternative actions are: (a) mark the root or any son already marked; (b_k) mark an unmarked son k; (c_k) mark a leaf of subtree (μ_k, j_k). For actions (b_k), we denote by μ'_i the marks of the sons after marking: $\mu'_i = \mu_i$ for $(i \neq k$ and $\mu'_k = \mu_k + 1$. Clearly, $\sum_{r=1}^m \mu'_r = 1 + \sum_{r=1}^m \mu_r$. For actions (c_k), which leaf is marked does not matter, so we ignore this information.

The right-hand side of the optimality equation in the cases (a), (b_k), and (c_k) are respectively:

$$\begin{aligned}
Q(t,a) &= \frac{m - \sum_{r=1}^m \mu_r}{m} + \sum_{r=1}^m \frac{1}{m}\tilde{f}(\mu_r;j_r) \\
Q(t,b_k) &= \frac{m - \sum_{r=1}^m \mu_r - 1}{m} + \sum_{r=1}^m \frac{1}{m}\tilde{f}(\mu'_r;j_r) \\
&= \frac{m - \sum_{r=1}^m \mu_r - 1}{m} - \sum_{r=1}^m \frac{1}{m}\left(\frac{1}{j_r} + \frac{\mathbb{H}_p}{p}\right) \\
&= 1 - \frac{\mathbb{H}_p}{p} - \frac{\sum_{r=1}^m \mu_r + 1}{m} - \frac{1}{m}\sum_{r=1}^m \frac{1}{j_r} = g + f(t) \qquad (33) \\
Q(t,c_k) &= \frac{m - \sum_{r=1}^m \mu_r}{m} + \frac{1}{m}\sum_{r=1|r\neq k}^m \tilde{f}(\mu_r;j_r) + \frac{1}{m}\tilde{f}(\mu_k;j_k^+)\,.
\end{aligned}$$

Then:

$$\begin{aligned}
Q(t,c_k) - Q(t,a) &= \frac{1}{m}(\tilde{f}(\mu_k;j_k^+) - \tilde{f}(\mu_k;j_k)) \\
Q(t,b_\ell) - Q(t,c_k) &= -\frac{1}{m} + \frac{1}{m}(\tilde{f}(\mu'_k;j_k) - \tilde{f}(\mu_k;j_k^+))\,.
\end{aligned}$$

Both differences are negative according to Lemma 10. This implies that action (b_k) dominates all actions (c_k), which in turn dominate action (a). It therefore realizes the minimum in the right-hand side of the optimality equation. With (33), the right-hand side and the left-hand side coincide.

Next, we consider Type 2 trees. According to the preliminary remark, we can focus our attention on trees with at most one son marked: if this tree is of Type 2 (all sons marked), then it has only one son. Let then $t = (\mu;(1,j))$, $j \in [1..p]$ be such a tree. The alternative actions are: (a) mark the root or the son; (b) mark leaf of the subtree (μ_k, j_k). The right-hand side of the optimality equation is respectively: $Q(t,a) = \tilde{f}(1;j)$ and $Q(t,b) = \tilde{f}(1;j^+)$. From Lemma 10, we know that action (b) dominates action (a). Further, from Definitions (29) and (32),

$$
\begin{aligned}
f(\mu;(1,j)) + g - Q(t,a) &= f(\mu;(1,j)) + 1 - \frac{\mathbb{H}_p}{p} - \tilde{f}(1;j^+) \\
&= -1 - \left(2 + \frac{\mathbb{H}_p - p}{p-1}\right) + 1 - \frac{\mathbb{H}_p}{p} + \left(1 + 2\frac{\mathbb{H}_p}{p} + \frac{\mathbb{H}_p - p}{p(p-1)}\right) \\
&= -1 + \frac{\mathbb{H}_p}{p} + \frac{\mathbb{H}_p - p}{p-1}\left(\frac{1}{p} - 1\right) = -1 + \frac{\mathbb{H}_p}{p} - \frac{\mathbb{H}_p - p}{p} = 0.
\end{aligned}
$$

□

Remark 4. *It was proven in [16] that, for any tree $t = (\mu;(s_1,\ldots,s_m))$, in the usable set, namely the set of trees with no sons marked, the finite-horizon total-cost optimal-value function is given by:*

$$
W_n^*(t) \;=\; n - \frac{1}{m} - \frac{n-2}{p}\,\mathbb{H}_p - \frac{1}{m}\sum_{r=1}^{m}\frac{1}{|s(s_r)|} \quad \textit{for} \quad 2 \le n \le N, \tag{34}
$$

and that this cost is realized by any greedy policy. From this result, the average cost of this policy in the infinite horizon is then:

$$
\lim_{N\to\infty}\frac{W_N^*(t)}{N} = \lim_{N\to\infty}\frac{1}{N}\left(N - \frac{1}{m} - \frac{N-2}{p}\,\mathbb{H}_p - \frac{1}{m}\sum_{r=1}^{m}\frac{1}{|s(s_r)|}\right) \;=\; 1 - \frac{\mathbb{H}_p}{p}.
$$

This matches with (28). Furthermore, it is compatible with the form of the function f in (29). The interpretation of $f(t) - f(t')$ is the difference in the total expected cost when starting from trees t or t'. According to (34), the tree-dependent cost for trees with unmarked sons would be:

$$
f(t) \;=\; -\frac{1}{|s(t)|}\left(1 + \sum_{r=1}^{|s(t)|}\frac{1}{|s(s_r)|}\right).
$$

This is indeed the value in (29) since $\sum_r \mu_r = 0$.

7. Trees of Depth $d = 2$ with Marking Budget $k = 2$

This section is devoted to the case where the marking budget is $k = 2$. In this case, we do not present general results, but we focus on the case of trees with depth 2. For small values of p, we describe the optimal policy, and we conjecture that this policy is optimal for general values of p.

We begin with some additional notation, then we introduce the definitions of the policies of interest. This allows us to formulate Conjecture 1. We then present numerical experiments made with small values of p supporting this conjecture.

7.1. Preliminary

Similarly as in Section 6, we argue that usable states are necessarily such that the marking of sons $\sum_{r=1}^{m} i_r$ takes values in $\{0,1,2\}$. Indeed, the sons of a tree are the leaves of a tree at the previous time step, and at most two leaves can be marked at any step.

7.2. Notation and Terminology

We first recall the representation of depth two trees of $\mathcal{M}_{2,p}^+$ from Section 6.2. Such a tree can be represented as $(\mu,(\mu_1,j_1),\cdots,(\mu_m,j_m))$ where $m \in [1,p]$, $\mu,\mu_r \in \{0,1\}$ and $1 \le j_r \le p$ for all $r \in [1,m]$. μ,μ_r are the markings of the root and the sons, m is the number of sons, and j_r are the number of leaves of the depth one subtree (μ_r,j_r). Based on the number of marked sons of the tree, we classify them into two types:

$$\text{Type 1: } \sum_{r=1}^{m} \mu_r \leq m-2, \qquad \text{Type 2: } \sum_{r=1}^{m} \mu_r \geq m-1.$$

Type 2 trees are further classified into three subtypes (remember that $\sum_r \mu_r$ cannot exceed 2):

$$\text{Type 2a: } \sum_{r=1}^{m} \mu_r = m-1, \quad \text{Type 2b: } m=2, \mu_1=\mu_2=1, \quad \text{Type 2c: } m=1, \mu_1=1.$$

We also introduce some shorthand notation for the different possible actions. Let $a(d1, d1)$ represent the action of marking two sons ($d1$ as in "depth 1"), $a(d1, l_{j_c})$ represent the action of marking a son and a leaf of the tree (μ_c, j_c) (l_j as in "leaf j"), and $a(l_{j_{c1}}, l_{j_{c2}})$ represent the action of marking a leaf in each of the subtrees (μ_{c1}, j_{c1}) and (μ_{c2}, j_{c2}). If $c1 = c2$, two leaves are marked in this subtree.

7.3. Policies

All policies of interest in our study are greedy in the sense of Definition 7. These policies do not specify what happens when some marking budget is left after marking all unmarked sons of a tree. We therefore specify the following variants of the greedy policy by their precise behavior in this situation. We begin with four simple rules; three of them rely on an order defined on the subtrees. The terms "first" and "second" used in the specification are relative to this order:

- Greedy depth 1: Ensure that only the sons of the tree are marked;
- Greedy smallest: Ensure that the sons of the tree are marked. If budget remains, then mark the first leaf of the smallest subtree. If budget still remains, mark the second leaf of the smallest subtree, if any. Otherwise, mark the first leaf of the second smallest subtree;
- Greedy largest: Ensure that the sons of the tree are marked. If budget remains, then mark the first leaf of the largest subtree. If budget still remains, mark the second leaf of the largest subtree, if any. Otherwise, mark the first leaf of the second largest subtree;
- Greedy leftmost: Ensure that the sons of the tree are marked. If budget remains, then mark the first leaf of the leftmost subtree. If budget still remains, mark the second leaf of the leftmost subtree, if any. Otherwise, mark the first leaf of the second leftmost subtree.

The cost of policy "greedy depth 1" is known by Lemma 7:

$$J_{\text{Greedy Depth 1}} = \frac{\mathbb{H}_{p2}}{p} = 1 + \frac{1 - 2\mathbb{H}_p}{p}. \tag{35}$$

Finally, we introduce the "greedy finite optimal" policy, a name that we use as a shorthand for "the policy that seems to emerge as optimal with the finite horizon criterion". Its behavior is specified in Table 2. It is explained in Appendix C how the features of this policy are extrapolated from the results obtained with the finite-horizon version of the MDP.

The behavior of the "greedy finite optimal" policy is obvious on Type 1 (more than two unmarked sons) and Type 2c (one son that is marked) trees. On Type 2a (one unmarked son) and Type 2b (two sons, both marked), it introduces a threshold of 3 or 4 on the size of the subtrees. When the tree is of Type 2a, one mark is left for marking subtrees. If all of them have a size less than 2, the largest one is marked. On the other hand, if some of them has a size larger than 3, the smallest of these is marked. When the tree is of Type 2b, two marks are left for marking subtrees. If both of them have a size larger than 4 ($j_2 \geq j_1 > 3$), the smallest subtree is marked. In the other case, both subtrees are marked.

Table 2. Specification of the greedy finite optimal policy.

Tree Type	Unmarked Sons	Specification of Subtree Size		Optimal Action
Type 1	≥ 2			$a(d1, d1)$
Type 2a	1	$j_r \geq 3$ for some r		$a(d1, l_{j_c})$, $j_c = \min_r j_r \geq 3$
		$j_r < 3$ for all r		$a(d1, l_{j_c})$, $j_c = \max_r j_r$
Type 2b	0	$j_1 > 3$	$j_2 \geq j_1$	$a(l_{j_1}, l_{j_1})$
		$j_1 \leq 3$	$j_2 \geq j_1$	$a(l_{j_1}, l_{j_2})$
Type 2c	0			$a(l_{j_1}, l_{j_1})$

A rationale for such a rule is as follows. When a tree has less than two unmarked sons, it can be made costless with the two marks of the budget. Therefore, when considering trees of Type 2a, a subtree that has less than two leaves has less priority than a subtree with more than three leaves, which is more "vulnerable". Among the vulnerable trees, it is better to mark the smallest ones, in order to reduce future expected costs. Trees of Type 2b have, so to speak, one round in advance since all sons are already marked. Subtrees of a size less than 3 can be "protected" by devoting one mark to them: if the surfer moves to them, the budget of the next round will be used to complete the protection. If both trees are too large to be fully protected, the budget is devoted again to the smallest one.

We can now state the conjecture that is the focus of this section.

Conjecture 1. *When $k = 2$ and $d = 2$, the "greedy finite optimal" policy is optimal.*

7.4. Numerical Experiments

We provide support to Conjecture 1 with results for small p. We implemented the policy improvement algorithm [15] (Chapter 8.6), starting with a particular greedy policy. Prior to the implementation of the algorithm, we evaluated the average cost of the four variants of the greedy policy introduced in Section 7.3, for $p = 3, 4, 5$. Of course, the greedy policies realize a zero cost for $p \leq 2$.

The average costs of the five greedy policies for small values of p are summarized in Table 3. The row concerning the "greedy depth 1" policy was evaluated using the exact formula (35), which gave respectively $1/9$, $5/24$, and $43/150$ for $p = 3, 4$, and 5. Several observations can be made from the data of this table. First, there is a substantial gain of marking subtrees after having marked all the sons: there is a larger gap between the performance of greedy depth one and the group of the other ones, than inside this group. Second, the best performance among the four simple policies introduced in Section 7.3 was achieved by marking the largest subtree, for all values of p tested. Picking the smallest subtree had the worst performance compared to picking the leftmost/largest subtree. Finally, choosing arbitrarily the leftmost (or, for that matter, the rightmost or a random) subtree resulted in a performance between these extremes.

Table 3. Average cost of different greedy policies.

Policy	Cost for $p = 3$	Cost for $p = 4$	Cost for $p = 5$
Greedy depth 1	0.111111	0.208333	0.286667
Greedy smallest	0.067912	0.161568	0.229741
Greedy leftmost	0.062802	0.160227	0.226289
Greedy largest	0.054369	0.156907	0.217443
Greedy finite optimal	0.054369	0.154401	0.208282

We used the greedy largest policy as a starting candidate policy for the policy iteration algorithm for $p = 3, 4, 5$, since it gave the best performance among the simple policies. In each case, the algorithm converged in a few iterations, and the resulting policy was the greedy finite optimal policy. The performance of the policy iteration algorithm is summarized in Table 4. The execution time figures correspond to an implementation in Python 3.9 running on a 1.4 GHz quad-core processor with 8 GB memory and a 1536 MB graphic card.

Table 4. Policy iteration performance.

Fanout p	No. of Iterations	Size of Matrices	Time to Converge
3	1	231×231	11 s
4	5	3336×3336	632 s
5	14	$57{,}860 \times 57{,}860$	7898 s

We could have also started the algorithm with the greedy finite optimal policy and checked that it solves the optimality equations. Selecting another policy was also a way of checking that our implementation of policy iteration works properly, as well as measuring "how far" from the optimum the greedy largest policy is.

Observe in Table 3 that the relative performance of the greedy finite optimal policy was more pronounced for $p = 5$ as compared to $p = 4$. The former MDP had a larger number of Type 2 trees. The greedy largest policy prescribed a suboptimal marking scheme for such trees, which explains the greater cost reduction by switching to the greedy finite optimal policy for $p = 5$. The greedy finite optimal policy for $p = 3$ coincided with the greedy largest policy, and hence, the costs were identical.

8. Discussion

We proposed a stochastic dynamic decision model for prefetching problems, which is simple in the sense that it has only three integer parameters, and yet can help conceive of optimal strategies in practical situations. The simplicity of the model lies in several assumptions that we discuss now.

We first observe that the modeling we proposed does not look practical for large values of parameters d and p. Indeed, Table 1 clearly shows that the state space sizes that could be handled numerically corresponded to small values of these parameters. On the other hand, the formal results obtained so far suggest that such a numerical solution would be needed in practice. We argue, consistent with our introduction, that large values of d are not desirable in practice: it may be better not to know the graph at a larger distance, as long as the complexity of the decision grows exponentially with the amount of information. A value $d = 2$ may be a good compromise between excessive shortsightedness and an excess of information. Concerning the parameters k and p, practical situations should involve cases where these values are not far from each other. Clearly, if $k \geq p$, the problem is easy, whereas if $k \ll p$, all policies will be bad because the controller is overwhelmed by the number of nodes to control. The modeling we propose is relevant if the network is a bottleneck of the system: in other words, if k is not very large. Therefore, p should not be very large either.

The next feature departing from practical cases is about the discovery process. In our model, we assumed a uniform distribution for the new generation of nodes. Practical graphs are known to have different node degree distributions. Here, since we identified this mechanism with a Galton–Watson branching process, it seems possible to use other distributions while conserving the possibility of characterizing the distribution of trees as we did in Section 4. Therefore, evaluating the performance of simple greedy policies and obtaining bounds might be possible with this generalization. Furthermore, the results we obtained with budget $k = 1$ or depth $d = 1$ were probably insensitive to the distribution of the number of sons.

The assumption we made about the movements of the "surfer" in the graph of documents may also be questioned. We assumed a uniform choice between neighboring documents. In addition to simplicity, we argue that this represents the most difficult situation for the controller, since the amount of information available to it is minimal. In the case where nonuniform movement probabilities are known, they can easily be integrated in the MDP, just as in the models reviewed in the Introduction. It is interesting to note that in such a case, models can be imagined where the optimal policy is *not* of the greedy type: consider just a case where the probability of moving to some son happens to be zero (or close to zero). The prefetching budget should then be devoted to marking the other sons and their subtrees. Apart from such obvious situations, it is difficult to imagine cases where an optimal policy would be formally identified, since this policy should weight the movement probabilities with the characteristics of the subtrees.

Going back to the simple model we proposed, the results of Section 7 (the case $d = 2$ and $k = 2$) and their interpretation suggest a general form for a heuristic policy. The first principle is that sons should be marked as priority. The issue is what to do with the remaining budget. On the one hand, the subtrees that have themselves less than k sons are easy to deal with and can be ignored. Among the remaining subtrees, those with the smallest number of sons should be marked first. If budget remains, the principle can be applied recursively.

9. Conclusions and Perspectives

Among the results of this paper, we proved that simple greedy policies are optimal in several situations, suggesting that optimal policies are always greedy. One first obvious step in future research will be to prove the following result, generalizing Proposition 4:

Conjecture 2. *When $k = 1$, any greedy policy is optimal.*

Next, our research will focus on the more challenging case $k = 2$. The first objective will be to prove Conjecture 1 and then determine how to adapt this policy to larger d. Some numerical investigation appears to be possible there for small values of p, despite the large size of the state space. Another line of research on formal solutions will focus on the analysis of Markov chains defined by simple policies with the purpose of providing bounds tighter than that of Proposition 3.

On the practical side, several issues need further investigation. The principal one is to efficiently identify the model and the optimal control from practical data. We plan to develop algorithms that would leverage the knowledge gained with the exact solution of simple models, with the objective of reducing learning time and learning errors.

Author Contributions: Formal analysis, K.K., A.J.-M., and S.A.; investigation, K.K., A.J.-M., and S.A.; supervision, A.J.-M. and S.A.; writing—original draft, K.K., A.J.-M., and S.A.; writing—review and editing, K.K., A.J.-M., and S.A. All authors have read and agreed to the published version of the manuscript.

Funding: This research received no external funding.

Institutional Review Board Statement: Not applicable.

Informed Consent Statement: Not applicable.

Data Availability Statement: Data is contained within the article.

Conflicts of Interest: The authors declare no conflict of interest.

Abbreviations

The following abbreviations are used in this manuscript:

CPU	Central Processing Unit
MDP	Markov Decision Process

Appendix A. Notation

Table A1. Main notation used in the paper.

Notation	Definition
d	depth of a tree
p	fanout of a tree; maximal number of sons of any node in a tree
k	prefetching budget; maximal number of nodes the controller can mark
$\textsc{Seq}_{1..p}(A)$	sequence of objects in set A, with length between 1 and p
$\mathcal{T}_{p,d}$	set of trees of depth d with fanout p
$T_{p,d}$	cardinal of $\mathcal{T}_{p,d}$
$\mathcal{M}_{p,d}$	set of trees of depth d with fanout p and marks in $\{0,1\}$
$M_{p,d}$	cardinal of $\mathcal{M}_{p,d}$
$\mathcal{M}^{+}_{p,d}$	set of trees of depth d with fanout p and marks in $\{0,1\}$ except for leaves that have mark 0
$M^{+}_{p,d}$	cardinal of $\mathcal{M}^{+}_{p,d}$
$\mu(t)$	mark (in $\{0,1\}$) of the root of tree t
$s(t)$	list of subtrees of tree t
$\text{leaves}(t)$	number of leaves in tree t
$\text{inode}(t)$	number of internal nodes in tree t (all nodes except leaves)
S	state space of MDP
t	state of MDP; it takes values in $\mathcal{M}^{+}_{p,d}$
A_t	action space of MDP when state is t
t_a	tree after the marking action; it takes values in $\mathcal{M}_{p,d}$
t_b	tree after the surfer's movement; it takes values in $\mathcal{M}_{p,d-1}$
$\mathcal{D}(t)$	set of trees that can be discovered from tree t
$\mathcal{SD}(t)$	set of trees after surfer movement and the discovery of new leaves when initial tree is t
$a(t)$	tree obtained after marking tree t according to action a
$P(t,a,t')$	transition probability of moving from a tree t to a tree t' under action a
$c(t,a)$	expected immediate cost incurred when applying action a to tree t
$\mathcal{U}$	set of usable states (states attainable through transitions given specific budget k)
HR	class of policies corresponding to history-dependent randomized strategies
γ	a marking policy
$\gamma_n(t)$	action prescribed by policy γ at time step n on tree t
$\gamma(t)$	stationary version of $\gamma_n(t)$
$\mathcal{M}_\gamma$	Markov chain generated by policy γ
$\sigma_{p,d}(t)$	shape of tree t; a tree with the same shape as tree t and all nodes' marks being 0
$\mathcal{M}$	Markov chain of tree shapes with depth d and fanout p
$\mathcal{M}_b$	Markov chain of tree shapes with depth $d-1$ and fanout p
$\pi_{p,d}$	stationary distribution of the Markov chain of tree shapes $\mathcal{M}$
$\mathbb{E}C$	upper bound on the optimal expected cost of any greedy policy
$\mathbb{H}_p$	harmonic number; finite partial sum of the harmonic series
g^*	optimal expected cost of the MDP
$N_{p,d}$	average number of nodes in trees distributed according to $\pi_{p,d}$
$L_{p,d}$	average number of leaves in trees distributed according to $\pi_{p,d}$
$C_{p,d}$	average number of nodes created in Markov chain $\mathcal{M}$
$D_{p,d}$	average number of nodes deleted in Markov chain $\mathcal{M}$
$g, f(s)$	constant and bounded function in optimality equation
$\overline{P}(t,a,t_b)$	transition probability of moving from a tree t to a tree t_b under action a
J_γ	expected cost for policy γ

Appendix B. MDP Facts

We borrow the below existential theorem from [17] (Theorem 2.1, Chapter V).

Theorem A1. *If there exists a bounded function $f(s)$ for every $s \in S$ and a constant g such that:*

$$g + f(s) = \max_a \left[r(s,a) + \sum_{s' \in S} P(s,a,s') f(s') \right] \tag{A1}$$

then there exists a stationary policy γ^* such that:

$$g = \max_\gamma \phi_\gamma(s) = \phi_{\gamma^*}(s). \tag{A2}$$

Theorem A1 guarantees the existence of an optimal policy given that there exists a bounded function f and constant g that satisfies (A1). We use the following theorem from [17] (Theorem 2.2, Chapter V), which proves the existence of such a function and constant.

Theorem A2. *Let $\{s_n\}_{n \geq 0}$ and $\{a_n\}_{n \geq 0}$ be the sequence of states and actions of the MDP when a policy γ_α is followed. Define:*

$$W^{\gamma_\alpha}(s) := E\left[\sum_{n=0}^{\infty} \alpha^n r(s_n, a_n) \right] \text{ where } 0 < \alpha < 1$$

and $W^{\gamma_\alpha^*}(s) := \max_{\gamma_\alpha} W^{\gamma_\alpha}(s)$. For some fixed $s_0 \in S$, if there exists a $B < \infty$ such that $|W^{\gamma_\alpha^*}(s) - W^{\gamma_\alpha^*}(s_0)| < B$ for all α and s, then there exist a bounded function f and a constant g satisfying (A1).

The uniform boundedness property on $W^{\gamma_\alpha^*}$ exists if the expected time to go from any state s to the fixed state s_0 is bounded by a finite value while using the optimal policy γ_α^*. The reader may refer to Theorem 2.4, Chapter V, from [17] for a proof of the same. A sufficient condition for the bounded expected time is that every stationary policy in the MDP yields a unichain.

Appendix C. Finite Horizon MDP for Trees of Depth $d = 2$ with Budget $k = 2$

This section is devoted to the findings from the study of the finite-horizon prefetching MDP, in the cases $d = 2$ and $k = 2$ and general p. These results are quoted from the unpublished report [16]. Other results for finite-horizon MDP are quoted in Remarks 3 and 4.

The optimal actions for all tree types for the finite horizons $n = 3, 4$ are specified in Table A2. The optimal actions for the $n = 3$ horizon were computed analytically and confirmed through numerical simulations. For the $n = 4$ horizon, the optimal actions in Table A2 are the numerical results. The same shorthands for marking actions from Section 7 are used here.

Table A2. Comparison of optimal actions at $n = 3$ and $n = 4$.

Tree Type	Specifications		Optimal Action at $n = 3$	Optimal Action at $n = 4$
Type 1			$a(d1, d1)$	$a(d1, d1)$
Type 2a	$j_r \geq 3$ for some r		$a(d1, l_{j_c})$, $j_c = \min_r j_r \geq 3$	$a(d1, l_{j_c})$, $j_c = \min_r j_r \geq 3$
	$j_r < 3$ for all r		$a(d1, l_{j_c})$, $j_c = \max_r j_r$	$a(d1, l_{j_c})$, $j_c = \max_r j_r$
Type 2b	$j_1 > 3$	$j_2 > 3$	$a(l_{j_1}, l_{j_1})$ assuming $j_1 \leq j_2$	$a(l_{j_1}, l_{j_1})$ assuming $j_1 \leq j_2$
	$j_1 = 3$	$j_2 \geq 3$	$a(l_{j_1}, l_{j_1})$ when (A3) holds true and $a(l_{j_1}, l_{j_2})$ otw.	$a(l_{j_1}, l_{j_2})$ for $p < 13$, $a(l_{j_1}, l_{j_1})$ when $p \geq 13$ for large j_2
	$j_1 = 2$	$j_2 > 3$	$a(l_{j_1}, l_{j_1})$ when (A4) holds true, $a(l_{j_1}, l_{j_2})$ when (A5) holds true, but (A4) does not, and $a(l_{j_2}, l_{j_2})$ otw.	$a(l_{j_1}, l_{j_2})$ for $p < 52$, $a(l_{j_1}, l_{j_1})$ when $p \geq 52$ for large j_2
		$j_2 = 3$	$a(l_{j_2}, l_{j_2})$	$a(l_{j_1}, l_{j_2})$
		$j_2 = 2$	$a(l_{j_1}, l_{j_2})$	$a(l_{j_1}, l_{j_2})$
	$j_1 = 1$	$j_2 \geq 4$	$a(l_{j_2}, l_{j_2})$ when (A6) holds true and $a(l_{j_1}, l_{j_2})$ otw.	$a(l_{j_1}, l_{j_2})$
		$j_2 = 3$	$a(l_{j_2}, l_{j_2})$ for $p < 5$, $a(l_{j_1}, l_{j_2})$ for $p \geq 5$	$a(l_{j_1}, l_{j_2})$
		$j_2 < 3$	$a(l_{j_1}, l_{j_2})$	$a(l_{j_1}, l_{j_2})$
Type 2c			$a(l_{j_1}, l_{j_1})$	$a(l_{j_1}, l_{j_1})$

The policy for Type 2 trees depends on the exact size of the subtrees, unlike the policy for Type 1 trees, where the sizes of the subtrees are irrelevant. There are thresholds that are functions of p, which decide the optimal action for Type 2*b* trees. The thresholds on j_2, with the obvious constraint $j_2 \leq p$, that decide the optimal action for certain specifications of Type 2b trees are below.

$$j_2 \geq \frac{6p^3}{6p^2(\mathbb{H}_p - 3) + 6p(5\mathbb{H}_p - 8) + 38\mathbb{H}_p - 15} \tag{A3}$$

$$j_2 \geq \frac{12p^2}{12p\mathbb{H}_p + 8\mathbb{H}_p - 39} \tag{A4}$$

$$j_2 \geq \frac{4p^2}{2p(2\mathbb{H}_p - 5) + 10\mathbb{H}_p - 7} \tag{A5}$$

$$j_2 \geq \frac{6p}{6\mathbb{H}_p - 11}. \tag{A6}$$

We observed a change in optimal actions for Type 2 trees for considerably large fanouts in the numerical experiments for $n = 4$. Since the results of $n = 4$ were obtained numerically, we could identify the critical p values after which there would be a change in the optimal action for some j_2. The precise threshold on j_2 for which the optimal action changes was not found due to the complex calculations involved. Comparing the optimal actions, we note a simplification when going from $n = 3$ to $n = 4$. The policy for trees of Types 1, 2a, and 2c is the same. For trees of Type 2b, the number of cases where a switch in optimal actions occurs is smaller in the $n = 4$ than in the $n = 3$ case. When a switch occurs in both cases, the threshold on the value j_2 is observed to be larger in the case $n = 4$. Naturally, we could expect that for a given specification of Type 2b trees, the optimal action would remain the same for most of the trees as the horizon increases. With this line of thought, we may conjecture that the threshold values disappear as $n \to \infty$ and the optimal actions are the same for all trees of a given type and specification. This is the principle that led to the definition of the greedy finite optimal policy in Table 2.

References

1. Grigoraş, R.; Charvillat, V.; Douze, M. Optimizing hypervideo navigation using a Markov decision process approach. In Proceedings of the ACM Multimedia, Juan-les-Pins, France, 1–6 December 2002; pp. 39–48.
2. Charvillat, V.; Grigoraş, R. Reinforcement learning for dynamic multimedia adaptation. *J. Netw. Comput. Appl.* **2007**, *30*, 1034–1058. [CrossRef]
3. Morad, O.; Jean-Marie, A. Prefetching Control for On-Demand Contents Distribution: A Markov Decision Process Model. In Proceedings of the 22nd IEEE International Symposium on Modeling, Analysis and Simulation of Computer and Telecommunication Systems (MASCOTS 2014), Paris, France, 9–11 September 2014; pp. 421–426. [CrossRef]
4. Morad, O.; Jean-Marie, A. On-Demand Prefetching Heuristic Policies: A Performance Evaluation. In Proceedings of the 29th International Symposium on Computer and Information Sciences (ISCIS 2014), Krakow, Poland, 27–28 October 2014; pp. 317–324. [CrossRef]
5. Pleşca, C.; Charvillat, V.; Ooi, W.T. Multimedia prefetching with optimal Markovian policies. *J. Netw. Comput. Appl.* **2016**, *69*, 40–53. [CrossRef]
6. Joseph, D.; Grunwald, D. Prefetching using Markov predictors. *IEEE Trans. Comput.* **1999**, *48*, 121–133. [CrossRef]
7. Dongshan, X.; Junyi, S. A new Markov model for Web access prediction. *Comput. Sci. Eng.* **2002**, *4*, 34–39. [CrossRef]
8. Jin, X.; Xu, H. An approach to intelligent Web pre-fetching based on hidden Markov model. In Proceedings of the 42nd IEEE International Conference on Decision and Control (IEEE Cat. No.03CH37475), Maui, HI, USA, 9–12 December 2003; Volume 3, pp. 2954–2958. [CrossRef]
9. Gellert, A.; Florea, A. Web Page Prediction Enhanced with Confidence Mechanism. *J. Web Eng.* **2014**, *13*, 507–524.
10. Kawazu, H.; Toriumi, F.; Takano, M.; Wada, K.; Fukuda, I. Analytical method of web user behavior using Hidden Markov Model. In Proceedings of the 2016 IEEE International Conference on Big Data (Big Data), Washington, DC, USA, 5–8 December 2016; pp. 2518–2524. [CrossRef]
11. Gupta, S.; Moharir, S. Request patterns and caching for VoD services with recommendation systems. In Proceedings of the 2017 9th International Conference on Communication Systems and Networks (COMSNETS), Bengaluru, India, 4–8 January 2017; pp. 31–38. [CrossRef]
12. Fomin, F.; Giroire, F.; Jean-Marie, A.; Mazauric, D.; Nisse, N. To satisfy impatient Web surfers is hard. *J. Theor. Comput. Sci.* **2014**, *526*, 1–17. [CrossRef]
13. Giroire, F.; Lamprou, I.; Mazauric, D.; Nisse, N.; Pérennes, S.; Soares, R. Connected surveillance game. *Theor. Comput. Sci.* **2015**, *584*, 131–143. [CrossRef]
14. Flajolet, P.; Sedgewick, R. *Analytic Combinatorics*; Cambridge University Press: Cambridge, UK, 2009.
15. Puterman, M. *Markov Decision Processes—Discrete Stochastic Dynamic Programming*; Wiley: New York, NY, USA, 1994.
16. Keshava, K. Prefetching: A Markov Decision Process Model. BS-MS Degree Dissertation, Indian Institute of Science Education and Research, Mohali, India, 2021.
17. Ross, S. *Introduction to Stochastic Dynamic Programming*; Academic Press: Cambridge, MA, USA, 1995.

Article

Convergence Bounds for Limited Processor Sharing Queue with Impatience for Analyzing Non-Stationary File Transfer in Wireless Network

Irina Kochetkova [1,2,*], Yacov Satin [3], Ivan Kovalev [3], Elena Makeeva [1], Alexander Chursin [1] and Alexander Zeifman [2,3]

1 Applied Probability and Informatics Department, Peoples' Friendship University of Russia (RUDN University), 6 Miklukho-Maklaya St., 117198 Moscow, Russia; elena-makeeva-96@mail.ru (E.M.); chursin-aa@rudn.ru (A.C.)
2 Institute of Informatics Problems, Federal Research Center "Computer Sciences and Control" of the Russian Academy of Sciences, 44-2 Vavilova St., 119333 Moscow, Russia; a_zeifman@mail.ru
3 Department of Applied Mathematics, Vologda State University, Lenina 15, 160000 Vologda, Russia; yacovi@mail.ru (Y.S.); kovalev.iv96@yandex.ru (I.K.)
* Correspondence: gudkova-ia@rudn.ru

Abstract: The data transmission in wireless networks is usually analyzed under the assumption of non-stationary rates. Nevertheless, they strictly depend on the time of day, that is, the intensity of arrival and daily workload profiles confirm this fact. In this article, we consider the process of downloading a file within a single network segment and unsteady speeds—for arrivals, file sizes, and losses due to impatience. To simulate the scenario, a queuing system with elastic traffic with non-stationary intensity is used. Formulas are given for the main characteristics of the model: the probability of blocking a new user, the average number of users in service, and the queue. A method for calculating the boundaries of convergence of the model is proposed, which is based on the logarithmic norm of linear operators. The boundaries of the rate of convergence of the main limiting characteristics of the queue length process were also established. For clarity of the influence of the parameters, a numerical analysis was carried out and presented.

Keywords: queuing system; elastic traffic; inpatient claim; non-stationary intensity; convergence analysis; bounds on the rate of convergence; wireless network; file transfer; daily traffic profile; blocking probability

Citation: Kochetkova, I.; Satin, Y.; Kovalev, I.; Makeeva, E.; Chursin, A.; Zeifman, A. Convergence Bounds for Limited Processor Sharing Queue with Impatience for Analyzing Non-Stationary File Transfer in Wireless Network. *Mathematics* **2022**, *10*, 30. https://doi.org/10.3390/math10010030

Academic Editor: Daniel-Ioan Curiac

Received: 17 November 2021
Accepted: 17 December 2021
Published: 22 December 2021

1. Introduction

The fifth generation (5G) networks will consist of different services with different specifications. Experts have identified network slicing as a key technology to enable 5G networks [1–4]. Within the framework of network slicing, many models are proposed with different principles for slicing radio resources [5]. In some cases, it is necessary to distribute slices between several users, and this distribution and the slices can be changed depending on time, for example, in [6–8], various options for re-slicing the network are shown. Earlier, we studied models for studying network slicing in the framework of papers [9,10], but we considered a model in the form of a queuing system with stationary intensities.

Given the need to re-slice the network and because all processes are non-stationary and depend on time. For example, a traffic profile—user activity is different depending on the time of day. Activity may also depend on the time of year; in the summer many users go on vacation and fly to other countries, and in winter most users are actively working and on weekends they sit at home watching movies, for example. Taking into account this dependence, it is necessary to consider models with non-stationary intensities. In our work, we consider an example of downloading user files depending on the time of day. For convenience, we consider one slice of the radio frequency channel taking into account the non-stationary nature, and look at the behavior of the characteristics. As a mathematical model, we take a queuing system with a non-stationary arrival intensity of user arrival in the system, as in [11,12], where such systems with non-stationary intensities are used for the joint service of radio frequencies.

This article discusses a special, but rather non-standard, heterogeneous birth and death process, for which, in principle, one can apply the methods described, for example, in articles [13,14], but due to the specifics of the model, in order to obtain acceptable estimates, the authors had to "trick". In particular, the rate of convergence in the example under consideration actually turns out to be rather slow, so that the existing (and constructed) limiting regime begins to adequately show the situation at sufficiently large times Usually, uniformization is used as a method for calculating the transition probabilities for Markov chains, as in [15,16]. However, the methods based on it work very poorly in the case of slow convergence, which takes place for the considered models. In addition, without a prior understanding of when the limiting regime is reached, significant computational efforts are required in order to be at least to some extent confident that the obtained solution is the required one [17].

The goal of the paper is to explore the nonstationary queue-length process based on file transferring in the wireless network and analyze the performance measures of this system model. The remainder of the paper is organized as follows. In Section 2, the queueing system and its performance measures are presented. The convergence analysis for large service rates and arrival rates of the mentioned queueing system are described in Section 3. Application for file transfer in the wireless network and numerical analysis of the considered system model are discussed in Section 4, followed by the conclusions in Section 5.

2. Queuing System

2.1. Overview and Assumptions

In this paper, we will consider a queuing system with elastic traffic and inpatient claims. To describe the flow of requests with a variable number of users, a Poisson flow of the first kind with the following parameters is suitable: the arrival intensity $\lambda(t)$, the minimum requirement for the resource b, and the length of the transmitted data block $\theta(t)$. Table 1 reflects the main parameters that describe the system model and the corresponding terms of the mathematical model. In the system under consideration, there is a resource of volume C, a storage device with a finite capacity r. Requests also have the property of impatience—they leave the queue with intensity $\gamma(t)$. We will assume that the block length is equal to some value $\theta(t)$. The entire volume C is divided equally between the orders, that is, if the number of customers is 1, then the entire resource is consumed by this customer, and the service rate is $\theta(t)/C$; if the number of customers is 2, then the service rate is $2\theta(t)/C$—the resource volume is divided in half. In the case when C cannot be divided equally between the customers with the provision of the minimum guaranteed threshold b, a new customer enters the queue.

Let $N(t) \in \{1,\dots,\lfloor C/b \rfloor\}$ be the number of processed orders at the moment $t \geq 0$. Hence, the number $\lfloor C/b \rfloor = N$ is the maximum number of requests that the device can process simultaneously. The state-space of the system looks like this:

$$X := \{n \in 0, ..., N, ..., N+r : c(n) \leq C\}. \tag{1}$$

Table 1. System model parameters.

Parameter	Description
$\lambda(t)$	Intensity of the flow of requests for the transfer of elastic data
b	Minimum guaranteed elastic data block transfer rate
$\theta(t)$	Average value of data block length
r	The queue of applications for the transfer of a block of elastic data
$\gamma(t)$	The rate of loss of impatient claims
C	Network bandwidth (service speed)

2.2. Continuous-Time Markov Chain

It is easy to see that the given model can be described by Markov process $X(t), t > 0$ where $X(t)$ denotes the number of customers in the system at time t (queue-length process). Denote by $p_n(t) = P(X(t) = n), n = 0, 1, 2, 3, \ldots$.

From the above assumptions, the resulting behavior of the state probabilities are described by a forward Kolmogorov system:

$$p_0'(t) = -\lambda(t)p_0(t) + \frac{C}{\theta(t)}p_1(t) \tag{2}$$

$$p_n'(t) = \lambda(t)p_{n-1}(t) - \left(\frac{C}{\theta(t)} + \lambda(t)\right)p_n(t) + \frac{C}{\theta(t)}p_{n+1}(t), 1 \le n < N \tag{3}$$

$$p_N'(t) = \lambda(t)p_{n-1}(t) - \left(\frac{C}{\theta(t)} + \lambda(t)\right)p_n(t) + \left(\frac{C}{\theta(t)} + \gamma(t)\right)p_{n+1}(t) \tag{4}$$

$$p_n'(t) = \lambda(t)p_{n-1}(t) - \left(\frac{C}{\theta(t)} + (n-N)\gamma(t) + \lambda(t)\right)p_n(t) + \left(\frac{C}{\theta(t)} + (n+1-N)\gamma(t)\right)p_{n+1}(t),$$
$$N < n < N + r \tag{5}$$

$$p_{N+r}'(t) = \lambda(t)p_{N+r-1}(t) - \left(\frac{C}{\theta(t)} + r\gamma(t) + \lambda(t)\right)p_{N+r}(t). \tag{6}$$

Now, we consider the corresponding nonstationary situation. Namely, we suppose that the queue-length process $\{X(t),\ t \ge 0\}$ is an inhomogeneous continuous-time Markov chain. All possible transition intensities say $q_{ij}(t)$, are supposed to be non-random functions of time. We suppose that all intensity functions are nonnegative and locally integrable on $[0, \infty)$.

Denote by $\mathbf{p}(t) = (p_0(t), p_1(t), p_2(t), \ldots, p_{N+r}(t))^T$ the vector of state probabilities at the moment t. Put $a_{ij}(t) = q_{ji}(t)$ for $j \ne i$ and $a_{ii}(t) = -\sum_{j \ne i} a_{ji}(t) = -\sum_{j \ne i} q_{ij}(t)$.

We can consider the forward Kolmogorov system (2)–(6) as a differential equation

$$\frac{d\mathbf{p}(t)}{dt} = A(t)\mathbf{p}(t), \tag{7}$$

in the space of sequences l_1, where $A(t)$ is a bounded for almost all $t \ge 0$ linear operator from l_1 to itself and it is generated be the corresponding transposed intensity matrix:

$$\begin{pmatrix}
-\lambda(t) & \frac{C}{\theta(t)} & 0 & \cdot & 0 & 0 & 0 & \cdot & 0 & 0 \\
\lambda(t) & -\left(\frac{C}{\theta(t)} + \lambda(t)\right) & \frac{C}{\theta(t)} & \cdots & 0 & 0 & 0 & \cdots & 0 & 0 \\
0 & \lambda(t) & -\left(\frac{C}{\theta(t)} + \lambda(t)\right) & \cdots & 0 & 0 & 0 & \cdots & 0 & 0 \\
\vdots & \vdots & \vdots & \vdots & \vdots & \vdots & \vdots & \vdots & \vdots & \vdots \\
0 & 0 & 0 & \cdots & -\left(\frac{C}{\theta(t)} + \lambda(t)\right) & \frac{C}{\theta(t)} & 0 & \cdots & 0 & 0 \\
0 & 0 & 0 & \cdots & \lambda(t) & -\left(\frac{C}{\theta(t)} + \lambda(t)\right) & \frac{C}{\theta(t)} + \gamma(t) & \cdots & 0 & 0 \\
0 & 0 & 0 & \cdots & 0 & \lambda(t) & -\left(\frac{C}{\theta(t)} + \gamma(t) + \lambda(t)\right) & \cdots & 0 & 0 \\
\vdots & \vdots & \vdots & \vdots & \vdots & \vdots & \vdots & \vdots & \vdots & \vdots \\
0 & 0 & 0 & \cdots & 0 & 0 & 0 & \cdots & -\left(\frac{C}{\theta(t)} + (r-1)\gamma(t) + \lambda(t)\right) & \frac{C}{\theta(t)} + \gamma(t)r \\
0 & 0 & 0 & \cdots & 0 & 0 & 0 & \cdots & \lambda(t) & -\left(\frac{C}{\theta(t)} + \gamma(t)r\right)
\end{pmatrix} \tag{8}$$

2.3. Performance Measures

For analysis of the system let us consider some characteristics of the devoted model. First, the probability of blocking an incoming application:

$$P_{block}(t) = p_{N+r}(t). \tag{9}$$

The average number of serviced applications:

$$\bar{C}(t) = \sum_{i=1}^{N} i p_i(t) + N \cdot \sum_{i=1}^{r} p_{N+i}(t). \tag{10}$$

The average number of applications in the queue:

$$Q(t) = \sum_{i=N+1}^{N+r} (i-N) p_i(t). \tag{11}$$

3. Convergence Analysis

3.1. Definitions of Terms

We denote the mathematical expectation (the mean) of $X(t)$ at the moment t if $X(0)=k$ as $E(t,k) = E\{X(t)|X(0)=k\}$.

The Markov chain $X(t)$ is called *weakly ergodic*, if $\lim_{t\to\infty} \|\mathbf{p}^1(t) - \mathbf{p}^2(t)\| = 0$ for any initial conditions $\mathbf{p}^1(0) = \mathbf{p}^1 \in \Omega$, $\mathbf{p}^2(0) = \mathbf{p}^2 \in \Omega$. For our situation *any* $\mathbf{p}^1(t)$ is considered as a *quasi-stationary distribution* of the chain $X(t)$.

The mentioned Markov chain $X(t)$ also has the limiting mean $\phi(t)$, if $|E(t;k) - \phi(t)| \to 0$ as $t \to \infty$ for any k.

We recall the logarithmic norm of operator function from l_1 to itself is calculated as (12):

$$\gamma(B(t))_1 = \sup_i \left(b_{ii}(t) + \sum_{j\neq i} |b_{ji}(t)| \right), \tag{12}$$

and the bound

$$\|U(t,s)\| \le e^{\int_s^t \gamma(B(\tau))\, d\tau}, \tag{13}$$

is valid for the Cauchy operator of the corresponding differential equation

$$\frac{d\mathbf{x}}{dt} = B(t)\mathbf{x}. \tag{14}$$

3.2. Preliminary Considerations

Let us put that $\mu(t) = \frac{C}{\theta(t)}$, and $\mu_n(t) = \mu(t) + \max(0, n-N)\gamma(t)$ for $n \ge 1$. Then

$$A(t) = \begin{pmatrix} -\lambda(t) & \mu_1(t) & 0 & \cdots & 0 & 0 \\ \lambda(t) & -(\mu_1(t)+\lambda(t)) & \mu_2(t) & \cdots & 0 & 0 \\ 0 & \lambda(t) & -(\mu_2(t)+\lambda(t)) & \cdots & 0 & 0 \\ \vdots & \vdots & \vdots & \vdots & \vdots & \vdots \\ 0 & 0 & 0 & \cdots & -(\mu_{N+r-1}(t)+\lambda(t)) & \mu_{N+r}(t) \\ 0 & 0 & 0 & \cdots & \lambda(t) & -\mu_{N+r}(t). \end{pmatrix} \tag{15}$$

As indicated above, the considered method is based on the concept of the logarithmic norm and the corresponding estimates for the Cauchy operator.

Assuming that $p_0 = 1 - \sum_{i=1}^{N+r} p_i(t)$, then from (7) we get the following equation:

$$\frac{d\mathbf{p}}{dt} = B(t)\mathbf{p} + \mathbf{g}(t), \quad t \ge 0, \tag{16}$$

where $\mathbf{g}(t) = (\lambda(t), 0, 0, \ldots, 0)^T$ and $B(t)$ equals (17).

$$B(t) = \begin{pmatrix} -(\mu_1(t)+2\lambda(t)) & \mu_2(t)-\lambda(t) & -\lambda(t) & -\lambda(t) & \cdots & -\lambda(t) & -\lambda(t) \\ \lambda(t) & -(\mu_2(t)+\lambda(t)) & \mu_3(t) & 0 & \cdots & 0 & 0 \\ 0 & \lambda(t) & -(\mu_3(t)+\lambda(t)) & \mu_4(t) & \cdots & 0 & 0 \\ \vdots & \vdots & \vdots & \vdots & \vdots & \vdots & \vdots \\ 0 & 0 & 0 & 0 & \cdots & -(\mu_{N+r-1}(t)+\lambda(t)) & \mu_{N+r}(t) \\ 0 & 0 & 0 & 0 & \cdots & \lambda(t) & -\mu_{N+r}(t). \end{pmatrix} \tag{17}$$

The solution to this equation can be represented in the following form:

$$\mathbf{p}(t) = U^*(t,0)\mathbf{p}(0) + \int_0^t U^*(t,\tau)\mathbf{g}(\tau)\,d\tau, \tag{18}$$

where $U^*(t,s)$ is the Cauchy operator of the corresponding homogeneous equation:

$$\frac{d\mathbf{x}}{dt} = B(t)\mathbf{x}. \tag{19}$$

Next, we will consider estimates in "weighted" norms. Suppose $d_1, d_2, \ldots, d_{N+r}$ are positive numbers. Then let

$$D = \begin{pmatrix} d_1 & d_1 & \cdots & d_1 \\ 0 & d_2 & \cdots & d_2 \\ & \ddots & \ddots & \vdots \\ 0 & 0 & \cdots & d_{N+r}. \end{pmatrix} \tag{20}$$

We will denote by $\|\mathbf{z}\|_{1\mathbf{D}} = \|\mathbf{Dz}\|_1$. Note that $B(t)$ is essentially non-negative, that is, all off-diagonal elements of $B(t)$ are non-negative for any $t \geq 0$. Then we get:

$$B^{**}(t) = DB(t)D^{-1} = \begin{pmatrix} (\mu_1(t)+\lambda(t)) & \frac{d_1}{d_2}\mu_1(t) & 0 & \cdots & 0 & 0 \\ \frac{d_2}{d_1}\lambda(t) & -(\mu_2(t)+\lambda(t)) & \frac{d_2}{d_3}\mu_2(t) & \cdots & 0 & 0 \\ 0 & \frac{d_3}{d_2}\lambda(t) & -(\mu_3(t)+\lambda(t)) & \cdots & 0 & 0 \\ 0 & 0 & \frac{d_4}{d_3}\lambda(t) & \cdots & 0 & 0 \\ \vdots & \vdots & \vdots & \vdots & \vdots & \vdots \\ 0 & 0 & 0 & \cdots & -\left(\mu_{N+r-1}(t)+\lambda(t)\right) & \frac{d_{N+r-1}}{d_{N+r}}\mu_{N+r-1}(t) \\ 0 & 0 & 0 & \cdots & \frac{d_{N+r}}{d_{N+r-1}}\lambda(t) & -(\mu_{N+r}(t)+\lambda(t)). \end{pmatrix} \tag{21}$$

Let us put the following:

$$\gamma_{**}(t) = \inf_i \left(|b_{ii}(t)| - \sum_{j \neq i} \frac{d_j}{d_i} b_{ji}(t) \right). \tag{22}$$

Then we will get:

$$\gamma(B(t))_{1\mathrm{D}} = \gamma\left(\mathrm{D}B(t)\mathrm{D}^{-1}\right) = \sup_i \left(b_{ii}(t) + \sum_{j \neq i} \frac{d_j}{d_i} b_{ji}(t) \right) = -\gamma_{**}(t). \tag{23}$$

For some positive δ we put $d_1 = 1$, $d_{k+1} = \delta d_k$, $k \geq 1$.

3.3. Bounds on the Rate of Convergence for Large Service Rates

First, we let $\delta > 1$. Then we will get the following:

$$\begin{gathered}\alpha_1(t) = \mu_1(t) + \lambda(t) - \delta\lambda(t) = \mu(t) - (\delta - 1)\lambda(t), \\ \alpha_k(t) = \mu_k(t) + \lambda(t) - \tfrac{1}{\delta}\mu_{k-1}(t) - \delta\lambda(t) \geq \left(1 - \tfrac{1}{\delta}\right)\mu_{k-1}(t) - (\delta - 1)\lambda(t) \geq \\ \left(1 - \tfrac{1}{\delta}\right)(\mu(t) - \delta\lambda(t)),\ 2 \leq k < N + r, \\ \alpha_{N+r}(t) = \mu_{N+r}(t) - \frac{1}{\delta}\mu_{N+r-1}(t) + \lambda(t) \geq \left(1 - \frac{1}{\delta}\right)(\mu(t) - \delta\lambda(t)).\end{gathered} \tag{24}$$

Therefore,

$$\gamma_{**}(t) = \min(\alpha_i(t)) = \left(1 - \frac{1}{\delta}\right)(\mu(t) - \delta\lambda(t)). \tag{25}$$

Theorem 1. *Let there be a positive number $delta > 1$ such that,*

$$\int_0^\infty (\mu(t) - \delta\lambda(t))dt = +\infty. \tag{26}$$

Then the Markov chain $X(t)$ is weakly ergodic and has the following convergence rate bounds:

$$\|\mathbf{p}^*(t) - \mathbf{p}^{**}(t)\|_{1D} \leq e^{-\int\limits_0^t \left(1-\frac{1}{\delta}\right)(\mu(t)-\delta\lambda(t))\,d\tau} \|\mathbf{p}^*(0) - \mathbf{p}^{**}(0)\|_{1D}, \tag{27}$$

$$\begin{aligned}\|\mathbf{p}^*(t) - \mathbf{p}^{**}(t)\| &\leq 4\delta^{N+r} e^{-\int\limits_0^t \left(1-\frac{1}{\delta}\right)(\mu(t)-\delta\lambda(t))\,d\tau} \|\mathbf{p}^*(0) - \mathbf{p}^{**}(0)\| \\ &\leq 8\delta^{N+r} e^{-\int\limits_0^t \left(1-\frac{1}{\delta}\right)(\mu(t)-\delta\lambda(t))\,d\tau},\end{aligned} \tag{28}$$

for any initial conditions $\mathbf{p}^*(0), \mathbf{p}^{**}(0)$ and any $t \geq 0$.

We let that $W = \min_{k \geq 1} \frac{d_k}{k} = \min_{k \geq 0} \frac{\delta^k}{k+1}$. Then we will get $W\|\mathbf{p}\|_{1E} \leq \|\mathbf{p}\|_{1D}$.

Corollary 1. *From the conditions of Theorem 1, $X(t)$ has a limit mean, then we say $\phi(t) = E(t, 0)$, and we obtain that the following estimate is true for any j and any $t \geq 0$:*

$$|E(t, j) - E(t, 0)| \leq \frac{1 + \delta^{j-1}}{W} e^{-\int\limits_0^t \left(1-\frac{1}{\delta}\right)(\mu(t)-\delta\lambda(t))\,d\tau}. \tag{29}$$

3.4. Bounds on the Rate of Convergence for Large Arrival Rates

Now consider the case $\delta < 1$ and assume that $\delta \in [\frac{r-1}{r}, 1)$. In this case, we have:

$$\begin{gathered}\alpha_1(t) \geq \mu(t) + (1 - \delta)\lambda(t), \\ \alpha_k(t) \geq (1 - \delta)\lambda(t) + \mu_k(t) - \tfrac{1}{\delta}\mu_{k-1}(t) \geq \\ (1 - \delta)\lambda(t) - \mu(t)\left(\tfrac{1}{\delta} - 1\right),\quad 2 \leq k \leq N + r - 1 \\ \alpha_{N+r}(t) \geq \lambda(t) - \left(\frac{1}{\delta} - 1\right)\mu(t).\end{gathered} \tag{30}$$

Then it follows that we have the same:

$$\gamma_{**}(t) = \min(\alpha_i(t)) = \left(\frac{1}{\delta} - 1\right)(\delta\lambda(t) - \mu(t)). \tag{31}$$

Theorem 2. *Let*

$$\int_0^\infty (\delta\lambda(t) - \mu(t))\, dt = +\infty, \tag{32}$$

for some $\delta \in [\frac{r-1}{r}, 1)$. *Then, the Markov chain* $X(t)$ *is weakly ergodic and has the following convergence rate bounds:*

$$\|\mathbf{p}^*(t) - \mathbf{p}^{**}(t)\|_{1D} \le e^{-\int\limits_0^t \left(\frac{1}{\delta}-1\right)(\delta\lambda(t)-\mu(t))\, d\tau} \|\mathbf{p}^*(0) - \mathbf{p}^{**}(0)\|_{1D}, \tag{33}$$

and

$$\|\mathbf{p}^*(t) - \mathbf{p}^{**}(t)\| \le 8\delta^{N+r} e^{-\int\limits_0^t \left(\frac{1}{\delta}-1\right)(\delta\lambda(t)-\mu(t))\, d\tau}, \tag{34}$$

for any initial conditions $\mathbf{p}^*(0), \mathbf{p}^{**}(0)$ *and any* $t \ge 0$. *In addition, there is also a marginal mean and bound (29).*

In addition, note the following: if the process is homogeneous (i.e., all intensities are constant), then the conditions of Theorems 1 and 2 are equivalent to the inequalities $\mu > \lambda$ and $\mu < \lambda$, respectively.

And it is also worth noting that when all intensities of the process are 1-periodic, then there is a weak ergodicity $X(t)$ and the estimates of Theorem 1 or Theorem 2 if:

$$\int_0^1 \lambda(t)\, dt \neq \int_0^1 \mu(t)\, dt. \tag{35}$$

3.5. Perturbed CTMC and Bounds

In this subsection, we will consider the application of the general perturbation bounding in the same way as in the work [13]) for the models under study. We will consider a "perturbed" queue-length process $\bar{X}(t), t \ge 0$ with the corresponding transposed intensity matrix $\bar{A}(t)$, where the "perturbing" matrix $\hat{A}(t) = A(t) - \bar{A}(t)$ is small. That is, we assume that the perturbed queue is of the same nature as the original one. Then the perturbed intensity matrix also has the same structure with the corresponding perturbed intensities $\bar{\theta}(t), \bar{\gamma}(t), \bar{\lambda}(t)$. We assume that $\bar{\mu}_n(t) = \frac{C}{\bar{\theta}(t)} + \max(0, n-N)\bar{\gamma}(t)$ for $n \ge 1$.

We suppose that:

$$\left|\frac{1}{\theta(t)} - \frac{1}{\bar{\theta}(t)}\right| = \left|\frac{1}{\hat{\theta}(t)}\right| < \hat{\epsilon}, |\gamma(t) - \bar{\gamma}(t)| = |\hat{\gamma}(t)| < \hat{\epsilon}, |\lambda(t) - \bar{\lambda}(t)| = |\hat{\lambda}(t)| < \hat{\epsilon}. \tag{36}$$

Hence, we will get the following:

$$\begin{aligned} |\mu_n(t) - \bar{\mu}_n(t)| = |\hat{\mu}_n(t)| &= |\tfrac{C}{\theta(t)} + \max(0, n-N)\gamma(t) - \tfrac{C}{\bar{\theta}(t)} \\ &\quad - \max(0, n-N)\bar{\gamma}(t)| \\ &\le \left|\tfrac{C}{\theta(t)} - \tfrac{C}{\bar{\theta}(t)}\right| + \max(0, n-N)|(\gamma(t) - \bar{\gamma}(t))| \\ &\le C\hat{\epsilon} + r\hat{\epsilon} = (C+r)\hat{\epsilon}. \end{aligned} \tag{37}$$

Then, from (8) we obtain the following bound:

$$\|\hat{A}(t)\| = 2\sup_k |\hat{a}_{kk}(t)| = 2\max\left(|\hat{\lambda}(t)|, |\hat{\lambda}(t)| + |\hat{\mu}_n(t)|, |\hat{\mu}_{N+r}(t)|\right) \le 2(C+r+1)\hat{\epsilon}. \tag{38}$$

Now from Theorem 1 and Corollary 1 in the paper [13] the following bounds of the perturbation follow.

Theorem 3. *Suppose that under the conditions of Theorem 1 or Theorem 2 the Markov chain $X(t)$ is exponentially ergodic, that is,*

$$e^{-\int\limits_s^t \gamma_{**}(\tau)\, d\tau} \leq K e^{-\gamma_0 (t-s)}, \tag{39}$$

for some positive K, γ_0. Then the following bounds of the perturbation take place:

$$\limsup_{t\to\infty} \|\mathbf{p}(t) - \bar{\mathbf{p}}(t)\| \leq \frac{2\hat{\epsilon}(C+r+1)(1+\log(4K)+(N+r)|\log(\delta)|)}{\gamma_0}, \tag{40}$$

and

$$\limsup_{t\to\infty} |E(t,0) - \bar{E}(t,0)| \leq \frac{2(N+r)\hat{\epsilon}(C+r+1)(1+\log(4K)+(N+r)|\log(\delta)|)}{\gamma_0}, \tag{41}$$

for any perturbed queue with the respectively closed intensities satisfying to (36).

4. File Transfer in Wireless Network

4.1. Multi-Service Network

The system model of this work has the form of a cell, in the coverage area of which mobile devices are located (Figure 1). Each of them has its MSISDN (Mobile Subscriber Integrated Services Digital Number)—the mobile subscriber number of digital network with the integration of services. Each user behaves as follows: he sends a request to download a file, then downloads it, and also the user can disappear from the system. Disappearance can be associated with leaving the coverage area of the cell, or with a change in the type of service, or with the end of the service. The conclusion follows from the description of the system model: if we sum up the flow from all users, then the duration of the intervals between requests will not depend on the number of users, which is described by the Poisson flow of the first kind.

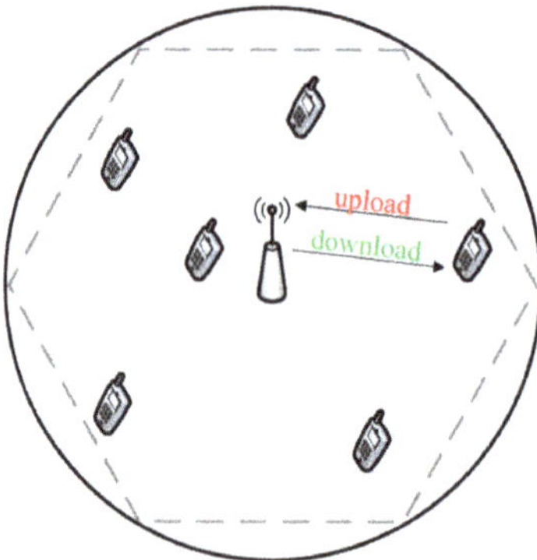

Figure 1. Scheme of System Model.

The users are provided with various services that belong to different categories of data transmission. A more detailed overview of the categories is illustrated in the Table 2. However, of these, we consider only those that can be described by elastic traffic, such as email, file transfer, and others.

Table 2. Description of service types.

N.	Service Name	Description
1	DB Transactions	Transaction in a database—for example, transfer of funds from a credit card through a mobile application.
2	File Systems	Work with remote file systems. This is how ATMs are arranged, for example. There is a machine with an operating system, and ATMs are connected to it and all information is stored there.
3	File Transfer	File transfer via FTP.
4	Games	Online game traffic.
5	Instant Messaging Applications	Instant messaging systems - messengers WhatsApp, Viber, Telegram, etc.
6	Legacy Protocols	Deprecated protocols.
7	Mail	Email.
8	Music Streaming	Streaming services for listening to music.
9	Network Operation	Network services.
10	Others	Others
11	P2P Applications	Applications where data transmission is based on the principles of peer-to-peer networks. The simplest example is Bittorrent applications.
12	Security	Online video cameras, data from alarm sensors, etc.
13	Streaming Applications	Streaming services for watching movies and video-chats.
14	Terminals	Mobile terminals for credit card payments.
15	VoIP	IP-telephony. For example, calls via WhatsApp or Skype.
16	Web Applications	Web applications—client-server applications in which the client interacts with the server using a browser (Microsoft Office Online, Google Documents).

4.2. Dataset Structure

The analysis of real traffic is of great importance, because the identification of the patterns of its arrival can make it possible to carry out studies of the system model described in the previous section in a non-stationary mode. Therefore, a task of this paper is formulated as follows: to analyze the traffic on one of the cell tower. There is a monitoring component to collect information about the system. Every hour, for each active device, the number of bits sent and received is added up. At the beginning of the next hour, the amount for the previous one hour is summed to the registration file. The principle of filling data during the monitoring is shown in Figure 2. There is a following designation $S_{d,t}^{up}$—the sum of bits sent by the device per day d, at hour t. The part of the log file is shown in the Table 3. There are the column START_HOUR contains the full date and hour when the data was transferred; and the column MASKED_MSISDN—masked device identifier of the user; APP_CLASS—the class of the application that transfers data; UPLOAD—the number of bits, which were sent as we denoted $S^{up}_{d,t}$; and the last DOWNLOAD—the number of received bits as we denoted $S^{down}_{d,t}$.

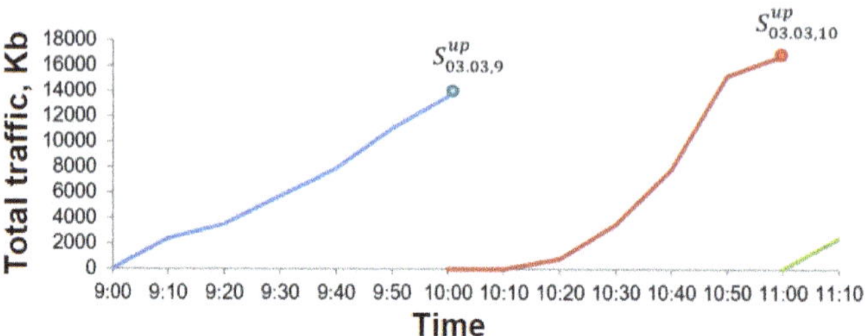

Figure 2. The principle of data accumulation during monitoring.

Table 3. Part of the log file.

START_HOUR	MASKED_MSISDN	APP_CLASS	UPLOAD	DOWNLOAD
12.02.2018 10:00	9BFA3001DE8C58C2453B6 9CE2E4A3704	Web Applications	111,339	78,208
12.02.2018 8:00	0C2C88351A593CD02727A 6207EB85E9E	Instant Messaging Applications	11,946	6741
12.02.2018 10:00	B9F45E1542162096408D9 4DD94499B89	Web Applications	67,143,832	4,230,675
12.02.2018 10:00	3CF6B81BC186E4C188D69 E1FE8919BB6	Web Applications	18,844	9769
11.02.2018 18:00	7EE1FCE60945D869A14EF 30E896E9131	Streaming Applications	17,219,335	403,361
11.02.2018 17:00	A746CCDCAC507B95C83 A3475C63C6BFD	Web Applications	235,455	131,712
11.02.2018 16:00	EA9D75DEB103F439A8C2 8300C2CC1757	Streaming Applications	136,767,000	4,078,817
12.02.2018 9:00	610FF30F1C9EC8939B4F BD259255BD63	Streaming Applications	3,330,234	112,395
12.02.2018 10:00	6DFD256C104B394A884E0 6A6EF7BE156	File Transfer	26,804,578	2,111,659
11.02.2018 17:00	6C9A6A5D444E7416C0C43 10E617EB611	Web Applications	587,786	87,897
...	...	...	...	...

4.3. Daily Traffic Profile

We consider elastic traffic, which is characterized by such a parameter as the length of the elastic data block. For the considered traffic model, monitoring data were taken with a class "File Transfer", which corresponds to the transfer of data using the FTP protocol.

To get the number of requests per second at a particular hour of the day for the selected application class, we perform aggregation of the form:

$$\lambda_{up}(t) = \frac{\sum_d^\infty S^{up}_{d,t}}{8 \cdot 1024 \cdot 1024 \cdot l} \tag{42}$$

$$\lambda_{down}(t) = \frac{\sum_d^\infty S^{down}_{d,t}}{8 \cdot 1024 \cdot 1024 \cdot l}, \tag{43}$$

where the received sum is determined in bits, d—date and l—average packet size, equal to 2 MB. Then, we perform the calculation on Formulas (42) and (43) and the results are shown in Table 4.

Table 4. Dependence of the number of requests on the time of day.

Times of Day, t	$\lambda_{up}(t)$	$\lambda_{down}(t)$
0	0.484847494	0.040711044
1	0.341865943	0.031079608
2	0.190823262	0.018468431
3	0.132143012	0.013982313
4	0.118697844	0.011683847
5	0.105803565	0.009923712
6	0.116474526	0.012585431
7	0.242942093	0.019329354
8	0.548677837	0.045828017
9	0.882484275	0.062668489
10	0.916328538	0.091028611
11	0.920243822	0.087240118
12	0.814816002	0.08172834
13	0.710077612	0.0853086
14	0.820103586	0.095698016
15	0.879327048	0.090266848
16	0.951204391	0.098226292
17	0.945908433	0.09180442
18	0.990004516	0.082148071
19	0.991890388	0.074312104
20	0.907643325	0.067218977
21	0.957584488	0.066221158
22	0.926637206	0.112652817
23	0.755175635	0.067874847

4.4. Fourier Series Approximation

From the found values it is necessary to obtain some continuous function of the dependence of the fluctuation of the upward and downward traffic flows on the time of day. To do this, we perform approximation by the Fourier series. For clarity, we will gradually increase the number of conditions in a row, which is to choose the most optimal option.

Consider the dependence of the intensity of the upward data flow λ_{up} from time, and we will carry out the approximation by the Fourier series with one, two, and three conditions. As a result, we get a functions of the form (44)–(46) respectively and for our example the parameters will take values, which are shown in Table 5:

$$a(t) = a_0 + a_1 cos(wt) + b_1 sin(wt) \tag{44}$$

$$a(t) = a_0 + a_1 cos(wt) + b_1 sin(wt) + a_2 cos(2wt) + b_2 sin(2wt) \tag{45}$$

$$\begin{aligned} a(t) = a_0 + a_1 cos(wt) + b_1 sin(wt) + a_2 cos(2wt) + b_2 sin(2wt) \\ + a_3 cos(3wt) + b_3 sin(3wt). \end{aligned} \tag{46}$$

Table 5. Parameters of Fourier Series Approximation for the upward data flow.

Condition	w	a_0	a_1	b_1	a_2	b_2	a_3	b_3
one	0.2519	0.6472	−0.168	−0.3675	–	–	–	–
two	0.2617	0.6521	−0.07656	−0.3992	0.1479	−0.1348	–	–
three	0.2603	0.6521	−0.08888	−0.3963	0.1384	−0.1411	−0.01446	0.05875

Using Matlab, we have built the plots of the considered approximation. It is easy to see that the plot of the obtained approximation for the one condition (Figure 3a) differs significantly from the initial data, therefore, so that is why we increased the number of conditions to two and the graph of the resulting function for two conditions (Figure 3b) significantly reflects more faithfully the relationship between real data. Then, according to the plot of the approximation for the three conditions (Figure 3c), it could be seen that, now, in the intervals with the highest network load, the function has become closer to the real data points. However, in general, the graph did not receive significant changes, and the variant of the approximation with two conditions can be considered the most optimal.

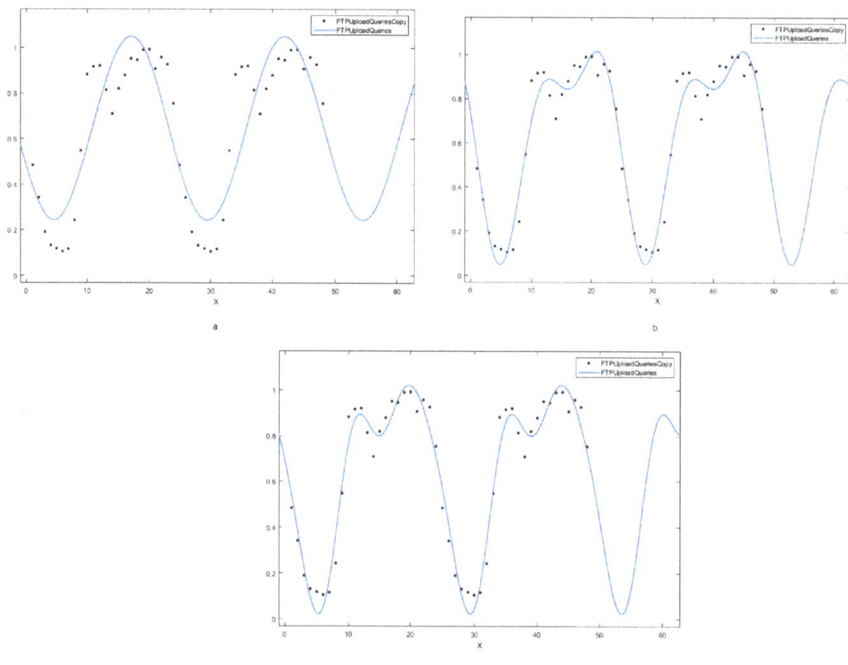

Figure 3. (**a**) Graph of the approximation function $a(t)$ for upward flow with one condition; (**b**) Graph of the approximation function $a(t)$ for upward flow with two conditions; (**c**) Graph of the approximation function $a(t)$ for upward flow with three conditions.

4.5. Numerical Analysis

For the numerical analysis, we will consider the following example: the volume of the resource block is $C = 100$ Mbps, and the finite capacity drive is $r = 100$. The size of the transferred file equals to $\theta(t) = \theta = 10$ MB, that is 80 Mb, and the minimum transfer rate is $b = 1$ Mbps. The arriving intensity of device equals to $\lambda(t) = \lambda \cdot a(t)$, where according to Section 4.4 $a(t)$ equals to (46), where parameters are indicated in the Table 5 on the line for the third condition. The intensity of the flow of leaving requests for the transfer of a block of elastic data due to "impatience" is $\gamma(t) = \gamma = 10^{-2}$.

Apply all our bounds for this specific situation.

For applying of Theorems 2 and 3 we put $\delta = 0.99$, $d_0 = 1$ and $d_{k+1} = \delta d_k$ for $k \geq 0$. Then, we will get:

$$\gamma_{**}(t) = \frac{1}{99}(2.97(a_0 + a_1\cos(wt) + b_1\sin(wt) + a_2\cos(2wt) + b_2\sin(2wt) + a_3\cos(3wt) + b_3\sin(3wt)) - 1.25), \quad (47)$$

$$e^{-\int_s^t \gamma_{**}(\tau)\,d\tau} \leq e^{-0.016(t-s)}; \quad (48)$$

therefore, one can get $K = 2 \cdot 10^6$ and $\gamma_0 = 0.016$ in (39).

Now we obtain the following bounds on the rate of convergence:

$$\|\mathbf{p}^*(t) - \mathbf{p}^{**}(t)\| \leq 3 \cdot 10^6 \cdot e^{-0.016t} \|\mathbf{p}^*(0) - \mathbf{p}^{**}(0)\|, \quad (49)$$

from Theorem 2;

$$\|\mathbf{p}^*(t) - \mathbf{p}^{**}(t)\|_{1D} \leq 2 \cdot 10^6 \cdot e^{-0.016t} \|\mathbf{p}^*(0) - \mathbf{p}^{**}(0)\|_{1D}, \quad (50)$$

$$|E(t,j) - E(t,0)| \leq \frac{1 + 0.99^{j-1}}{W} e^{-0.016t}, \quad (51)$$

from Theorem 2 and Corollary 1.

The corresponding perturbation bounds are:

$$\limsup_{t \to \infty} \|\mathbf{p}(t) - \bar{\mathbf{p}}(t)\| \leq 5 \cdot 10^5 \hat{\varepsilon}, \quad (52)$$

and

$$\limsup_{t \to \infty} |E(t,0) - \bar{E}(t,0)| \leq 10^8 \hat{\varepsilon}, \quad (53)$$

from Theorem 3.

To solve the Cauchy problem, the 4th order Adams–Multon method was used with the use of IntelliJ IDEA software, JDK and the JFreeChart library, the functionality of which is used for plotting. The convergence plots were built for the characteristics specified in Section 2.3, shown in Figures 4–6. To analyze the characteristics, two scenarios were chosen: 1. At the initial moment of time, our system is empty: $X(0) = 0$; 2. At the initial moment, the system is completely occupied, that is, all devices are occupied and there are no free places in the queue $X(0) = 200$. Figure 4 shows graphs for the blocking probability. Note that, since we are considering two scenarios, then according to Figure 4a. for the first scenario, the starting value of the probability is 0, since the system is empty and it is logical that there will be no locks, and for the second scenario, the starting value is 1 since the system is completely busy. We can notice that the blocking probability for the two scenarios converges at time $t = 500$ and according to Figure 4b. We see that its values fluctuate within the range of 0.000–0.013, and the period of fluctuation is 24. Further, in Figure 5, we see that the average number of service requests converges at 100; this may indicate a high load of devices in our system. However, the probability of blocking is small. Let us look at the average number of applications in the queue shown in Figure 6. As shown in Figure 6a. for the two scenarios, the graphs converge and the mean value is no less than 64 and no more than 75, as we can see in Figure 6b.

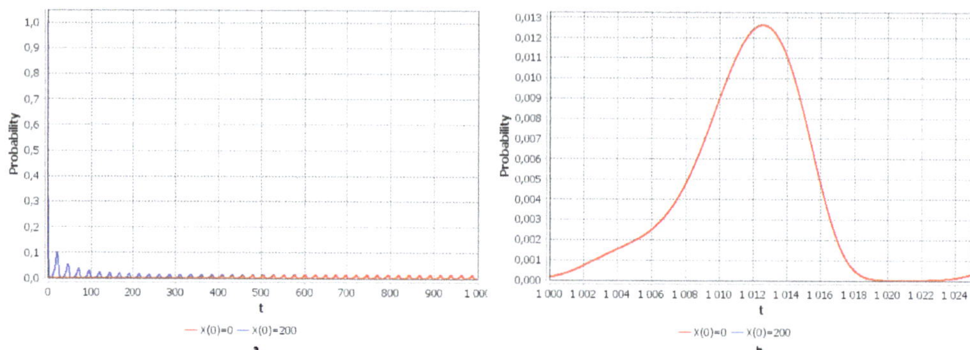

Figure 4. (**a**) Probability of blocking for $t \in [0, 1304]$. (**b**) Approximation of the limiting probability of blocking for $t \in [1304, 1329]$.

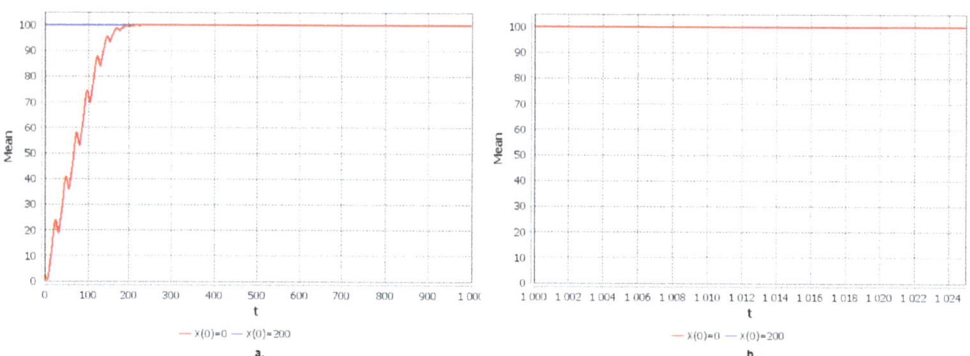

Figure 5. (**a**) The mean of applications served $E(t, k)$ for $t \in [0, 1304]$. (**b**) Approximation of the limiting mean of applications served $E(t, k)$ for $t \in [1304, 1329]$.

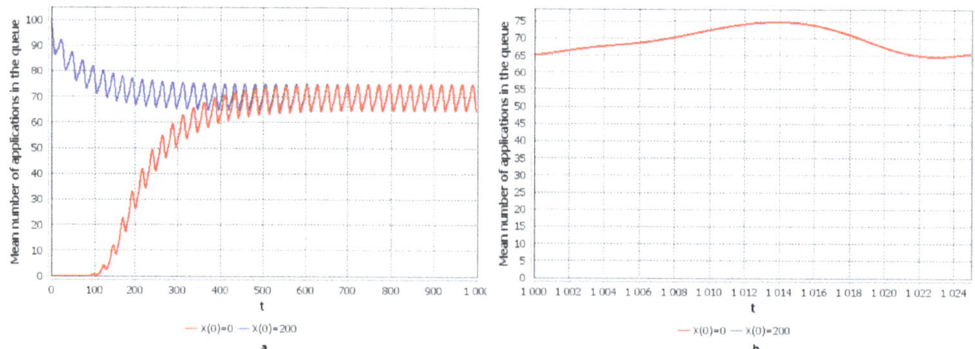

Figure 6. (**a**) The mean of applications in the queue $Q(t)$ for $t \in [0, 1304]$. (**b**) Approximation of the limiting mean of applications in the queue $Q(t)$ for $t \in [1304, 1329]$.

5. Conclusions

In this paper, we have investigated the process of downloading user files depending on the time in the form of a queuing system with elastic traffic and non-stationary intensities. We have developed a method for calculating the limits of convergence of such a model. Estimates of the rate of convergence are obtained based on the logarithmic norm of linear operators. As a result, it was found that the rate of convergence turned out to be low enough for an adequate representation of the situation in the limiting mode to begin at sufficiently long times. We evaluated the characteristics of such a model, namely, the probability of blocking new users, the average number of users downloading data, and the average number of users waiting for the download to start. We obtained the upper and lower bounds of their values.

We considered the model in the form of a single network slice and, in the framework of further tasks, we can consider the model in the form of several slices. If we want to scale our system, for example, to add another incoming stream, it will be necessary to solve the problem of redistribution; we will have to adapt our method for the new case. In addition, in our case, we have a one-dimensional random process, and for a new system where it will be multidimensional, we will first need to identify a function—a mapping, to go to the one-dimensional case to apply our method. This complexity can be viewed as a challenge for future research.

Author Contributions: Conceptualization, supervision, A.Z. and I.K. (Irina Kochetkova); methodology, Y.S. and I.K. (Irina Kochetkova); software, validation, visualization I.K. (Ivan Kovalev) and E.M.; investigation, writing, I.K. (Irina Kochetkova), Y.S., Ivan Kovalev, E.M., A.C., A.Z. All authors have read and agreed to the published version of the manuscript.

Funding: This research was funded by Russian Science Foundation, grant number 19-11-00020 (recipients I.K. (Irina Kochetkova), Y.S., I.K. (Ivan Kovalev), A.Z., Sections 3–5). This paper has been supported by the RUDN University Strategic Academic Leadership Program (recipients A.C., E.M., Sections 1 and 2).

Institutional Review Board Statement: Not applicable.

Informed Consent Statement: Not applicable.

Data Availability Statement: Not applicable.

Acknowledgments: The authors are sincerely grateful to Luis M. Correia (IST/INESC-ID, University of Lisbon) for providing the dataset for Section 4.

Conflicts of Interest: The authors declare no conflict of interest.

References

1. Kochetkov, D.; Almaganbetov, M. Using patent landscapes for technology benchmarking: A case of 5g networks. *Adv. Syst. Sci. Appl.* **2021**, *21*, 20–28. [CrossRef]
2. Kochetkov, D.; Vuković, D.; Sadekov, N.; Levkiv, H. Smart cities and 5G networks: An emerging technological area? *J. Geogr. Inst. Jovan Cvijic SASA* **2019**, *69*, 289–295. [CrossRef]
3. Barakabitze, A.A.; Ahmad, A.; Mijumbi, R.; Hines, A. 5G network slicing using SDN and NFV: A survey of taxonomy, architectures and future challenges. *Comput. Netw.* **2020**, *167*, 106984. [CrossRef]
4. Khan, L.U.; Yaqoob, I.; Tran, N.H.; Han, Z.; Hong, C.S. Network Slicing: Recent Advances, Taxonomy, Requirements, and Open Research Challenges. *IEEE Access* **2020**, *8*, 36009–36028. [CrossRef]
5. Muhizi, S.; Ateya, A.A.; Muthanna, A.; Kirichek, R.; Koucheryavy, A. A novel slice-oriented network model. *Commun. Comput. Inf. Sci.* **2018**, *919*, 421–431. [CrossRef]
6. Ageev, K.A.; Sopin, E.S.; Yarkina, N.V.; Samouylov, K.E.; Shorgin, S.Y. Analysis of the network slicing mechanisms with guaranteed allocated resources for various traffic types. *Inform. Primen.* **2020**, *14*, 94–100. [CrossRef]
7. Yarkina, N.; Gaidamaka, Y.; Correia, L.M.; Samouylov, K. An analytical model for 5G network resource sharing with flexible SLA-oriented slice isolation. *Mathematics* **2020**, *8*, 1177. [CrossRef]
8. Kochetkova, I.; Vlaskina, A.; Burtseva, S.; Savich, V.; Hosek, J. Analyzing the effectiveness of dynamic network slicing procedure in 5g network by queuing and simulation models. *Lect. Notes Comput. Sci.* **2020**, *12525*, 71–85. [CrossRef]
9. Vlaskina, A.; Polyakov, N.; Gudkova, I. Modeling and Performance Analysis of Elastic Traffic with Minimum Rate Guarantee Transmission under Network Slicing. *Lect. Notes Comput. Sci.* **2019**, *11660*, 621–634. [CrossRef]

10. Vlaskina, A.S.; Polyakov, N.A.; Gudkova, I.A.; Gaidamaka, Y.V. Performance analysis of elastic traffic with minimum bit rate guarantee transmission in wireless network under network slicing, Izvestiya of Saratov University. *Informatics* **2020**, *20*, 378–387. [CrossRef]
11. Gudkova, I.; Korotysheva, A.; Zeifman, A.; Shilova, G.; Korolev, V.; Shorgin, S.; Razumchik, R. Modeling and analyzing licensed shared access operation for 5G network as an inhomogeneous queue with catastrophes. In Proceedings of the International Congress on Ultra Modern Telecommunications and Control Systems and Workshops, Lisbon, Portugal, 18–20 October 2016; pp. 282–287. [CrossRef]
12. Markova, E.; Satin, Y.; Kochetkova, I.; Zeifman, A.; Sinitcina, A. Queuing system with unreliable servers and inhomogeneous intensities for analyzing the impact of non-stationarity toperformance measures of wireless network under licensed shared access. *Mathematics* **2020**, *8*, 800. [CrossRef]
13. Zeifman, A.; Korolev, V.; Satin, Y. Two approaches to the construction of perturbation bounds for continuous-time Markov chains. *Mathematics* **2020**, *8*, 253. [CrossRef]
14. Zeifman, A.; Satin, Y.; Kryukova, A.; Razumchik, R.; Kiseleva, K.; Shilova, G. On the Three Methods for Bounding the Rate of Convergence for some Continuous-time Markov Chains. *arXiv* **2020**, arXiv:1911.04086.
15. Arns, M.; Buchholz, P.; Panchenko, A. On the numerical analysis of inhomogeneous continuous-time Markov chains. *Informs J. Comput.* **2010**, *22*, 416–432. [CrossRef]
16. Andreychenko, A.; Sandmann, W.; Wolf, V. Approximate adaptive uniformization of continuous-time Markov chains. *Appl. Math. Model.* **2018**, *61*, 561–576. [CrossRef]
17. Katoen, J.-P.; Zapreev, I.S. Safe on-the-fly steady-state detection for time-bounded reachability. In Proceedings of the 3rd International Conference on the Quantitative Evaluation of Systems, California, CA, USA, 11–14 September 2016.

Article

The Estimators of the Bent, Shape and Scale Parameters of the Gamma-Exponential Distribution and Their Asymptotic Normality

Alexey Kudryavtsev [1,2,*] and Oleg Shestakov [1,2,3,*]

[1] Faculty of Computational Mathematics and Cybernetics, M. V. Lomonosov Moscow State University, 119991 Moscow, Russia
[2] Moscow Center for Fundamental and Applied Mathematics, 119991 Moscow, Russia
[3] Federal Research Center "Computer Science and Control" of the Russian Academy of Sciences, 119333 Moscow, Russia
* Correspondence: aakudryavtsev@cs.msu.ru (A.K.); oshestakov@cs.msu.ru (O.S.)

Abstract: When modeling real phenomena, special cases of the generalized gamma distribution and the generalized beta distribution of the second kind play an important role. The paper discusses the gamma-exponential distribution, which is closely related to the listed ones. The asymptotic normality of the previously obtained strongly consistent estimators for the bent, shape, and scale parameters of the gamma-exponential distribution at fixed concentration parameters is proved. Based on these results, asymptotic confidence intervals for the estimated parameters are constructed. The statements are based on the method of logarithmic cumulants obtained using the Mellin transform of the considered distribution. An algorithm for filtering out unnecessary solutions of the system of equations for logarithmic cumulants and a number of examples illustrating the results obtained using simulated samples are presented. The difficulties arising from the theoretical study of the estimates of concentration parameters associated with the inversion of polygamma functions are also discussed. The results of the paper can be used in the study of probabilistic models based on continuous distributions with unbounded non-negative support.

Keywords: parameter estimation; gamma-exponential distribution; mixed distributions; generalized gamma distribution; generalized beta distribution; method of moments; cumulants; asymptotic normality

Citation: Kudryavtsev, A.; Shestakov, O. The Estimators of the Bent, Shape and Scale Parameters of the Gamma-Exponential Distribution and Their Asymptotic Normality. *Mathematics* **2022**, *10*, 619. https://doi.org/10.3390/math10040619

Academic Editors: Alexander Zeifman, Victor Korolev, Alexander Sipin and Stelios Psarakis

Received: 22 December 2021
Accepted: 14 February 2022
Published: 17 February 2022

1. Introduction

Gamma and beta classes of distributions play an important role in applied probability theory and mathematical statistics and have proven to be convenient and effective tools for modeling many real processes. The generalized gamma distribution and generalized beta distribution of the second kind are quite wide classes, including distributions that have such useful properties as, for example, infinite divisibility and stability, which makes it possible to use distributions from these classes as asymptotic approximations in various limit theorems. The article discusses the distribution proposed in the Ref. [1], that is closely related to the listed popular distributions.

Definition 1. *We say that the random variable ζ has the gamma-exponential distribution $GE(r,\nu,s,t,\delta)$ with the parameters of bent $0 \le r < 1$, shape $\nu \ne 0$, concentration $s,t > 0$, and scale $\delta > 0$, if its density at $z > 0$ is*

$$g_E(z) = \frac{|\nu| z^{t\nu-1}}{\delta^{t\nu}\Gamma(s)\Gamma(t)} \mathrm{Ge}_{r,\,tr+s}\left(-(z/\delta)^{\nu}\right), \tag{1}$$

where $E = (r,\nu,s,t,\delta)$ and $\mathrm{Ge}_{\alpha,\,\beta}(z)$ are the gamma-exponential function [2]:

$$\mathrm{Ge}_{\alpha,\beta}(z) = \sum_{k=0}^{\infty} \frac{z^k}{k!}\Gamma(\alpha k + \beta), \ z \in \mathbb{R}, \ 0 \leq \alpha < 1, \ \beta > 0. \tag{2}$$

Function (2) generalizes to the case $\beta \neq 1$, which is the transformation introduced by Le Roy [3] to study generating functions of a special form. In addition, Function (2) can be considered (under some assumptions) as a special case of the Srivastava–Tomovski function [4], that generalizes the Mittag–Leffler function [5].

In the Ref. [1] it was shown that the distribution (1) adequately describes Bayesian balance models [6]. This is primarily due to the fact that the distribution with the density (1) can be represented as a scaled mixture of two random variables with generalized gamma distributions.

In turn, the generalized gamma distribution $GG(v,q,\theta)$ with the density

$$f(x) = \frac{|v| x^{vq-1} e^{-(x/\theta)^v}}{\theta^{vq}\Gamma(q)}, \ v \neq 0, \ q > 0, \ \theta > 0, \ x > 0, \tag{3}$$

proposed in 1925 by the Italian economist Amoroso [7], has proven its validity in many applied problems that use continuous distributions with an unbounded non-negative support for modeling. The class of distributions (3) is wide enough and includes exponential distribution; χ^2-distribution; Erlang distribution; gamma distribution; half-normal distribution (the distribution of the maximum of the Brownian motion process); Rayleigh distribution; Maxwell–Boltzmann distribution; χ-distribution; Nakagami m-distribution; Wilson–Hilferty distribution; Weibull–Gnedenko distribution, and many others, including scaled and inverse analogs of the above.

In addition, it was shown in the Ref. [8] that the distribution (1) when $r \to 1$ gives in, is the limit of the generalized beta distribution of the second kind $GB2(\nu,s,t,\delta)$ with the density

$$f(x) = \frac{|\nu|(x/\delta)^{t\nu-1}}{\delta B(s,t)\left(1+(x/\delta)^{\nu}\right)^{t+s}}, \ \nu \neq 0, \ s > 0, \ t > 0, \ \delta > 0, \ x > 0, \tag{4}$$

proposed in 1984 by McDonald [9]. The distribution (4), used primarily in econometrics and regression analysis, includes the Burr distribution (or Singh–Maddala distribution); Dagum distribution; Pearson distribution; Pareto distribution; Lomax distribution; the Fisher–Snedecor F-distribution, and others.

Traditionally, an important place in problems of applied mathematical statistics is occupied by the problem of estimating unknown distribution parameters. At the same time, in order to improve the consistency between mathematical models and analyzed real processes, researchers are considering increasingly complex mathematical abstractions. The relevance of the statistical analysis of the distributions (3) and (4), and their particular types and mixtures is evidenced by a large number of publications on this topic, for example, the Refs. [10–19].

In the Ref. [1], it was shown that the gamma-exponential distribution has the following properties.

Lemma 1. *1. Let the independent random variables λ and μ have the distributions $GG(v,q,\theta)$ and $GG(u,p,\alpha)$, $uv > 0$, respectively. Then the distribution of λ coincides with $GE(0,v,\cdot,q,\theta)$; the distribution of λ/μ for $|u| > |v|$ coincides with $GE(v/u,v,p,q,\theta/\alpha)$; and the distribution of λ/μ for $|v| > |u|$ coincides with $GE(u/v,-u,q,p,\theta/\alpha)$.*

2. For $0 < r < 1$, the density $g_E(x)$, $E = (r,\nu,s,t,\delta)$ coincides with the density of the ratio of independent random variables with generalized gamma distributions $GG(\nu,t,\delta)$ and $GG(\nu/r,s,1)$.

The possibility of representing the gamma-exponential distribution as a ratio of random variables having a generalized gamma distribution allows it to be used in a wide range of applied problems [6,20]. In addition, the five-parameter gamma-exponential distribution can be used to model a wide range of real phenomena, due to the wide variety of its possible densities [20]. Besides, for a random variable ζ with a distribution (1), the following representation is valid [8]:

$$\zeta \overset{d}{=} \delta\left(\frac{\lambda}{\mu^r}\right)^{1/\nu}, \tag{5}$$

where the independent random variables λ and μ have gamma distributions $GG(1,t,1)$ and $GG(1,s,1)$, respectively. Moreover, if we put $r = 0$, the right-hand part of the ratio (5) will have the distribution $GG(\nu,t,\delta)$ [8], and for $r = 1$, the right-hand part of (5) will have the distribution $GB2(\nu,s,t,\delta)$ [19]. Consequently, the gamma-exponential distribution can be viewed as the distribution connecting and generalizing distributions from the gamma and beta classes.

In practice, the researcher deals with observable quantities that reflect the evolution of the analyzed real process. In relation to these quantities, some model assumptions are made about the form of their distribution. The problem of estimating unknown parameters from real data also arises in the case of modeling a real process using the gamma-exponential distribution. Due to the representation of the density (1) in terms of a special gamma-exponential Function (2), the maximum likelihood method seems to be too complicated. The same can be said about the direct method of moments, since the moments of distribution (1) can be represented as a product of non-monotone gamma functions [1]:

$$\mathsf{E}\zeta^m = \frac{\delta^m\Gamma(t+m/\nu)\Gamma(s-mr/\nu)}{\Gamma(t)\Gamma(s)}, \quad t+\frac{m}{\nu}>0, \quad s-\frac{mr}{\nu}>0. \tag{6}$$

For this reason, In the Refs. [20,21] it was proposed to estimate the parameters of the gamma-exponential distribution using a modified method based on logarithmic moments. In this paper, we consider the estimators for three out of five parameters of the gamma-exponential distribution, constructed by the method of logarithmic cumulants.

The paper is organized as follows. Section 2 is devoted to the description of the method based on logarithmic cumulants; it provides an explicit form of theoretical logarithmic cumulants, their connection with logarithmic moments, as well as the form of strongly consistent estimators obtained using this method. Section 3 contains auxiliary relations necessary for formulating the main results. Section 4 contains the main results of the paper on the asymptotic normality of estimators for unknown parameters. In Section 5, a numerical analysis of the obtained results is carried out using the generated samples. The paper also contains the sections of discussions and conclusions.

2. Estimators for the Parameters of the Gamma-Exponential Distribution

This section defines the estimators for the parameters of bent r, shape ν, and scale δ of the gamma-exponential distribution (1) for fixed values of the concentration parameters s and t. These estimators were obtained by equating the sample and theoretical cumulants of the gamma-exponential distribution.

Let us introduce the polygamma functions

$$\psi(z) = \frac{d}{dz}\ln\Gamma(z), \quad \psi^{(m)}(z) = \frac{d^{m+1}}{dz^{m+1}}\ln\Gamma(z), \quad m = 1,2,\ldots$$

To obtain an explicit form of theoretical logarithmic cumulants, consider the Mellin transform

$$\mathcal{M}_\zeta(z) = \int\limits_0^\infty x^z\, dF_\zeta(x), \quad z \in \mathbb{C}.$$

We use Lemma 1 and the representation $\zeta \overset{d}{=} \lambda/\mu$, where the independent random variables λ and μ have distributions $GG(\nu,t,\delta)$ and $GG(\nu/r,s,1)$, respectively. For $\lambda \sim GG(\nu,t,\delta)$, the Mellin transform has the form

$$\mathcal{M}_{\lambda}(z) = \frac{\delta^z}{\Gamma(t)}\Gamma\Big(t+\frac{z}{\nu}\Big), \ t+\frac{\mathrm{Re}(z)}{\nu} > 0.$$

Hence, for the ratio of $\lambda \sim GG(\nu,t,\delta)$ to $\mu \sim GG(\nu/r,s,1)$

$$\mathcal{M}_{\lambda/\mu}(z) = \frac{\delta^z}{\Gamma(t)\Gamma(s)}\Gamma\Big(t+\frac{z}{\nu}\Big)\Gamma\Big(s-\frac{rz}{\nu}\Big), \ t+\frac{\mathrm{Re}(z)}{\nu} > 0, \ s-\frac{r\mathrm{Re}(z)}{\nu} > 0,$$

from where we get the characteristic function of the logarithm of ζ:

$$\mathsf{E}e^{iy\ln\zeta} = \frac{\delta^{iy}}{\Gamma(t)\Gamma(s)}\Gamma\Big(t+\frac{iy}{\nu}\Big)\Gamma\Big(s-\frac{iry}{\nu}\Big), \ y \in \mathbb{R}.$$

Thus, the cumulants of the random variable $\ln\zeta$ for fixed s and t have the form

$$\kappa_1(r,\nu,\delta) = \mathsf{E}\ln\zeta = \frac{\nu\ln\delta + \psi(t) - r\psi(s)}{\nu};$$

$$\kappa_m(r,\nu) = (-i)^m \frac{d^m}{dy^m}\ln \mathsf{E}e^{iy\ln\zeta}\Big|_{y=0} = \frac{\psi^{(m-1)}(t) + (-r)^m\psi^{(m-1)}(s)}{\nu^m}, \ m > 1. \tag{7}$$

The moments of the random variable $\ln\zeta$ can be represented as [22]

$$\mu_m(r,\nu,\delta) \equiv \mathsf{E}\ln^m\zeta = B_m(\kappa_1(r,\nu,\delta),\kappa_2(r,\nu),\ldots,\kappa_m(r,\nu)), \tag{8}$$

where B_m is a complete exponential Bell polynomial that can be recurrently defined as

$$B_{m+1}(x_1,\ldots,x_{m+1}) = \sum_{k=0}^{m} C_m^k B_{m-k}(x_1,\ldots,x_{m-k})x_{k+1}, \ B_0 = 1.$$

An explicit form of the necessary relations connecting moments and cumulants can be found in the Ref. [22].

In addition, we will need the following moment characteristics of the logarithm of a random variable with a gamma-exponential distribution, calculated using the Formula (8):

$$\sigma_m^2(r,\nu,\delta) \equiv \mathsf{D}\ln^m\zeta = \mu_{2m}(r,\nu,\delta) - \mu_m^2(r,\nu,\delta); \tag{9}$$

$$\sigma_{ml}(r,\nu,\delta) \equiv \mathsf{cov}(\ln^m\zeta,\ln^l\zeta) = \mu_{m+l}(r,\nu,\delta) - \mu_m(r,\nu,\delta)\mu_l(r,\nu,\delta). \tag{10}$$

To define the sample logarithmic cumulants, we introduce a notation for the sample logarithmic moments of the random variable ζ:

$$L_m(X) = \frac{1}{n}\sum_{i=1}^{n}\ln^m X_i, \tag{11}$$

where $X = (X_1,\ldots,X_n)$ is a sample from the distribution of ζ.

Let us denote $l = (l_1,l_2,l_3,l_4)$. Consider the functions

$$K_1(l) \equiv K_1(l_1) = (\psi(s))^{-1}l_1;$$

$$K_2(l) \equiv K_2(l_1,l_2) = (\psi'(s))^{-1}(l_2 - l_1^2);$$

$$K_3(l) \equiv K_3(l_1,l_2,l_3) = (\psi''(s))^{-1}(l_3 - 3l_2l_1 + 2l_1^3);$$

$$K_4(l) \equiv K_4(l_1,l_2,l_3,l_4) = (\psi'''(s))^{-1}(l_4 - 4l_3l_1 - 3l_2^2 + 12l_2l_1^2 - 6l_1^4).$$

Consider the statistics

$$K_1(X) \equiv K_1(L_1(X));$$

$$K_2(X) \equiv K_2(L_1(X), L_2(X)); \tag{12}$$

$$K_3(X) \equiv K_3(L_1(X), L_2(X), L_3(X));$$

$$K_4(X) \equiv K_4(L_1(X), L_2(X), L_3(X), L_4(X)); \tag{13}$$

$$K(X) = (K_1(X), K_2(X), K_3(X), K_4(X)).$$

Note that the statistics $\psi^{(m-1)}(s)K_m(X)$ are the m-th sample logarithmic cumulants of the gamma-exponential distribution.

The method for estimating unknown parameters considered in the paper is based on solving the system for logarithmic cumulants:

$$\kappa_m(r, \nu, \delta) = \psi^{(m-1)}(s)K_m(X), \quad m = 1, 2, 3, 4. \tag{14}$$

To describe the solution of this system, we introduce a number of functions of sample logarithmic cumulants with the arguments $k = (k_1, k_2, k_3, k_4)$:

$$\phi_m = \frac{\psi^{(m)}(t)}{\psi^{(m)}(s)}; \quad \tau(k) \equiv \tau(k_2, k_4) = \phi_1^2 k_4 + \phi_3\left(k_4 - k_2^2\right); \tag{15}$$

$$R_\pm(k) \equiv R_\pm(k_2, k_4) = \sqrt{\frac{\phi_1 k_4 \pm k_2\sqrt{\tau(k)}}{k_2^2 - k_4}}; \tag{16}$$

$$V_\pm(k) \equiv V_\pm(k_2, k_4) = \sqrt{\frac{\phi_1 k_2 \pm \sqrt{\tau(k)}}{k_2^2 - k_4}}; \tag{17}$$

$$D_\pm(k) \equiv D_\pm(k_1, k_2, k_4) = \exp\left\{\psi(s)k_1 + \frac{\psi(s)R_\pm(k) - \psi(t)}{V_\pm(k)}\right\}; \tag{18}$$

$$\Delta_\pm(k) \equiv \Delta_\pm(k_2, k_3, k_4) = \left|\frac{\phi_2 - R_\pm^3(k)}{V_\pm^3(k)} - k_3\right|;$$

$$\Delta(k) \equiv \Delta(k_2, k_3, k_4) = \min\{\Delta_+(k), \Delta_-(k)\};$$

$$R_\Delta(k) \equiv R_\Delta(k_2, k_3, k_4) = \sqrt{\frac{\phi_1 k_4 - \mathrm{sgn}(\Delta_+(k) - \Delta_-(k))k_2\sqrt{\tau(k)}}{k_2^2 - k_4}};$$

$$V_\Delta(k) \equiv V_\Delta(k_2, k_3, k_4) = \sqrt{\frac{\phi_1 k_2 - \mathrm{sgn}(\Delta_+(k) - \Delta_-(k))\sqrt{\tau(k)}}{k_2^2 - k_4}};$$

$$D_\Delta(k) \equiv D_\Delta(k_1, k_2, k_3, k_4) = \exp\left\{\psi(s)k_1 + \frac{\psi(s)R_\Delta(k) - \psi(t)}{V_\Delta(k)}\right\}.$$

The system (14) has several solutions. It was shown in the Ref. [23] that the estimators for the parameters of bent r, shape ν, and scale δ have the form

$$\hat{r}(X) = R_\Delta(K(X)); \tag{19}$$

$$\hat{\nu}(X) = V_\Delta(K(X)); \tag{20}$$

$$\hat{\delta}(X) = D_\Delta(K(X)), \tag{21}$$

and the following statement holds.

Lemma 2. *For fixed parameters s and t of the distribution $GE(r, \nu, s, t, \delta)$, the estimators (19)–(21) for the parameters $0 \leq r < 1$, $\nu > 0$ and $\delta > 0$ are strongly consistent.*

Remark 1. *If it is known that $v < 0$, one should consider the estimator $\hat{v}(X) = -V_\Delta(K(X))$ instead of $V_\Delta(K(X))$, and the estimator*

$$\hat{\delta}(X) = \exp\left\{ \psi(s)K_1(X) - \frac{\psi(s)R_\Delta(K(X)) - \psi(t)}{V_\Delta(K(X))} \right\} \tag{22}$$

instead of $D_\Delta(K(X))$.

3. Auxiliary Relations

In what follows, we will need the derivatives of Functions (16)–(18) expressed in terms of the functions ϕ_m and τ, defined in (15). Note that

$$\begin{aligned}
R_{k_2,\pm}(k) &\equiv \frac{\partial R_\pm}{\partial k_2}(k_2,k_4) = \mp \frac{k_4\left(\phi_1^2 k_2^2 + \tau(k) \pm 2\phi_1 k_2\sqrt{\tau(k)}\right)}{2(k_2^2 - k_4)^{3/2}\sqrt{\tau(k)}\sqrt{\phi_1 k_4 \pm k_2\sqrt{\tau(k)}}}; \\
R_{k_4,\pm}(k) &\equiv \frac{\partial R_\pm}{\partial k_4}(k_2,k_4) = \pm \frac{k_2\left(\phi_1^2 k_2^2 + \tau(k) \pm 2\phi_1 k_2\sqrt{\tau(k)}\right)}{4(k_2^2 - k_4)^{3/2}\sqrt{\tau(k)}\sqrt{\phi_1 k_4 \pm k_2\sqrt{\tau(k)}}}; \\
V_{k_2,\pm}(k) &\equiv \frac{\partial V_\pm}{\partial k_2}(k_2,k_4) = \mp \frac{k_2(\phi_1^2 k_4 + \tau(k)) \pm \phi_1(k_2^2 + k_4)\sqrt{\tau(k)}}{2(k_2^2 - k_4)^{3/2}\sqrt{\tau(k)}\sqrt{\phi_1 k_2 \pm \sqrt{\tau(k)}}}; \\
V_{k_4,\pm}(k) &\equiv \frac{\partial V_\pm}{\partial k_4}(k_2,k_4) = \pm \frac{\phi_1^2 k_2^2 + \tau(k) \pm 2\phi_1 k_2\sqrt{\tau(k)}}{4(k_2^2 - k_4)^{3/2}\sqrt{\tau(k)}\sqrt{\phi_1 k_2 \pm \sqrt{\tau(k)}}}; \\
D_{k_1,\pm}(k) &\equiv \frac{\partial D_\pm}{\partial k_1}(k_1,k_2,k_4) = \psi(s)\exp\left\{ \psi(s)k_1 + \frac{\psi(s)R_\pm(k) - \psi(t)}{V_\pm(k)} \right\}; \\
D_{k_2,\pm}(k) &\equiv \frac{\partial D_\pm}{\partial k_2}(k_1,k_2,k_4) = \exp\left\{ \psi(s)k_1 + \frac{\psi(s)R_\pm(k) - \psi(t)}{V_\pm(k)} \right\} \times \\
&\quad \times \frac{\psi(t)V_{k_2,\pm}(k) + \psi(s)R_{k_2,\pm}(k)V_\pm(k) - \psi(s)R_\pm(k)V_{k_2,\pm}(k)}{V_\mp^2(k)}; \\
D_{k_4,\pm}(k) &\equiv \frac{\partial D_\pm}{\partial k_4}(k_1,k_2,k_4) = \exp\left\{ \psi(s)k_1 + \frac{\psi(s)R_\pm(k) - \psi(t)}{V_\pm(k)} \right\} \times \\
&\quad \times \frac{\psi(t)V_{k_4,\pm}(k) + \psi(s)R_{k_4,\pm}(k)V_\pm(k) - \psi(s)R_\pm(k)V_{k_4,\pm}(k)}{V_\pm^2(k)}.
\end{aligned} \tag{23}$$

Using the formula for the derivative of a composite function, we obtain

$$\begin{aligned}
\frac{\partial R_\pm}{\partial l_1}(l) &= -\frac{2l_1}{\psi'(s)}R_{k_2,\pm}(K_2(l),K_4(l)) - \frac{4l_3 - 24l_2l_1 + 24l_1^3}{\psi'''(s)}R_{k_4,\pm}(K_2(l),K_4(l));\\
\frac{\partial R_\pm}{\partial l_2}(l) &= \frac{1}{\psi'(s)}R_{k_2,\pm}(K_2(l),K_4(l)) - \frac{6l_2 - 12l_1^2}{\psi'''(s)}R_{k_4,\pm}(K_2(l),K_4(l));\\
\frac{\partial R_\pm}{\partial l_3}(l) &= -\frac{4l_1}{\psi'''(s)}R_{k_4,\pm}(K_2(l),K_4(l));\\
\frac{\partial R_\pm}{\partial l_4}(l) &= \frac{1}{\psi'''(s)}R_{k_4,\pm}(K_2(l),K_4(l));\\
\frac{\partial V_\pm}{\partial l_1}(l) &= -\frac{2l_1}{\psi'(s)}V_{k_2,\pm}(K_2(l),K_4(l)) - \frac{4l_3 - 24l_2l_1 + 24l_1^3}{\psi'''(s)}V_{k_4,\pm}(K_2(l),K_4(l));\\
\frac{\partial V_\pm}{\partial l_2}(l) &= \frac{1}{\psi'(s)}V_{k_2,\pm}(K_2(l),K_4(l)) - \frac{6l_2 - 12l_1^2}{\psi'''(s)}V_{k_4,\pm}(K_2(l),K_4(l));\\
\frac{\partial V_\pm}{\partial l_3}(l) &= -\frac{4l_1}{\psi'''(s)}V_{k_4,\pm}(K_2(l),K_4(l)); \qquad (24)\\
\frac{\partial V_\pm}{\partial l_4}(l) &= \frac{1}{\psi'''(s)}V_{k_4,\pm}(K_2(l),K_4(l));\\
\frac{\partial D_\pm}{\partial l_1}(l) &= \frac{1}{\psi(s)}D_{k_1,\pm}(K_1(l),K_2(l),K_4(l)) - \frac{2l_1}{\psi'(s)}D_{k_2,\pm}(K_1(l),K_2(l),K_4(l)) -\\
&\quad - \frac{4l_3 - 24l_2l_1 + 24l_1^3}{\psi'''(s)}D_{k_4,\pm}(K_1(l),K_2(l),K_4(l));\\
\frac{\partial D_\pm}{\partial l_2}(l) &= \frac{1}{\psi'(s)}D_{k_2,\pm}(K_1(l),K_2(l),K_4(l)) - \frac{6l_2 - 12l_1^2}{\psi'''(s)}D_{k_4,\pm}(K_1(l),K_2(l),K_4(l));\\
\frac{\partial D_\pm}{\partial l_3}(l) &= -\frac{4l_1}{\psi'''(s)}D_{k_4,\pm}(K_1(l),K_2(l),K_4(l));\\
\frac{\partial D_\pm}{\partial l_4}(l) &= \frac{1}{\psi'''(s)}D_{k_4,\pm}(K_1(l),K_2(l),K_4(l)),
\end{aligned}$$

where the partial derivatives of the functions $R_\pm(k)$, $V_\pm(k)$ and $D_\pm(k)$ are defined in the relations (23).

4. Asymptotic Normality of the Estimators for the Parameters of the Gamma-Exponential Distribution

Further arguments are based on the following statements [24].

Lemma 3. *In $\mathbb{R}^n$, the random vector X_n converges in distribution to the random vector X if and only if each linear combination of the components of X_n converges in a distribution to the same linear combination of the components of X.*

Lemma 4. *Suppose that in $\mathbb{R}^m$,*

$$\sqrt{n}(T_{n1},\ldots,T_{nm}) \Longrightarrow N(\mu,\Sigma), \quad n\to\infty,$$

with Σ a covariance matrix. Let $g(t) = g(t_1,\ldots,t_m)$ be a real-valued function with a nonzero differential at $t=\mu$. Put

$$d = \left(\frac{\partial g}{\partial t_1}\Big|_{t=\mu},\ldots,\frac{\partial g}{\partial t_m}\Big|_{t=\mu}\right).$$

Then $\sqrt{n}g(T_{n1},\ldots,T_{nm}) \Longrightarrow N(g(\mu),d\Sigma d^T)$.

Let us formulate the statements about the asymptotic normality of the estimators (19)–(21) with fixed concentration parameters s and t.

Denote

$$\Sigma = \begin{pmatrix} \sigma_1^2(r,\nu,\delta) & \sigma_{12}(r,\nu,\delta) & \sigma_{13}(r,\nu,\delta) & \sigma_{14}(r,\nu,\delta) \\ \sigma_{12}(r,\nu,\delta) & \sigma_2^2(r,\nu,\delta) & \sigma_{23}(r,\nu,\delta) & \sigma_{24}(r,\nu,\delta) \\ \sigma_{13}(r,\nu,\delta) & \sigma_{23}(r,\nu,\delta) & \sigma_3^2(r,\nu,\delta) & \sigma_{34}(r,\nu,\delta) \\ \sigma_{14}(r,\nu,\delta) & \sigma_{24}(r,\nu,\delta) & \sigma_{34}(r,\nu,\delta) & \sigma_4^2(r,\nu,\delta) \end{pmatrix}; \tag{25}$$

$$d_{R_\pm} = \left(\frac{\partial R_\pm}{\partial l_1}(l)\Big|_{l=\mu}, \frac{\partial R_\pm}{\partial l_2}(l)\Big|_{l=\mu}, \frac{\partial R_\pm}{\partial l_3}(l)\Big|_{l=\mu}, \frac{\partial R_\pm}{\partial l_4}(l)\Big|_{l=\mu} \right), \tag{26}$$

where the variances $\sigma_m^2(r,\nu,\delta)$ and the covariances $\sigma_{ml}(r,\nu,\delta)$ are defined in (9) and (10), respectively, and partial derivatives $\partial R_\pm/\partial l_k(l)$ are defined in (24).

Let ϕ_m, $m = 1,3$ be defined in (15); $R_\pm(K_2(X), K_4(X))$ be defined in (16); $K_m(X)$, $m = 2,4$, be defined in (12) and (13). The following statement holds.

Theorem 1. *Suppose that $r^2 \neq (\phi_3 - \phi_1^2)/(2\phi_1)$.*

1. Let $r^2 > \phi_3/\phi_1$. Then the estimator $\hat{r}(X)$ for the unknown parameter r has the form $\hat{r}(X) = R_+(K_2(X), K_4(X))$, and when $n \to \infty$ has the property of asymptotic normality:

$$\sqrt{n}\frac{\hat{r}(X) - r}{\sqrt{d_{R_+}\Sigma d_{R_+}^T}} \Longrightarrow N(0,1).$$

2. Let $r^2 < \phi_3/\phi_1$. Then the estimator $\hat{r}(X)$ for the unknown parameter r has the form $\hat{r}(X) = R_-(K_2(X), K_4(X))$ and when $n \to \infty$ has the property of asymptotic normality:

$$\sqrt{n}\frac{\hat{r}(X) - r}{\sqrt{d_{R_-}\Sigma d_{R_-}^T}} \Longrightarrow N(0,1).$$

Proof of Theorem 1. The sample logarithmic moments $L_m(X)$ defined in (11) are sums of independent identically distributed random variables with means $\mu_m(r,\nu,\delta)$ defined in (8) and variances $\sigma_m^2(r,\nu,\delta)/n$ defined in (9). Therefore, when $n \to \infty$, the statistics $L_m(X), k = 1,2,3,4$, together with any of their linear combinations, have the property of asymptotic normality with the corresponding limit means depending on $\mu_m(r,\nu,\delta)$, and variances determined by the covariance matrix Σ given in (25).

In addition, under the conditions of the theorem, the components of the vector $d_{R_\pm}$ defined in (26) are finite and the function $R_\pm(K_2(l), K_4(l))$ has a nonzero differential at the point $\mu = (\mu_1(r,\nu,\delta), \ldots, \mu_4(r,\nu,\delta))$.

Thus, all conditions of Lemmas 3 and 4 are satisfied. Hence,

$$\sqrt{n}R_\pm(K_2(X), K_4(X)) \Longrightarrow N(R_\pm(K_2(\mu), K_4(\mu)), d_{R_\pm}\Sigma d_{R_\pm}^T).$$

Consider the limiting mean $R_\pm(K_2(\mu), K_4(\mu))$. Note that when $n \to \infty$

$$K_2(X) \longrightarrow \frac{\phi_1 + r^2}{\nu^2} \text{ a. s.;}$$

$$K_4(X) \longrightarrow \frac{\phi_3 + r^4}{\nu^4} \text{ a. s.}$$

Therefore, for the function $\tau(k)$ defined in (15),

$$\tau(K_2(X), K_4(X)) \longrightarrow \frac{(\phi_1 r^2 - \phi_3)^2}{\nu^4} \text{ a. s.}$$

when $n \to \infty$.

Let $r^2 > \phi_3/\phi_1$. Then

$$R_+(K_2(X), K_4(X)) \longrightarrow R_+(K_2(\mu), K_4(\mu)) = r \text{ a. s.,}$$

and the statistic $\hat{r}(X) = R_+(K_2(X), K_4(X))$ is a strongly consistent estimator for r.

Since

$$R_-(K_2(\mu), K_4(\mu)) = \sqrt{\frac{2\phi_1\phi_3 + (\phi_3 - \phi_1^2)r^2}{\phi_1^2 - \phi_3 + 2\phi_1 r^2}},$$

the statistic $R_-(K_2(X), K_4(X))$ estimates the function $R_-(K_2(\mu), K_4(\mu)) \neq r$ and does not satisfy the statement of Lemma 2.

If $r^2 < \phi_3/\phi_1$, we similarly conclude that the statistic $\hat{r}(X) = R_-(K_2(X), K_4(X))$ is a strongly consistent estimator for r and the statistic $R_+(K_2(X), K_4(X))$ does not satisfy the statement of Lemma 2. □

Let

$$d_{V_\pm} = \left(\frac{\partial V_\pm}{\partial l_1}(l)\Big|_{l=\mu}, \frac{\partial V_\pm}{\partial l_2}(l)\Big|_{l=\mu}, \frac{\partial V_\pm}{\partial l_3}(l)\Big|_{l=\mu}, \frac{\partial V_\pm}{\partial l_4}(l)\Big|_{l=\mu}\right);$$

$$d_{D_\pm} = \left(\frac{\partial D_\pm}{\partial l_1}(l)\Big|_{l=\mu}, \frac{\partial D_\pm}{\partial l_2}(l)\Big|_{l=\mu}, \frac{\partial D_\pm}{\partial l_3}(l)\Big|_{l=\mu}, \frac{\partial D_\pm}{\partial l_4}(l)\Big|_{l=\mu}\right),$$

where the partial derivatives $\partial V_\pm/\partial l_k(l)$ and $\partial D_\pm/\partial l_k(l)$ are defined in (24).

Let ϕ_m, $m = 1, 3$ be defined in (15); $V_\pm(K_2(X), K_4(X))$ be defined in (17); $D_\pm(K_2(X), K_4(X))$ be defined in (18); $K_m(X)$, $m = 2, 4$, be defined in (12) and (13); and the matrix Σ be defined in (25).

Theorems 2 and 3 are proved in a completely similar way to Theorem 1.

Theorem 2. *Suppose that $r^2 \neq (\phi_3 - \phi_1^2)/(2\phi_1)$, $\nu > 0$.*

1. Let $r^2 > \phi_3/\phi_1$. Then the estimator $\hat{\nu}(X)$ for the unknown parameter ν has the form $\hat{\nu}(X) = V_+(K_2(X), K_4(X))$, and when $n \to \infty$ has the property of asymptotic normality:

$$\sqrt{n}\frac{\hat{\nu}(X) - \nu}{\sqrt{d_{V_+}\Sigma d_{V_+}^T}} \Longrightarrow N(0,1).$$

2. Let $r^2 < \phi_3/\phi_1$. Then the estimator $\hat{\nu}(X)$ for the unknown parameter ν has the form $\hat{\nu}(X) = V_-(K_2(X), K_4(X))$, and when $n \to \infty$ has the property of asymptotic normality:

$$\sqrt{n}\frac{\hat{\nu}(X) - \nu}{\sqrt{d_{V_-}\Sigma d_{V_-}^T}} \Longrightarrow N(0,1).$$

Theorem 3. *Suppose that $r^2 \neq (\phi_3 - \phi_1^2)/(2\phi_1)$, $\nu > 0$.*

1. Let $r^2 > \phi_3/\phi_1$. Then the estimator $\hat{\delta}(X)$ for the unknown parameter δ has the form $\hat{\delta}(X) = D_+(K_2(X), K_4(X))$, and when $n \to \infty$ has the property of asymptotic normality:

$$\sqrt{n}\frac{\hat{\delta}(X) - \delta}{\sqrt{d_{D_+}\Sigma d_{D_+}^T}} \Longrightarrow N(0,1).$$

2. Let $r^2 < \phi_3/\phi_1$. Then the estimator $\hat{\delta}(X)$ for the unknown parameter δ has the form $\hat{\delta}(X) = D_-(K_2(X), K_4(X))$ and when $n \to \infty$ has the property of asymptotic normality:

$$\sqrt{n}\frac{\hat{\delta}(X) - \delta}{\sqrt{d_{D_-}\Sigma d_{D_-}^T}} \Longrightarrow N(0,1).$$

Remark 2. *By analogy with the arguments of the Ref. [23] concerning the statement of Lemma 2, if it is known that $\nu < 0$, it is easy to show that the statements of the Theorems 2 and 3 hold for the statistics $\hat{\nu}(X) = -V_\Delta(K(X))$ and $\hat{\delta}(X)$ defined in (22) with the corresponding modification of the vectors of derivatives $d_{V_\pm}$ and $d_{D_\pm}$.*

Denote

$$s_{mm}(X) \equiv \sigma_m^2(\hat{r}(X), \hat{v}(X), \hat{\delta}(X)); \tag{27}$$

$$s_{ml}(X) = s_{lm}(X) \equiv \sigma_{ml}(\hat{r}(X), \hat{v}(X), \hat{\delta}(X)); \tag{28}$$

$$d_r^{[m]}(X) \equiv \frac{\partial \hat{r}(X)}{\partial l_m}; \; d_v^{[m]}(X) \equiv \frac{\partial \hat{v}(X)}{\partial l_m}; \; d_\delta^{[m]}(X) \equiv \frac{\partial \hat{\delta}(X)}{\partial l_m}, \tag{29}$$

where $\sigma_m^2(r, v, \delta)$ and $\sigma_{ml}(r, v, \delta)$ are defined in (9) and (10), respectively, and the estimators $\hat{r}(X)$, $\hat{v}(X)$ and $\hat{\delta}(X)$ satisfy the conditions of Theorems 1–3.

Corollary 1. *Suppose that the conditions of Theorems 1–3 are met; then,*

$$\sqrt{n}\frac{\hat{r}(X) - r}{\sqrt{\sum_{m=1}^4 \sum_{l=1}^4 d_r^{[m]}(X) s_{ml}(X) d_r^{[l]}(X)}} \Longrightarrow N(0,1);$$

$$\sqrt{n}\frac{\hat{v}(X) - v}{\sqrt{\sum_{m=1}^4 \sum_{l=1}^4 d_v^{[m]}(X) s_{ml}(X) d_v^{[l]}(X)}} \Longrightarrow N(0,1);$$

$$\sqrt{n}\frac{\hat{\delta}(X) - \delta}{\sqrt{\sum_{m=1}^4 \sum_{l=1}^4 d_\delta^{[m]}(X) s_{ml}(X) d_\delta^{[l]}(X)}} \Longrightarrow N(0,1),$$

when $n \to \infty$, where $s_{ml}(X)$, $d_r^{[m]}(X)$, $d_v^{[m]}(X)$, $d_\delta^{[m]}(X)$ are defined in (27)–(29).

Proof of Corollary 1. Due to the strong consistency of the estimators $\hat{r}(X)$, $\hat{v}(X)$, and $\hat{\delta}(X)$, the quadratic form

$$\sum_{m=1}^4 \sum_{l=1}^4 d_r^{[m]}(X) s_{ml}(X) d_r^{[l]}(X)$$

converges almost surely to the normalizing function from the Theorem 1. Therefore, by Slutsky's theorem, we obtain the statement of the Corollary 1 for the estimator of the parameter r. Similarly, we obtain statements for estimators of the parameters v and δ. □

Based on Corollary 1, it is possible to construct asymptotic confidence intervals for unknown parameters of the gamma-exponential distribution.

By u_γ, we denote the $(1+\gamma)/2$-quantile of the standard normal distribution.

Corollary 2. *Suppose that the conditions of Theorems 1–3 are met; then, asymptotic confidence intervals with the confidence level γ based on the estimators $\hat{r}(X)$, $\hat{v}(X)$, $\hat{\delta}(X)$ for the unknown parameters r, v, δ have the form*

$$(A_r(X), B_r(X)) = \left(\hat{r}(X) - \frac{u_\gamma}{\sqrt{n}} C_r(X), \hat{r}(X) + \frac{u_\gamma}{\sqrt{n}} C_r(X)\right);$$

$$(A_v(X), B_v(X)) = \left(\hat{v}(X) - \frac{u_\gamma}{\sqrt{n}} C_v(X), \hat{v}(X) + \frac{u_\gamma}{\sqrt{n}} C_v(X)\right);$$

$$(A_\delta(X), B_\delta(X)) = \left(\hat{\delta}(X) - \frac{u_\gamma}{\sqrt{n}} C_\delta(X), \hat{\delta}(X) + \frac{u_\gamma}{\sqrt{n}} C_\delta(X)\right),$$

where

$$C_r(X) = \sqrt{\sum_{m=1}^4 \sum_{l=1}^4 d_r^{[m]}(X) s_{ml}(X) d_r^{[l]}(X)};$$

$$C_v(X) = \sqrt{\sum_{m=1}^4 \sum_{l=1}^4 d_v^{[m]}(X) s_{ml}(X) d_v^{[l]}(X)};$$

$$C_\delta(X) = \sqrt{\sum_{m=1}^{4}\sum_{l=1}^{4} d_\delta^{[m]}(X)s_{ml}(X)d_\delta^{[l]}(X)},$$

and $s_{ml}(X)$, $d_r^{[m]}(X)$, $d_\nu^{[m]}(X)$, $d_\delta^{[m]}(X)$ *are defined in (27)–(29).*

Proof of Corollary 2. The proof is based on the relation

$$\mathsf{P}\left(|\hat{r}(X) - r| < \frac{u_\gamma}{\sqrt{n}}C_r(X)\right) = \mathsf{P}\left(\frac{\sqrt{n}}{C_r(X)}|\hat{r}(X) - r| < u_\gamma\right) \asymp 2\Phi(u_\gamma) - 1 = \gamma,$$

from which we obtain the form of the confidence interval $(A_r(X), B_r(X))$. Similarly, asymptotic confidence intervals for the parameters ν and δ are obtained. □

5. Numerical Analysis of Theoretical Results

Let us consider the problem of obtaining numerical values of estimates for the parameters of bent r, shape ν, and scale δ of the gamma-exponential distribution $GE(r,\nu,s,t,\delta)$ for fixed values of concentration parameters s and t.

The method for obtaining estimators for the parameters r and ν is based on solving the system of equations [23], where the theoretical logarithmic cumulants (7) of the second and fourth orders are equated to their sample counterparts:

$$\frac{\phi_1 + r^2}{\nu^2} = K_2(X); \tag{30}$$

$$\frac{\phi_3 + r^4}{\nu^4} = K_4(X). \tag{31}$$

Note that the solutions [23]

$$\hat{r}^2_\pm(X) = \frac{\phi_1 K_4(X) \pm K_2(X)\sqrt{\tau(K(X))}}{K_2^2(X) - K_4(X)}; \tag{32}$$

$$\hat{\nu}^2_\pm(X) = \frac{\phi_1 K_2(X) \pm \sqrt{\tau(K(X))}}{K_2^2(X) - K_4(X)}$$

do not uniquely determine the estimators for the parameters r and ν, and whereas the sign of r is always known, ν can be either positive or negative. In addition, numerical experiments show that for a fixed sample, the expression (32) can give non-controversial estimates for any sign before the radical. For this reason, when processing real data, one should use the algorithm for filtering out unnecessary system solutions.

The algorithm for choosing the "correct" solution $(\hat{r}(X), \hat{\nu}(X))$ of the system is as follows.

At the first stage, one should try to determine the sign before the radical in the relation (32), using the domain of the parameter $r \in [0,1)$. According to the condition of Theorem 1, it is necessary to compare the value of ϕ_3/ϕ_1 with one. If $\phi_3/\phi_1 \geq 1$, one should choose $\hat{r}(X) = R_-(K(X))$, where the function $R_-(k)$ is defined in (16). If $\phi_3/\phi_1 < 1$, the values of the right-hand sides of (32) are calculated for the given sample. If the value of $\hat{r}^2_\pm(X)$ for some signs before the radical does not belong to the interval $[0,1)$, the corresponding solution is eliminated and the solution with the opposite sign before the radical is chosen as the estimate of r.

At the second stage, one should additionally use an equation similar to (30) and (31):

$$\frac{\phi_2 - r^3}{\nu^3} = K_3(X).$$

Since the estimators $\hat{r}(X)$ and $\hat{\nu}(X)$, along with the statistics $K_3(X)$, are continuous functions of sample logarithmic moments,

$$\left|\frac{\phi_2 - \hat{r}^3(X)}{\hat{v}^3(X)} - K_3(X)\right| \longrightarrow 0 \text{ a. s.}$$

when $n \to \infty$.

For a fixed sample size n, this relation makes it possible to determine the "correct" solution of the system using the values

$$\Delta_{\pm,+} = \left|\frac{\phi_2 - R^3_\pm(K(X))}{V^3_\pm(K(X))} - K_3(X)\right| \text{ and } \Delta_{\pm,-} = \left|\frac{\phi_2 - R^3_\pm(K(X))}{-V^3_\pm(K(X))} - K_3(X)\right|$$

based on the following criteria:

- If the result $\hat{r}(X) = \hat{r}_+(X)$ is obtained at the first stage, one should choose the pair $(R_+(K(X)), V_+(K(X)))$ as the solution $(\hat{r}(X), \hat{v}(X))$ if $\Delta_{+,+} < \Delta_{+,-}$, and the pair $(R_+(K(X)), -V_+(K(X)))$ if $\Delta_{+,-} < \Delta_{+,+}$;
- If the result $\hat{r}(X) = \hat{r}_-(X)$ is obtained at the first stage, one should choose the pair $(R_-(K(X)), V_-(K(X)))$ as the solution $(\hat{r}(X), \hat{v}(X))$ if $\Delta_{-,+} < \Delta_{-,-}$, and the pair $(R_-(K(X)), -V_-(K(X)))$ if $\Delta_{-,-} < \Delta_{-,+}$;
- If the values (32) obtained at the first stage are non-controversial for any sign before the radical, the solution $(\hat{r}(X), \hat{v}(X))$ should be chosen from four possible pairs $(R_\pm(K(X)), \pm V_\pm(K(X)))$, using the minimum of the values $\Delta_{+,+}, \Delta_{+,-}, \Delta_{-,+}, \Delta_{-,-}$ to determine the "correct" set of signs.

The estimate $\hat{\delta}(X)$ of the scale parameter, δ is found from the equation for the first logarithmic cumulant

$$\ln \delta + \frac{\psi(t) - r\psi(s)}{v} = L_1(X),$$

substituting the solution $(\hat{r}(X), \hat{v}(X))$ is found by the above algorithm instead of (r, v), and has the form

$$\hat{\delta}(X) = \exp\left\{L_1(X) + \frac{\psi(s)\hat{r}(X) - \psi(t)}{\hat{v}(X)}\right\}. \tag{33}$$

Let us present some numerical results illustrating the method of choosing the "correct" estimates for the parameters of bent r, shape v, and scale δ at fixed concentration parameters s and t of the gamma-exponential distribution (1).

Tables 1–8 provide examples of numerical values of parameter estimates obtained using the algorithm for eliminating unnecessary solutions and constructed from the samples of the size n from the model distribution (1) with a set of parameters $E = (r, v, s, t, \delta)$ and examples of the boundary values of asymptotic confidence intervals with a confidence level $\gamma = 0.95$ for these estimates.

A simulation of pseudo-random samples from the gamma-exponential distribution is based on the relation (5) and is carried out using standard tools in any programming language that has the ability to generate samples from the gamma distribution.

Table 1 shows the values of the estimates of the parameters r, v and δ, obtained from the sample of the size n from a model distribution with a set of parameters $E = (0.5; 2.5; 2.4; 1.9; 1.0)$. For this set of parameters, the inequality $\phi_3/\phi_1 \geq 1$ holds, and therefore the first stage of the algorithm for eliminating unnecessary solutions gives the estimate $\hat{r}(X) = R_-(K(X))$. At the second stage, since $\Delta_{-,+} < \Delta_{-,-}$, the pair $(R_-(K(X)), V_-(K(X)))$ is selected as a solution $(\hat{r}(X), \hat{v}(X))$. The estimate $\hat{\delta}(X)$ is obtained from the relation (33). Table 2 shows the values of the estimates $(\hat{r}(X), \hat{v}(X), \hat{\delta}(X))$ obtained in the previous step and the boundaries of corresponding confidence intervals.

Table 1. Examples of estimates for the parameters of the model distribution $GE(0.5;2.5;2.4;1.9;1.0)$.

n	$R_-(K(X))$	$\Delta_{-,+}$	$\Delta_{-,-}$	$\pm V_-(K(X))$	$\hat{\delta}(X)$
10^5	0.5096	0.0008	0.2048	±2.5179	1.0018
5×10^5	0.5040	0.0009	0.2089	±2.5038	1.0005
10^6	0.4980	0.0001	0.2101	±2.4985	0.9990
5×10^6	0.5008	0.0001	0.2091	±2.5013	1.0002

Table 2. Examples of estimates for the parameters and boundaries of confidence intervals for a model distribution $GE(0.5;2.5;2.4;1.9;1.0)$.

n	$\hat{r}(X)$	$A_r(X)$	$B_r(X)$	$\hat{v}(X)$	$A_v(X)$	$B_v(X)$	$\hat{\delta}(X)$	$A_\delta(X)$	$B_\delta(X)$
10^5	0.5096	0.3935	0.6258	2.5179	2.4197	2.6161	1.0018	0.9711	1.0325
5×10^5	0.5040	0.4519	0.5560	2.5038	2.4603	2.5472	1.0005	0.9866	1.0143
10^6	0.4980	0.4611	0.5349	2.4985	2.4680	2.5290	0.9990	0.9891	1.0088
5×10^6	0.5008	0.4843	0.5173	2.5013	2.4877	2.5150	1.0002	0.9958	1.0046

Table 3 shows the values of the estimates of the parameters r, v and δ obtained from the sample of the size n from a model distribution with a set of parameters $E=(0.7;-1.8;3.6;3.9;1.5)$. For this set of parameters, the inequality $\phi_3/\phi_1<1$ holds, while the values of the statistics $R_+(K(X))$ are outside the domain of the parameter r, and therefore the first stage of the algorithm for eliminating unnecessary solutions gives the estimate $\hat{r}(X)=R_-(K(X))$. At the second stage, since $\Delta_{-,-}<\Delta_{-,+}$, the pair $(R_-(K(X)),-V_-(K(X)))$ is selected as a solution $(\hat{r}(X),\hat{v}(X))$. The estimate $\hat{\delta}(X)$ is obtained from the relation (33). Table 4 shows the values of the estimates $(\hat{r}(X),\hat{v}(X),\hat{\delta}(X))$ obtained in the previous step, and the boundaries of corresponding confidence intervals.

Table 3. Examples of estimates for the parameters of the model distribution $GE(0.7;-1.8;3.6;3.9;1.5)$.

n	$R_+(K(X))$	$R_-(K(X))$	$\Delta_{-,+}$	$\Delta_{-,-}$	$\pm V_-(K(X))$	$\hat{\delta}(X)$
10^5	1.1037	0.7556	0.1394	0.0132	±1.8551	1.4415
5×10^5	1.1254	0.7405	0.1526	0.0138	±1.8347	1.4579
10^6	1.1940	0.6962	0.1696	0.0013	±1.7980	1.5036
5×10^6	1.1828	0.7031	0.1681	0.0016	±1.8018	1.4965

Table 4. Examples of estimates for the parameters and boundaries of confidence intervals for a model distribution $GE(0.7;-1.8;3.6;3.9;1.5)$.

n	$\hat{r}(X)$	$A_r(X)$	$B_r(X)$	$\hat{v}(X)$	$A_v(X)$	$B_v(X)$	$\hat{\delta}(X)$	$A_\delta(X)$	$B_\delta(X)$
10^5	0.7556	0.4082	1.1031	−1.8551	−2.1847	−1.5255	1.4415	1.0839	1.7991
5×10^5	0.7405	0.5998	0.8813	−1.8347	−1.9662	−1.7033	1.4579	1.3090	1.6069
10^6	0.6962	0.6182	0.7742	−1.7980	−1.8683	−1.7277	1.5036	1.4152	1.5920
5×10^6	0.7031	0.6670	0.7392	−1.8018	−1.8345	−1.7691	1.4965	1.4560	1.5370

Table 5 shows the values of the estimates of the parameters r, v and δ obtained from the sample of the size n from a model distribution with a set of parameters $E=(0.8;1.3;0.3;1.4;2.5)$. For this set of parameters, the expression under the outer radical in $R_-(K(X))$ is negative, so the first stage of the algorithm for eliminating unnecessary solutions gives the estimate $\hat{r}(X)=R_+(K(X))$. At the second stage, since $\Delta_{+,+}<\Delta_{+,-}$, the pair $(R_+(K(X)),V_+(K(X)))$ is selected as a solution $(\hat{r}(X),\hat{v}(X))$. The estimate $\hat{\delta}(X)$ is obtained from the relation (33). Table 6 shows the values of the estimates $(\hat{r}(X),\hat{v}(X),\hat{\delta}(X))$ obtained in the previous step, and the boundaries of corresponding confidence intervals.

Table 5. Examples of estimates for the parameters of the model distribution $GE(0.8; 1.3; 0.3; 1.4; 2.5)$.

n	$R_+(K(X))$	$\Delta_{+,+}$	$\Delta_{+,-}$	$\pm V_+(K(X))$	$\hat{\delta}(X)$
10^5	0.8254	0.0006	0.4652	±1.3310	2.4735
5×10^5	0.8146	0.0007	0.4565	±1.3213	2.4893
10^6	0.7958	0.0002	0.4525	±1.2944	2.4996
5×10^6	0.8009	0.0001	0.4548	±1.3007	2.5035

Table 6. Examples of estimates for the parameters and boundaries of confidence intervals for a model distribution $GE(0.8; 1.3; 0.3; 1.4; 2.5)$.

n	$\hat{r}(X)$	$A_r(X)$	$B_r(X)$	$\hat{v}(X)$	$A_v(X)$	$B_v(X)$	$\hat{\delta}(X)$	$A_\delta(X)$	$B_\delta(X)$
10^5	0.8254	0.6302	1.0207	1.3310	1.0545	1.6076	2.4735	2.2936	2.6535
5×10^5	0.8146	0.7302	0.8989	1.3213	1.2015	1.4411	2.4893	2.4088	2.5698
10^6	0.7958	0.7397	0.8519	1.2944	1.2149	1.3738	2.4996	2.4426	2.5566
5×10^6	0.8009	0.7754	0.8264	1.3007	1.2646	1.3369	2.5035	2.4779	2.5290

Table 7 shows the values of the estimates of the parameters r, v and δ obtained from the sample of the size n from a model distribution with a set of parameters $E = (0.6; -2.9; 2.1; 3.9; 0.5)$. For this set of parameters, the inequality $\phi_3/\phi_1 < 1$ holds, while the values of both statistics $R_+(K(X))$ and $R_-(K(X))$ lie in the interval $[0, 1)$. Therefore, at the second stage, since $\Delta_{+,-} = \min\{\Delta_{+,+}, \Delta_{+,-}, \Delta_{-,+}, \Delta_{-,-}\}$, the pair $(R_+(K(X)), -V_+(K(X)))$ is selected as a solution $(\hat{r}(X), \hat{v}(X))$. The estimate $\hat{\delta}(X)$ is obtained from the relation (33). Table 8 shows the values of the estimates $(\hat{r}(X), \hat{v}(X), \hat{\delta}(X))$ obtained in the previous step, and the boundaries of corresponding confidence intervals.

Table 7. Examples of estimates for the parameters of the model distribution $GE(0.6; -2.9; 2.1; 3.9; 0.5)$.

n	$R_+(K(X))$ $R_-(K(X))$	$\Delta_{+,+}$ $\Delta_{-,+}$	$\Delta_{+,-}$ $\Delta_{-,-}$	$\pm V_+(K(X))$ $\pm V_-(K(X))$	$\hat{\delta}(X)$
10^5	0.6125 0.3854	0.0009 0.0120	0.0004 0.0106	±2.9217 ±2.5055	0.4979
5×10^5	0.5871 0.4075	0.0021 0.0109	0.0006 0.0094	±2.8789 ±2.5487	0.5023
10^6	0.6047 0.3921	0.0013 0.0117	0.0001 0.0103	±2.9081 ±2.5181	0.4991
5×10^6	0.6017 0.3948	0.0015 0.0116	0.0001 0.0100	±2.9047 ±2.5249	0.4995

Table 8. Examples of estimates for the parameters and boundaries of confidence intervals for a model distribution $GE(0.6; -2.9; 2.1; 3.9; 0.5)$.

n	$\hat{r}(X)$	$A_r(X)$	$B_r(X)$	$\hat{v}(X)$	$A_v(X)$	$B_v(X)$	$\hat{\delta}(X)$	$A_\delta(X)$	$B_\delta(X)$
10^5	0.6125	0.4204	0.8046	−2.9217	−3.3210	−2.5224	0.4979	0.4604	0.5355
5×10^5	0.5871	0.4833	0.6909	−2.8789	−3.0904	−2.6674	0.5023	0.4814	0.5232
10^6	0.6047	0.5408	0.6686	−2.9081	−3.0401	−2.7760	0.4991	0.4865	0.5117
5×10^6	0.6017	0.5724	0.6309	−2.9047	−2.9650	−2.8445	0.4995	0.4938	0.5053

Remark 3. *In some cases, when processing real data using the above methods, the lengths of confidence intervals may unboundedly increase. This indicates that the conditions of Theorems 1–3 are violated, that is, either* $r^2 = (\phi_3 - \phi_1^2)/(2\phi_1)$ *or* $r^2 = \phi_3/\phi_1$.

6. Discussion

The majority of models of real processes using continuous distributions with unbounded non-negative support operate with special cases of the generalized gamma distribution, proposed in the 1920s by the Italian economist Amoroso in the framework of the study of the dynamic equilibrium theory [7], and special cases of the generalized beta distribution of the second kind, proposed in the 1980s by McDonald as a generalization of the well-known beta-type distributions used to model profitability [9].

The study of probabilistic and statistical properties of distributions from the gamma and beta classes is very important. For example, in the Refs. [10–14], it was proposed to use the generalized gamma distribution and its particular cases in problems of processing radar signals and images, evaluating the concentration of harmful gases in industrial areas, studying the periods of remission of cancer patients, analyzing neurotransmission and anorexia. The results of the Refs. [15–19] concerning the generalized beta distribution of the second kind and its representatives are used for meteorological research, analysis of infectious diseases, climatic phenomena and profitability, the study of physiological characteristics, and consumer price indexes, and can also be used in the theory of reliability when modeling the time of failure.

This article considers the gamma-exponential distribution, a generalization of the Amoroso distribution that gives the McDonald distribution in the limit. Thus, it can be argued that the results of the article will be in demand when studying various models that allow descriptions of real processes using continuous distributions with non-negative unbounded support.

The main problem in the study of the gamma-exponential distribution is the representation of the density (1) in terms of a special gamma-exponential function (2). This fact makes it difficult to study the probabilistic and statistical properties of the distribution using classical methods such as, for example, the maximum likelihood method. In addition, the moments (6) of the gamma-exponential distribution may not exist for some parameter values and are the products of nonmonotonic gamma functions, with arguments depending on several parameters at once. It significantly complicates not only the application of the method of moments, but also the interpretation of the parameters as characteristics of the mean, spread, asymmetry, and so forth. The latter cannot be considered as a disadvantage of this distribution, since in practice, the value of each characteristic is influenced by many factors of a different nature.

The results of this paper concern the estimation of the bent, shape and scale parameters of the gamma-exponential distribution under the assumption that the concentration parameters are known and fixed. This formulation naturally arises in the case of using the gamma-exponential distribution to study scale mixtures of Rayleigh, Maxwell–Boltzmann, Fréchet (Weibull–Gnedenko), Lévy (with zero bias) distributions, and some others. However, a natural question arises about the form of statistical estimates in the case when all five parameters are unknown.

The difficulty in applying the considered method, based on equating theoretical logarithmic cumulants to their sample counterparts, lies in the fact that the concentration parameters enter the equations as arguments of polygamma functions.

There are many papers related to the study of polygamma functions. For example, the Refs. [25,26] provide the estimates of polygamma functions and inverse polygamma functions in terms of elementary functions, Riemann and Hurwitz zeta functions, and Bernoulli numbers, and also investigate the monotonicity properties of expressions associated with polygamma functions. However, the usefulness of these results for the statistical estimation of the arguments of polygamma functions is not obvious yet.

Thus, the problem of developing effective theoretical methods of inverting polygamma functions is an urgent and, apparently, unsolved problem. However, due to the strict monotonicity and continuity of the polygamma functions, the possibility of numerical inversion is obvious, which will allow for an estimation of all five parameters of the gamma-exponential distribution using computer technologies. The solution to this problem is a direction of further research of the authors.

7. Conclusions

The paper considers the problem of estimating the parameters of the gamma-exponential distribution, which is a generalization and an intermediate link between the generalized gamma distribution and the generalized beta distribution of the second kind. A method for estimating unknown parameters based on logarithmic cumulants is discussed. An algorithm for eliminating unnecessary solutions obtained by solving a system based on logarithmic cumulants is described. The asymptotic normality of the strongly consistent estimators for the bent, shape and scale parameters of the gamma-exponential distribution at fixed concentration parameters is proved. Based on this result, asymptotic confidence intervals for the estimated parameters are constructed. The results are illustrated by numerical examples constructed on the basis of model samples from the gamma-exponential distribution, implemented using the representation of the gamma-exponential distribution as a fractional-scale mixture of gamma distributions. Possible applications of the results in the analysis of processes using continuous distributions with a non-negative unbounded support are discussed.

Author Contributions: Conceptualization, A.K. and O.S.; methodology, A.K. and O.S.; formal analysis, A.K. and O.S.; investigation, A.K. and O.S.; writing—original draft preparation, A.K. and O.S.; writing—review and editing, A.K. and O.S.; supervision, A.K. and O.S.; funding acquisition, O.S. All authors have read and agreed to the published version of the manuscript.

Funding: This research was supported by the Ministry of Science and Higher Education of the Russian Federation, project No. 075-15-2020-799.

Institutional Review Board Statement: Not applicable.

Informed Consent Statement: Not applicable.

Data Availability Statement: Not applicable.

Conflicts of Interest: The authors declare no conflict of interest.

References

1. Kudryavtsev, A.A. On the representation of gamma-exponential and generalized negative binomial distributions. *Inform. Appl.* **2019**, *13*, 78–82.
2. Kudryavtsev, A.A.; Titova, A.I. Gamma-exponential function in Bayesian queueing models. *Inform. Appl.* **2017**, *11*, 104–108.
3. Le Roy, É. Sur les séries divergentes et les fonctions définies par un développement de Taylor. *Ann. Fac. Sci. Toulouse* **1900**, *2*, 317–384. [CrossRef]
4. Srivastava, H.M.; Tomovski, Ž. Fractional calculus with an integral operator containing a generalized Mittag-Leffler function in the kernel. *Appl. Math. Comput.* **2009**, *211*, 198–210. [CrossRef]
5. Gorenlo, R.; Kilbas, A.A.; Mainardi, F.; Rogosin, S.V. *Mittag-Leffler Functions, Related Topics and Applications*; Springer: Berlin, Germany, 2014.
6. Kudryavtsev, A.A. Bayesian balance models. *Inform. Appl.* **2018**, *12*, 18–27.
7. Amoroso, L. Ricerche intorno alla curva dei redditi. *Ann. Mat. Pura Appl.* **1925**, *21*, 123–159. [CrossRef]
8. Vorontsov, M.O.; Kudryavtsev, A.A.; Shestakov, O.V. Some Probability-Statistical Properties of the Gamma-Exponential Distribution. *Syst. Means Inform.* **2021**, *29*, 18–35.
9. McDonald, J.B. Some Generalized Functions for the Size Distribution of Income. *Econometrica* **1984**, *52*, 647–665. [CrossRef]
10. Gao, G.; Ouyang, K.; Luo, Y.; Liang, S.; Zhou, S. Scheme of Parameter Estimation for Generalized Gamma Distribution and Its Application to Ship Detection in SAR Images. *IEEE Trans. Geosci. Remote Sens.* **2017**, *55*, 1812–1832. [CrossRef]
11. Zhou, Y.; Zhu, H. Image Segmentation Using a Trimmed Likelihood Estimator in the Asymmetric Mixture Model Based on Generalized Gamma and Gaussian Distributions. *Math. Probl. Eng.* **2018**, *2018*, 3468967. [CrossRef]
12. Iriarte, Y.A.; Varela, H.; Gómez, H.J.; Gómez, H.W. A Gamma-Type Distribution with Applications. *Symmetry* **2020**, *12*, 870. [CrossRef]
13. Barranco-Chamorro, I.; Iriarte, Y.A.; Gómez, Y.M.; Astorga, J.M.; Gómez, H.W. A Generalized Rayleigh Family of Distributions Based on the Modified Slash Model. *Symmetry* **2021**, *13*, 1226. [CrossRef]
14. Combes, C.; Ng, H.K.T. On parameter estimation for Amoroso family of distributions. *Math. Comp. Sim.* **2021**, *191*, 309–327. [CrossRef]
15. López-Rodríguez, F.; García-Sanz-Calcedo, J.; Moral-García, F.J.; García-Conde, A.J. Statistical Study of Rainfall Control: The Dagum Distribution and Applicability to the Southwest of Spain. *Water* **2019**, *11*, 453. [CrossRef]

16. Hassan, N.J.; Mahdi Hadad, J.; Hawad Nasar, A. Bayesian Shrinkage Estimator of Burr XII Distribution. *Int. J. Math. Sci.* **2020**, *2020*, 7953098. [CrossRef]
17. Bantan, R.A.R.; Elgarhy, M.; Chesneau, C.; Jamal, F. Estimation of Entropy for Inverse Lomax Distribution under Multiple Censored Data. *Entropy* **2020**, *22*, 601. [CrossRef]
18. Shi, X.; Shi, Y. Inference for Inverse Power Lomax Distribution with Progressive First-Failure Censoring. *Entropy* **2021**, *23*, 1099. [CrossRef]
19. Sarabia, J.M.; Jordá, V.; Prieto, F.; Guillén, M. Multivariate Classes of GB2 Distributions with Applications. *Mathematics* **2021**, *9*, 72. [CrossRef]
20. Kudryavtsev, A.; Shestakov, O. Asymptotically Normal Estimators for the Parameters of the Gamma-Exponential Distribution. *Mathematics* **2021**, *9*, 273. [CrossRef]
21. Kudryavtsev, A.A.; Shestakov, O.V. Method of logarithmic moments for estimating the gamma-exponential distribution parameters. *Inform. Appl.* **2020**, *14*, 49–54.
22. Kendall, M.G.; Stuart, A. *The Advanced Theory of Statistics*, 3rd ed.; Griffin: London, UK, 1969; Volume 1.
23. Kudryavtsev, A.A.; Shestakov, O.V. A Method for Estimating Bent, Shape and Scale Parameters of the Gamma-Exponential Distribution. *Inform. Appl.* **2021**, *15*, 57–62.
24. Serfling, R.J. *Approximation Theorems of Mathematical Statistics*; John Wiley & Sons, Inc.: New York, NY, USA, 2002.
25. Guo, B.-N.; Qi, F.; Zhao, J.-L.; Luo, Q.-M. Sharp inequalities for polygamma functions. *Math. Slovaca* **2015**, *65*, 103–120. [CrossRef]
26. Batir, N. Inequalities for the inverses of the polygamma functions. *Arch. Math.* **2018**, *110*, 581–589. [CrossRef]

 mathematics

MDPI

Article

Monte Carlo Algorithms for the Extracting of Electrical Capacitance

Andrei Kuznetsov * **and Alexander Sipin**

Department of Applied Mathematics, Vologda State University, Lenina 15, 160000 Vologda, Russia; cac1909@mail.ru
* Correspondence: pm_kan@mail.ru or kuznetsovan@vogu35.ru

Abstract: We present new Monte Carlo algorithms for extracting mutual capacitances for a system of conductors embedded in inhomogeneous isotropic dielectrics. We represent capacitances as functionals of the solution of the external Dirichlet problem for the Laplace equation. Unbiased and low-biased estimators for the capacitances are constructed on the trajectories of the Random Walk on Spheres or the Random Walk on Hemispheres. The calculation results show that the accuracy of these new algorithms does not exceed the statistical error of estimators, which is easily determined in the course of calculations. The algorithms are based on mean value formulas for harmonic functions in different domains and do not involve a transition to a difference problem. Hence, they do not need a lot of storage space.

Keywords: capacitance; dirichlet boundary value problem; monte carlo method; unbiased estimator; von-neumann-ulam scheme

Citation: Kuznetsov, A.; Sipin, A. Monte Carlo Algorithms for the Extracting of Electrical Capacitance. *Mathematics* **2021**, *9*, 2922. https://doi.org/10.3390/math9222922

Academic Editor: Tuan Phung-Duc

Received: 28 October 2021
Accepted: 14 November 2021
Published: 17 November 2021

1. Introduction

The problems of finding potentials and mutual capacitances for complex three-dimensional objects have become widespread with the development of high-frequency electrical engineering. In the case of one or two conductors, they can still be solved analytically, but solving problems for systems of a large number of conductors of complex shape causes significant difficulties. The more the operation frequency is, the more impact on the system of parasitic capacitance and induction. This is true for radio frequency communication devices, as well as very large-scale integration circuits and multilayer printed-circuit boards [1,2].

In inhomogeneous media with permittivity $\varepsilon(x)$ the electrostatic potential $\varphi(x)$ satisfies the boundary value problem:

$$\left.\begin{array}{l}\Delta\varphi = 0,\ x \in \mathbf{R}^3 \setminus (\Gamma_i \cup \Gamma_d);\\ \varphi(x) \underset{|x|\to\infty}{\longrightarrow} 0;\\ \varphi|_{\Gamma_i} = \varphi_i,\ \varphi_i = \text{const};\\ \varphi^+(x) = \varphi^-(x),\ x \in \Gamma_d;\\ \varepsilon^+ \frac{\partial\varphi^+(x)}{\partial n} = \varepsilon^- \frac{\partial\varphi^-(x)}{\partial n},\ x \in \Gamma_d;\\ \oint_{\Gamma_i} \varepsilon \frac{\partial\varphi}{\partial n} dS = -q_i.\end{array}\right\} \qquad (1)$$

Here Γ_i denotes the conductor surfaces, Γ_d is the union of the dielectric interfaces, n is the external normal to Γ_d, $|x|$ is Euclidean length of x, φ^+ and φ^- are the values of the potential on different sides dielectric interfaces, ε^+ and ε^- are the permittivity constants on different sides dielectric interfaces, φ_i are values of the potential on the Γ_i, dS is the differential element of area, and q_i is the charge on Γ_i (Figure 1).

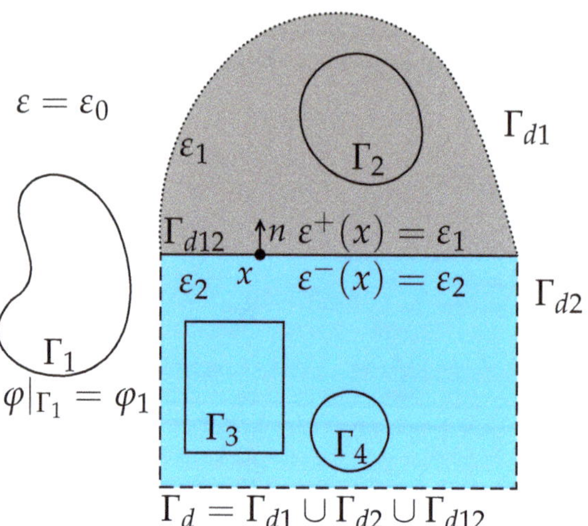

Figure 1. Domain for the boundary value problem.

Charges linearly depend on potentials [3]: $q_i = \sum_{j=1}^{m} C_{ij}\varphi_j$. Here, C_{ij} is mutual electrostatic capacitance for the conductors i and j, it is known that $C_{ij} = C_{ji}$. Hence, C_{ij} is equal to the charge q_i, when all potential $\varphi_k = 0$, if $k \neq j$, and $\varphi_j = 1$.

The analytical solution is only available for simple geometries [3], and could not be used for real-life tasks. Another way is to use pattern-matching algorithms, but there is dependency on available patterns and the quality of geometry approximation with patterns [2,4].

The methods used most for computing capacitances in complicated three-dimensional geometries are the boundary-element technique (for example, [5–7]) and Monte Carlo methods (for example, [8–11]).

The boundary element method is used to solve the system of integral equations of potential theory for the charge density on the surfaces of conductors. The charge on the conductor is then calculated by integrating the density. The main drawbacks of these methods are the necessity of approximation of the conductor's surface, high random access memory requirements, and additional computational error when equations are solved using the iterative technique.

The Monte Carlo method is used to solve the Dirichlet boundary value problem (1). Capacitance is calculated using the Gaussian formula through the normal derivative of the potential. Monte Carlo algorithms for a boundary value problem are based on the representation of its solution in the form of the mathematical expectation of some random variable, which in mathematical statistics is called an unbiased estimator. A common drawback for Monte Carlo methods isthe necessity of a large number of simulations, but usually they are highly parallelizable and have low random access memory requirements.

There are various formulas for the average value for the potential, which determine both the estimate itself and the type of Random Walk along the trajectories of which it is calculated.

One of the first works on using Monte Carlo method for real-life capacitance extraction is [8]. This article describes Random Walk on Cubes methods for rectilinear conductors in a homogeneous medium. The proposed algorithm uses the mean value theorem for the potential at the center of a cube. To simplify the procedure for modeling a Random Walk, the problem was discretized. The development of the method of Random Walk on Cubes in various directions (multiple dielectrics, non-Manhattan polygonal shapes, optimizations) can be found, for example, in [12–14]. Besides the statistical error of Monte

Carlo approximations, Walk on Cubes have additional bias because of the approximation of Green's function for cubes using the Fourier series.

In [15], Random Walk on Boundary was described for calculating conductor's capacitance in free space, in [9] Random Walk on Spheres and Walk on Boundary were used for estimating electrostatic properties of molecules, including cases for different (constant) permittivities. These methods were extended in [16] for analysis on multidielectric integrated circuits of arbitrary geometry from a scanning electron microscopy image. Besides the statistical error of Monte Carlo approximations, there is additional bias, because of various discretizations, that could not be estimated along with the calculations.

In this article we discuss algorithms of the Monte Carlo method that do not require discretization of the boundary value problem. Consequently, there is no approximation error in them. Due to this, it is possible to estimate the error of the approximate solution of the problem during the calculations. Furthermore, Random Walks in unbounded regions may not reach the boundary of the conductors in a finite time with a positive probability. Forced completion of the trajectory leads to a bias in the estimate of the potential, which authors usually do not take into account. Our proposed algorithms are free from this drawback.

In our previous work [10,11], we developed algorithms for mutual capacitance calculation in homogeneous media on trajectories of a Walk on Spheres and in inhomogeneous media on trajectories of a Walk on Hemispheres, when dielectric interfaces are polyhedral. We summarize the main results of these works here.

In this paper, we also consider a new version of the Walk on Hemispheres and its application to the calculation of electrostatic capacitances for systems with various dielectric interfaces, including non-Manhattan geometries.

Using the examples of conductor systems for which the capacitances are calculated analytically [3], it is shown that the accuracy of the Monte Carlo approximation is within the statistical error. In more complex examples, the simulation results are compared with the results of calculating these capacitances using the programs FastCap2 and FFTCap [6,17,18].

The paper is organized as follows. Section 2 introduces a description of the problem. Section 3 describes different kinds of unbiased estimators for the capacitance. It begins with a description of previously proposed algorithms in Sections 3.1 and 3.2, followed by the description of a new version of the Walk on Hemispheres in Section 3.3, and finishes with a description of the generic algorithm for capacitance extraction with these methods in Section 3.4. Section 4 contains the numerical results for capacitance extraction, where we compare the results of the proposed algorithms with analytical solutions or other programs. Section 5 concludes the paper.

2. Integral Representation for the Capacitance

Using Gauss's theorem we have

$$C_{ij} = -\oint_{\Gamma} \varepsilon \frac{\partial \varphi}{\partial n} dS, \tag{2}$$

where Γ is the surface containing the i-th conductor inside and separating it from others conductors and interfaces.

Using the Poisson formula, we obtain the following representation for the normal derivative of the potential on the shell Γ:

$$\frac{\partial \varphi(x)}{\partial n} = \frac{1}{4\pi r^2} \oint_{S_r} \frac{3}{r^2}(y - x, n)\varphi(y) d_y S, \tag{3}$$

where x is a point on the shell around the i-th conductor, r is distance from point x to the nearest conductor or interface, S_r is sphere of radius r centered at point x, y is a point on S_r (Figure 2).

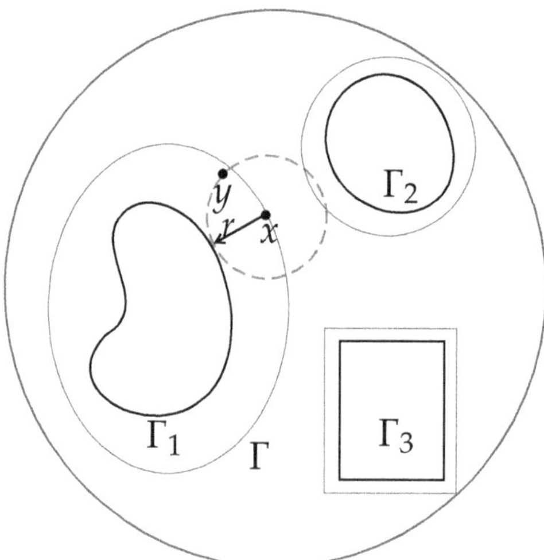

Figure 2. First steps of Random Walk on Spheres for estimation of C_{1j}. Here Γ_i are conductor's surfaces, Γ is a shell around first conductor.

Finally, replacing the normal derivative $\partial\varphi/\partial n$ by its integral representation in the ball, which lies entirely in the region with the dielectric constant ε, we obtain an integral representation of the mutual capacitance of the i-th and j-th conductors:

$$C_{ij} = -\frac{1}{\sigma_\Gamma} \oint_\Gamma \frac{\varepsilon}{r^2} \oint_{S_r} \frac{3\sigma_\Gamma}{4\pi r^2}(y-x,n)\varphi(y) d_y S d_x S, \tag{4}$$

where σ_Γ is a surface area Γ.

3. Unbiased Estimators for the Capacitance

Using Formula (4) we have unbiased estimator for capacitance C_{ij}

$$\xi = \frac{3\varepsilon(X)\sigma_\Gamma}{r}(\omega, n)\varphi(X + r\omega). \tag{5}$$

Here, a random point X is uniformly distributed on Γ, $r = r(X)$, and ω is an isotropic vector (random unit vector). It remains to estimate the potential at the point $Y = X + r(X)\omega$. This can be done using the mean value formula

$$\varphi(x) = \int_Q \varphi(y) P(x, dy), x \in Q, \tag{6}$$

where $Q = \mathbf{R}^3 \setminus D$, and D is the set of interior points of all conductors. The unbiased estimators for $\varphi(Y)$ are constructed on trajectories of Random Walk $\{Y_k\}_{k=0}^{\infty}$, $(Y_0 = Y)$, in space Q. The kernel $P(x, dy)$ must be stochastic or sub-stochastic. It determines the distribution of the next point of the Random Walk over the current point.

Let

$$\xi_0 = \frac{3\varepsilon(X)\sigma_\Gamma}{r}(\omega, n). \tag{7}$$

If at time k the "weight" $W_k = P(Y_k, Q) < 1$, then the current value of the estimator is multiplied by the "weight": $\xi_{k+1} = W_k \xi_k$. The Random Walk stops at time ν, when it reaches the δ-boundary of the conductors, that is, when the distance $\text{dist}(Y_k, \partial\overline{D})$ from point Y_k to the boundary of the conductors becomes less than the δ. Hence, it must satisfy the condition $P\{\nu < \infty\} = 1$. We define estimator $\xi_\delta = \xi_\nu$, if $\text{dist}(Y_\nu, \Gamma_j) < \delta$, and zero, otherwise. If the boundary ∂D is smooth enough, then $|C_{ij} - E\xi_\delta| < c\delta$, for some constant c. In practice, the estimator ξ_δ is simulated in a reasonable time, only if $E\xi_\nu < \infty$. Having received a sufficient number of realizations of the estimator ξ_δ and calculating their arithmetic mean, we obtain an approximate value of the capacitance C_{ij}.

We will now describe some of the types of Random Walks used to calculate the capacitances of conductors.

3.1. Random Walk on Spheres for the External Dirichlet Problem

Random Walk on Spheres (WoS) is used to solve the external Dirichlet problem for the Laplace equation [19], and allows for the calculation of the capacitances of conductors in a homogeneous medium [10]. Let all the conductors lie inside a sphere S_R of radius R centered at the origin. Let $\rho(y)$ be a continuous function such that $c \cdot \mathrm{dist}(y, \partial\overline{D}) \leq \rho(y) \leq \mathrm{dist}(y, \partial\overline{D})$ for some constant $c > 0$.

By the mean value theorem for harmonic functions, we obtain $\varphi(x) = E\varphi(x + \rho(x)\omega)$ for $x \in \mathbf{R}^3 \setminus \overline{D}$. Let $\{\omega_k\}_{k=1}^{\infty}$ be a sequence of independent isotropic vectors. Then we get a Random Walk

$$\begin{aligned} Y_{k+1} &= Y_k + \rho(Y_k)\omega_{k+1}, \\ Y_0 &= Y, \\ W_k &= 1, \\ k &= 0, 1, 2, \ldots. \end{aligned}$$

To restrict the region of the Random Walk, we use the Poisson formula for $|x| > R$:

$$\varphi(x) = \frac{1}{4\pi R} \oint_{S_R} \frac{|x|^2 - R^2}{|x - y|^3} d_y S.$$

Namely, if $|Y_k| > R$, then "weight" $W_k = R/|Y_k|$, and Y_{k+1} is distributed on the sphere S_R with density

$$p(Y_k, y) = \frac{|Y_k|^2 - R^2}{|Y_k - y|^3} \cdot \frac{|Y_k|}{4\pi R^2} \tag{8}$$

(Figure 3).

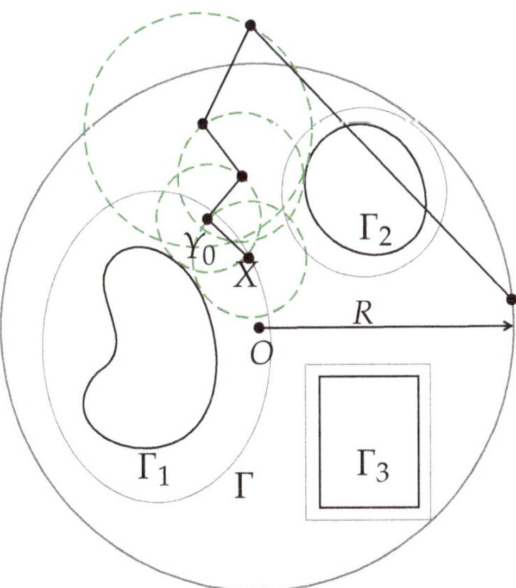

Figure 3. Random Walk on Spheres. Return on external sphere S_R.

It is proved [19] that the Random Walk on Spheres reaches the δ-neighborhood of the boundary of the conductors in a finite time. The formulas for simulating the Random Walk are also given.

3.2. Random Walk on Hemispheres

The Random Walk on Hemispheres (WoH) algorithm was proposed in [20] for solving various boundary value problems for the Laplace and Poisson equations. It allows for the calculation of capacitances when dielectric interfaces are polyhedral [11]. In cases when surfaces of the conductors are also polyhedral, the algorithm gives unbiased statistical estimators of the capacitances. We will now briefly describe this algorithm.

Let all the conductors and dielectric interfaces lie inside a sphere S_R of radius R centered at the origin. If $|Y_k| > R$, then $Y_{k+1} \in S_R$ and has a distribution density (8). "Weight" $W_k = R/|Y_k|$.

Now, let $Y_k \in \Gamma_{d_l}$, where Γ_{d_l} is a component of the dielectric interface. Next, we choose the maximum r, such, that $0 < r < \text{dist}(Y_k, \overline{D} \cup \Gamma_d \setminus \Gamma_{d_l})$ and part of the Γ_{d_l}, lying in sphere $S_r(Y_k)$, is plane. The sphere is divided into two parts $S_r^+(Y_k)$ and $S_r^-(Y_k)$ lying in media with permittivity constants ε^+ and ε^-, respectively. The point Y_{k+1} is uniformly distributed in $S_r^+(Y_k)$ or in $S_r^-(Y_k)$ with probability $\varepsilon^+/(\varepsilon^+ + \varepsilon^-)$ and $\varepsilon^-/(\varepsilon^+ + \varepsilon^-)$ respectively (Figure 4).

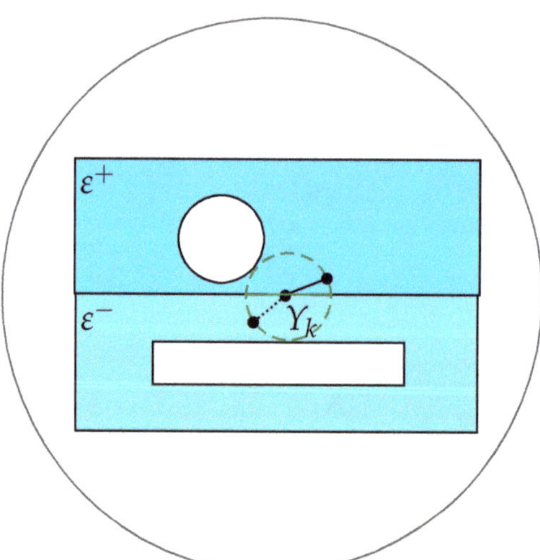

Figure 4. Random Walk on Hemispheres. Exit from interface.

If $Y_k \notin \Gamma_d$ and $|Y_k| \leq R$, then Y_{k+1} is distributed on a sphere or hemisphere. The center of the hemisphere $\widehat{Y}_k$ must be in a plane containing a face of the conductor surface or interface and is the orthogonal projection of Y_k onto this plane.

Hemisphere radius $r_k = |Y_k - \widehat{Y}_k|/\beta$, where $0 < \beta < 1$ is a fixed constant. The hemisphere must be contained in a medium with a dielectric constant $\varepsilon(Y_k)$. The distribution density of the point Y_{k+1} on the hemisphere is the normal derivative of the Green's function for the half of the ball

$$p(Y_k, y) = \begin{cases} \frac{2r_k\beta}{4\pi}\left(\frac{1}{|Y_k - y|^3} - \frac{1}{(\beta|\overline{Y_k^*} - y|)^3}\right), & y \in H, \\ \frac{r_k}{4\pi}(1-\beta^2)\left(\frac{1}{|Y_k - y|^3} - \frac{1}{|\overline{Y_k} - y|^3}\right), & y \in S. \end{cases} \quad (9)$$

Here H is the plane part, and S is the spherical part of the hemisphere. The point $\overline{Y_k}$ is symmetric to Y_k relative to plane H. The point Y_k^* lies outside of the sphere S and it is inverse to the point Y_k ($|\widehat{Y}_k - Y_k| \cdot |\widehat{Y}_k - Y_k^*| = r^2$) (Figure 5).

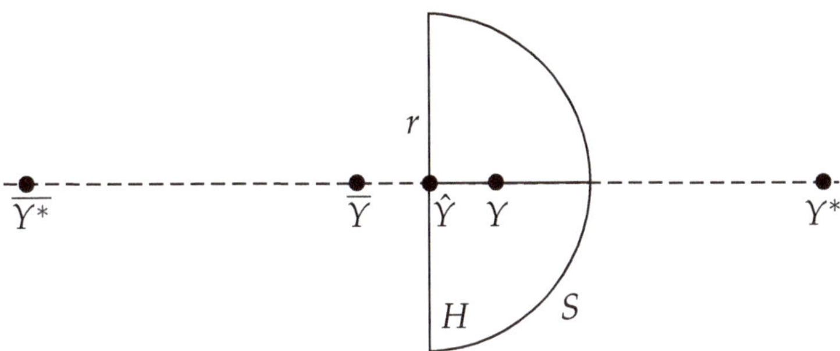

Figure 5. Random Walk on Hemispheres. Symmetrical points on hemisphere.

If it is impossible to construct such a hemisphere, then Y_{k+1} is distributed uniformly on a sphere of radius $r_k = \text{dist}(Y_k, \overline{D} \cup \Gamma_d)$, centered at Y_k.

Von Neumann's Acceptance-Rejection Method can be used to simulate density (9). To do this, we write the density in the form

$$p(Y_k, y) = \frac{1}{4\pi} \frac{\cos \varphi_{Y_k y}}{|Y_k - y|^2} \cdot k_1(Y_k, y), \tag{10}$$

where $\varphi_{Y_k y}$ is the angle between the vector $y - Y_k$ and the external normal to the surface of the hemisphere at the point y.

The first factor in this formula is the distribution density of the point Z on the surface of the hemisphere and the vector $\omega = (Z - Y_k)/|Z - Y_k|$ is isotropic one. The second factor does not exceed the constant $M = \max(M_1, M_2)$, where

$$M_1 = \frac{2}{\beta}\sqrt{1+\beta^2}\left(1 - \left(\frac{\beta}{\sqrt{1+\beta^2}}\right)^3\right),$$

$$M_2 = \sqrt{1+\beta^2}\frac{1+\beta}{1-\beta}\left(1 - \left(\frac{1-\beta}{1+\beta}\right)^3\right).$$

To select the next point of the random walk, we simulate an isotropic vector ω and a random variable α with uniform distribution on $[0, 1]$. Then we define the point Z, in which the ray emerging from the Y_k in the direction ω crosses the hemisphere. If $\alpha M < k_1(Y_k, Z)$, then $Y_{k+1} = Z$. Otherwise, it is necessary to repeat the simulation until the inequality is true (Figure 6).

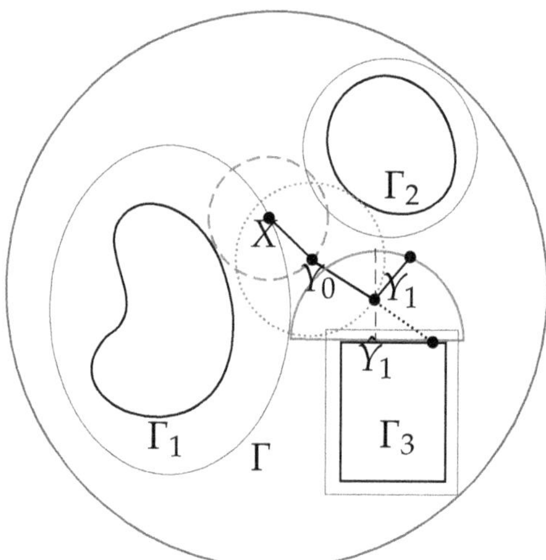

Figure 6. Random Walk on Hemispheres. Jump on hemisphere.

3.3. Random Walk on Hemispheres for a Convex Dielectric Interfaces (RWHC)

Let γ be a connected convex part of some dielectric interface Γ_{d_l} lying inside a sphere $S_r(x)$ of radius r, centered at point x. For all $y \in \Gamma_{d_l}$ we choose the direction of the normal vector n_y so, that the surface Γ_{d_l} lies in the half-space $(z-y, n_y) \leq 0$. Surface γ divides the sphere into two parts S^+ and S^-, lying in media with permittivity constants ε^+ and ε^-, respectively, and $(z-y, n_y) \leq 0$ for all $z \in S^-$.

The potential $\varphi(x)$ is a harmonic function for the part of the ball bounded by surfaces S^-, γ and part of the ball bounded by surfaces S^+, γ also. Using the second Green's formula for a harmonic function in a bounded domain, we obtain the following Theorem.

Theorem 1. *Let $\lambda = \varepsilon^+/\varepsilon^-$ and let φ_{xy} be the angle between vectors $n_y, y-x$. Then the potential $\varphi(y)$ satisfies the mean value formulas:*

$$\varphi(x) = \frac{1}{1+\lambda} \cdot \frac{1}{2\pi r^2} \int_{S^-} \varphi(y) d_y S + \frac{\lambda}{1+\lambda} \cdot \frac{1}{2\pi r^2} \int_{S^+} \varphi(y) d_y S + \\ + \frac{1-\lambda}{1+\lambda} \cdot \frac{1}{2\pi} \int_\gamma \frac{\cos \varphi_{xy}}{|x-y|^2} \varphi(y) d_y S, \quad x \in \gamma, \quad (11)$$

$$\varphi(x) = \frac{1}{4\pi r^2} \int_{S^-} \varphi(y) d_y S + \lambda \cdot \frac{1}{4\pi r^2} \int_{S^+} \varphi(y) d_y S + \\ + (1-\lambda) \cdot \frac{1}{4\pi} \int_\gamma \frac{\cos \varphi_{xy}}{|x-y|^2} \varphi(y) d_y S, \quad x \notin \gamma, \quad \varepsilon(x) = \varepsilon^-, \quad (12)$$

$$\varphi(x) = \frac{1}{4\pi r^2} \int_{S^+} \varphi(y) d_y S + \frac{1}{\lambda} \cdot \frac{1}{4\pi r^2} \int_{S^-} \varphi(y) d_y S - \\ - \left(1 - \frac{1}{\lambda}\right) \cdot \frac{1}{4\pi} \int_\gamma \frac{\cos \varphi_{xy}}{|x-y|^2} \varphi(y) d_y S, \quad x \notin \gamma, \quad \varepsilon(x) = \varepsilon^+. \quad (13)$$

If $\lambda < 1$, the Formula (11) defines the stochastic kernel. To simulate the transition from the surface γ, we chose with the probability $\lambda/(1+\lambda)$ a random direction ω, that satisfies the condition $(\omega, n_x) > 0$, and define $Y = x + r\omega$. With probability $1/(1+\lambda)$, we simulate a random direction ω satisfying the condition $(\omega, n_x) < 0$. We calculate $Y = x + r\omega$. If $Y \notin S^-$, we change Y to a point $Z \in \gamma$, which is visible from x in the direction ω (Figure 7).

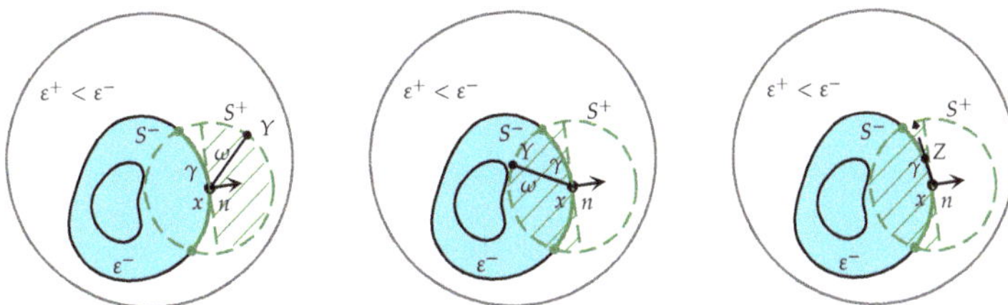

Figure 7. RWHC. Jump from convex part of interface.

If $\lambda < 1$, the Formula (12) defines the stochastic kernel also. To simulate the transition from x, we simulate a random direction ω and calculate $Y = x + r\omega$. If $Y \notin S^-$, than with probability $1 - \lambda$ we change Y to a point $Z \in \gamma$, which is visible from x in the direction ω (Figure 8).

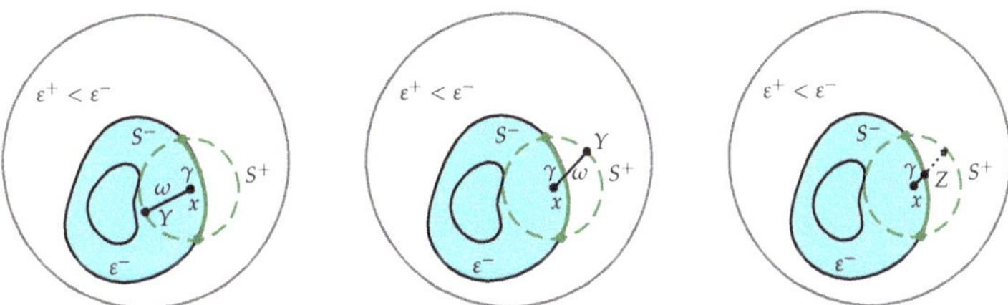

Figure 8. RWHC. Jump from dielectric with higher permittivity.

If $\lambda > 1$, the Formula (13) defines the stochastic kernel, if any ray outgoing from point x intersects γ at no more than one point. The modeling procedure is similar to the algorithm for the Formula (12).

Thus, Formulas (11)–(13) make it possible to simulate transitions from a region with a higher dielectric constant to a region with a lower dielectric constant. To pass from point x through the interface Γ_{d_l}, it is sufficient to take such $r \leq \text{dist}(x, \overline{D} \cup \Gamma_d \setminus \Gamma_{d_l})$ that $S_r(x) \cap \Gamma_{d_l} \neq \emptyset$. Reverse transitions can be provided using, for example, formulas for solving external and internal Dirichlet problems for standard domains. The exit from the "bad" point x can be done by Random Walk on Spheres or Hemispheres in the set $Q(x) = \{y | \varepsilon(y) = \varepsilon(x)\}$. As always, from distant points of the external medium there is a transition to the sphere S_R.

3.4. *Algorithm for Mutual Capacitance Calculation*

On this basis we could describe algorithm for capacitance estimation as follows:

1. For each conductor i select shell Γ_i, that separate it from other conductors and dielectric interfaces.
2. Select radius R for, centered at the origin, "outer" sphere, that contains all conductors and shells.
3. Select point X uniformly on Γ_i and Y uniformly on sphere of radius $r = r(X) = \text{dist}(X, \cup_{k=1,k\neq i}^{n} \Gamma_k)$ centered at X. Set ξ_0 as shown in (7).
4. From point Y started appropriate kind of Walk on Hemispheres (see Sections 3.2 and 3.3).
5. If at some step n process exit outside of sphere S_R, next point is selected at sphere S_R and "weight" updated, as described in Section 3.1.

6. Otherwise, if at some step n, Y_n located at flat surface of k-th conductor or at δ-neighborhood of non-flat surface of k-th conductor, estimation ξ_n included into C_{ik} accumulator and number of trajectories, that used this conductor for evaluation is incremented by 1. (Because $C_{ij} = C_{ji}$, we could use the same accumulator and counter for both of them.) Values in other accumulators C_{ij} are not changed, but the number of trajectories for j-th conductor is also incremented by 1 with the exception of nested conductors: if j-th conductor is inside i-th, its number of trajectories is not changed.
7. If the number of evaluated trajectories is not sufficient, return to step 3.
8. The approximation for capacitance C_{ik} is calculated as value stored in the corresponded accumulator divided by the stored number of trajectories for this accumulator. If the system contains nested conductors, the self-capacitance of external conductor m is updated as $C_{mm} = C_{mm} - \sum_{j:D_j \subset D_m} C_{jm}$ (here sum is taken by numbers of conductors that are located inside m-th).

4. Results

4.1. Mutual Capacitance of Two Spheres in Free Space

Mutual capacitance for two spheres could be calculated analytically [3]. When spheres are not nested:

$$
\begin{aligned}
C_{1,1} &= 4\pi\varepsilon r_1 r_2 \sinh\alpha \sum_{n=1}^{\infty} \frac{1}{r_2 \sinh(n\alpha) + r_1 \sinh[(n-1)\alpha]};\\
C_{1,2} &= -4\pi\varepsilon \frac{r_1 r_2 \sinh\alpha}{d} \sum_{n=1}^{\infty} \frac{1}{\sinh(n\alpha)};\\
C_{2,2} &= 4\pi\varepsilon r_1 r_2 \sinh\alpha \sum_{n=1}^{\infty} \frac{1}{r_1 \sinh(n\alpha) + r_2 \sinh[(n-1)\alpha]};\\
\cosh\alpha &= \frac{d^2 - r_1^2 - r_2^2}{2 r_1 r_2}
\end{aligned}
$$

where r_1 and r_2 are radii, d—distance between sphere centers. For nested spheres ($r_2 > r_1$):

$$
\begin{aligned}
C_{1,1} &= 4\pi\varepsilon r_1 r_2 \sinh\alpha \sum_{n=1}^{\infty} \frac{1}{r_2 \sinh(n\alpha) - r_1 \sinh[(n-1)\alpha]};\\
C_{1,2} &= -C_{1,1};\\
\cosh\alpha &= -\frac{d^2 - r_1^2 - r_2^2}{2 r_1 r_2}.
\end{aligned}
$$

In Table 1 results of mutual capacitance estimation for two non-nested spheres using Walk on Spheres are presented. Here and below Δ is error estimation, calculated as triple square root of ratio of sample variance to number of trajectories, *Time* is "wall time" of calculation, and *Memory* is a peak memory usage. Calculations were performed on one personal computer (PC) with central processing unit (CPU) "AMD Ryzen 7 2700 Eight-Core Processor 3.20 GHz". Monte Carlo simulations were performed in parallel by eight worker processes on one PC using the Message Passing Interface, and memory usage is at its peak for one worker process. FastCap2 and FFTCap are 32-bit single-threaded applications, so no parallel execution were used for them. It should also be noted that we have not used additional optimizations, so the calculation time could be improved for the Monte Carlo case, for example, by using a different pseudo random number generator or using optimizations in distance calculations.

1st sphere radius: 5;
1st sphere center: (1, 2, 3);
1st sphere shell radius: 8;
2nd sphere radius: 3;
2nd sphere center: (10, 13, 12);
2nd sphere shell radius: 8;
"External" sphere radius: 31.155;
δ: 10^{-8}.

Table 1. Mutual capacitance estimation for two non-nested spheres, $C_{i,j}/4\pi\varepsilon_0$.

Method	1,1	Δ	1,2	Δ	2,2	Δ	Time	Memory
Analytical	5.29133	–	−0.94883	–	3.18564	–	–	–
WoS, 10^5	5.14975	$2.711 \cdot 10^{-1}$	−0.93335	$5.799 \cdot 10^{-2}$	3.24719	$1.268 \cdot 10^{-1}$	–	11 Mb
WoS, 10^7	5.29655	$2.717 \cdot 10^{-2}$	−0.94894	$5.805 \cdot 10^{-3}$	3.19233	$1.263 \cdot 10^{-2}$	40 s	11 Mb

Usually, the bias order is the same as the order of δ, so, in this and following examples, error of methods is equal to statistical error. We will say that results of the estimation are *matched*, when the modulus of difference between the Monte Carlo estimation and the reference solution are not more than the statistical error ($|ref - est| \leq \Delta$). As we can see in the Table 1, the analytical solution and our estimation are matched, so the algorithm is working correctly.

In Table 2, results of mutual capacitance estimation for two *nested* spheres using Walk on Spheres (WoS) are presented. There is no formula for C_{22} in the case of two nested spheres in [3], so we do not show the estimation results for this value.

1st sphere radius: 3;
1st sphere center: (10, 13, 12);
1st sphere shell radius: 5;
2nd sphere radius: 31;
2nd sphere center: (1, 2, 3);
2nd sphere shell radius: 35;
"External" sphere radius: 42.616;
δ: 10^{-8}.

Table 2. Mutual capacitance estimation for two nested spheres, $C_{i,j}/4\pi\varepsilon_0$.

Method	1,1	Δ	1,2	Δ	Time	Memory
Analytical	3.47735	–	−3.47735	–	–	–
WoS, 10^5	3.48643	$1.531 \cdot 10^{-1}$	−3.53438	$1.292 \cdot 10^{-1}$	–	11 Mb
WoS, 10^7	3.47455	$1.531 \cdot 10^{-2}$	−3.47807	$1.289 \cdot 10^{-2}$	30 s	11 Mb

Estimation results in Table 2 are within statistical error, so analytical solution and our estimation are matched.

4.2. Capacitance of "Coated" Sphere

In Tables 3 and 4, results of capacitance estimation for the conductive sphere of radius a encased in concentric spherical dielectric of radius b with relative permittivity ε using RWHC are presented. In this example, the external sphere radius is equal to the dielectric shell radius. Analytical solutions for this case is $\dfrac{16\pi^2\varepsilon\varepsilon_0 ab}{\varepsilon a + b - a}$ [3]. Also here we compare results with FastCap2 (FC2) [18] (correspondent sphere discretization was made using spheregen tool from [21] with refine depth 5).

Table 3. Capacitance estimation for coated sphere, $C/4\pi\varepsilon_0$.

$a = 1, b = 3$	Anal.	FC2	RWHC, 10^6	Δ	RWHC, 10^8	Δ
$\varepsilon = 2$	1.5000	1.5018	1.4813	$3.111 \cdot 10^{-2}$	1.4994	$3.114 \cdot 10^{-3}$
$\varepsilon = 10$	2.5000	2.5872	2.4403	$1.889 \cdot 10^{-1}$	2.5094	$1.889 \cdot 10^{-2}$
$\varepsilon = 100$	2.9411	4.3926	2.7198	2.055	2.8966	$2.055 \cdot 10^{-1}$

Comparing values in columns 2, 4, 5 of Table 3 we conclude that the analytical solution and RWHC estimation are matched in all cases. However, comparing columns 2 and 3, we can see that the estimation error for FC2 grows up with ε (as it is stated in [7]). So we can state, that RWHC working correctly with high permittivities too, but more number of simulation may be needed to get estimation with desired statistical error.

Table 4. Capacitance estimation for coated sphere. Time and memory usage.

	FC2	RWHC, 10^8
	$a = 1, b = 3, \varepsilon = 2$	
Time, s	9	158
Memory, Mb	899	11
	$a = 1, b = 3, \varepsilon = 10$	
Time, s	9	148
Memory, Mb	899	11
	$a = 1, b = 3, \varepsilon = 100$	
Time, s	9	147
Memory, Mb	900	11

Table 4 shows that RWHC is used more often than boundary-element technique-based methods, such as FC2, but use much less memory.

4.3. Mutual Capacitance of Two Spheres in Spherical Dielectric

In the Tables 5 and 6 results of mutual capacitance estimation using FC2 (correspondent sphere discretization was made using spheregen tool from [21] with refined depth 5) and RWHC for two conductive spheres in spherical dielectric (Figure 9) are presented. In this case external sphere radius is set to the dielectric shell radius.

1st sphere radius:	5;
1st sphere center:	(1, 2, 3);
1st sphere shell radius:	6;
2nd sphere radius:	3;
2nd sphere center:	(10, 3, 11);
2nd sphere shell radius:	4;
Dielectric ball radius:	20;

δ: 10^{-8}.

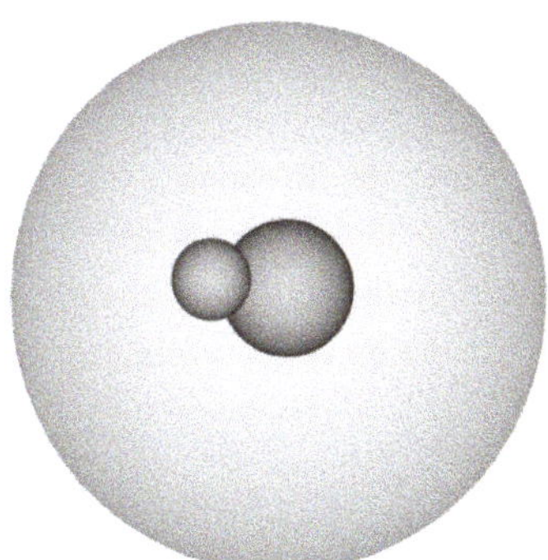

Figure 9. Two spheres in dielectrical shell.

Table 5. Mutual capacitance estimation for two spheres in dielectric shell, $C_{i,j}/4\pi\varepsilon_0$.

i,j	FC2	RWHC, 10^6	Δ	RWHC, 10^8	Δ
			$\varepsilon = 2$		
1,1	9.8444	9.9718	$1.597 \cdot 10^{-1}$	9.8192	$1.598 \cdot 10^{-2}$
1,2	−3.3396	−3.3463	$2.254 \cdot 10^{-2}$	−3.3440	$2.260 \cdot 10^{-3}$
2,2	6.0965	6.1454	$6.760 \cdot 10^{-2}$	6.0968	$6.755 \cdot 10^{-3}$
			$\varepsilon = 10$		
1,1	33.262	33.440	$8.264 \cdot 10^{-1}$	32.890	$8.270 \cdot 10^{-2}$
1,2	−21.038	−21.045	$1.317 \cdot 10^{-1}$	−21.118	$1.318 \cdot 10^{-2}$
2,2	25.380	25.424	$3.495 \cdot 10^{-1}$	25.276	$3.494 \cdot 10^{-2}$

In this case we have no analytical solutions for reference, so we compare our results with FC2. But we also have no error estimation for FC2 result, so we could not guarantee that the difference will be within statistical margin of error. Comparing results in columns 2 and 3 of Table 5 we could say that the estimations are matched, but it is not true for columns 2 and 4. In previous cases we ascertained that RWHC is matched with the analytical solution. Also, in this case, we can see that results in columns 3 and 4 are matched. So we can state that in this case the estimation error with FC2 is more than with RWHC on 10^8 trajectories.

Table 6. Mutual capacitance estimation for two spheres in dielectric shell. Time and memory usage.

	FC2	RWHC, 10^8
	$\varepsilon = 2$	
Time, min	13	3
Memory, Mb	1710	11
	$\varepsilon = 10$	
Time, min	14	3
Memory, Mb	1712	11

In Table 6 we can see that RWHC is better both in time and memory. This is due to the fact that for FC2 spheres should be approximated with a large number of panels. Also, we can see that with the refinement depth we used, we have almost reached the memory limit for the original 32-bit fastcap application, so we cannot compare to these results with better discretization.

4.4. Mutual Capacitance of Two Spheres in Spherical Dielectrics

In Tables 7 and 8, results of mutual capacitance estimation using FC2 (correspondent sphere discretization was made using the spheregen tool from [21] with refined depth 5) and RWHC for two conductive spheres, each in its own spherical dielectric, (Figure 10) are presented.

1st sphere radius:	5;
1st sphere center:	$(1, 2, 3)$;
1st dielectric radius:	9;
1st dielectric center:	$(2, 3, 4)$;
Relative permittivity of 1st dielectric:	2;
2nd sphere radius:	3;
2nd sphere center:	$(10, -3, 20)$;
2nd dielectric radius:	6;
2nd dielectric center:	$(10, -4, 20)$;
Relative permittivity of 2nd dielectric:	5;
External sphere radius:	30;
δ: 10^{-8}.	

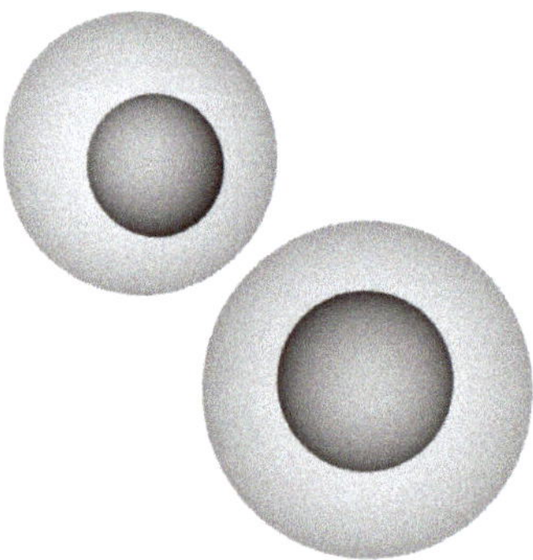

Figure 10. Two spheres in dielectrical shells.

Table 7. Mutual capacitance estimation for two spheres in dielectric shells, $C_{i,j}/4\pi\varepsilon_0$.

i, j	FC2	RWHC, 10^6	Δ	RWHC, 10^8	Δ
1, 1	7.0277	7.0672	$1.655 \cdot 10^{-1}$	7.0255	$1.654 \cdot 10^{-2}$
1, 2	−1.8095	−1.7865	$2.002 \cdot 10^{-2}$	−1.8003	$1.997 \cdot 10^{-3}$
2, 2	5.5841	5.6409	$1.866 \cdot 10^{-1}$	5.4919	$1.866 \cdot 10^{-2}$

There are no reference analytical solutions in this case either. Using the results from Table 7, we have reached the same conclusion as in the previous case.

Table 8. Mutual capacitance estimation for two spheres in dielectric shells. Time and memory usage.

	FC2	RWHC, 10^8
Time, s	17	296
Memory, Mb	1775	11

4.5. *Mutual Capacitance of Parallel "Pins"*

In Tables 9 and 10, the results of the mutual capacitance estimation using FFT-Cap [17,18] (correspondent discretization was made using cubegen tool from [18] with 5 panels per side) and Walk on Hemispheres for 81 conductive "pins" placed in uniform lattice points (Figure 11) are presented. The full capacitance matrix has dimensions of 81×81, so we show only a few values in the table.

Pin size:	$1 \times 1 \times 10$;
Cell size:	2×2;
Shell offset:	0.05;
β:	0.5;
"External" sphere radius:	14.786;
δ:	10^{-16}.

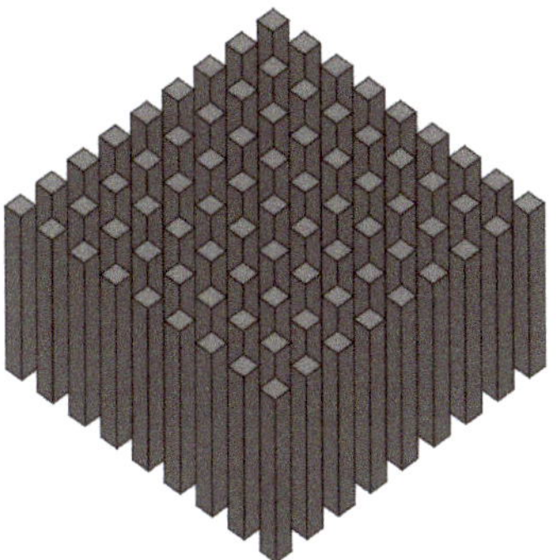

Figure 11. 9×9 conductive pins.

Table 9. Mutual capacitance of 9×9 conductive pins, $C_{i,j}/4\pi\varepsilon_0$.

i,j	FFTCap	WoH, 10^5	Δ	WoH, 10^7	Δ
1, 1	3.9865	3.9447	$2.874 \cdot 10^{-1}$	4.0079	$2.855 \cdot 10^{-2}$
1, 2	-1.3384	-1.3553	$7.000 \cdot 10^{-2}$	-1.3545	$7.048 \cdot 10^{-3}$
1, 81	$-5.9305 \cdot 10^{-3}$	$-7.2766 \cdot 10^{-3}$	$3.514 \cdot 10^{-3}$	$-5.9480 \cdot 10^{-3}$	$3.455 \cdot 10^{-4}$

In this case objects have flat faces, so this task is "good" for FFTCap and we can take this solution as reference. The results from Table 9 shows that the estimations are matched. Also, we could see that, besides matching statistical error, the results for WoH estimation when $i = 1, j = 81$ have a larger statistical error than in other cases (about 6%). This is due to the fact that these conductors are located in opposite corners of the lattice, so only a small number of trajectories started near the first conductor will end on 81st. In this case, to get an estimation with the desired statistical error, more simulations may be required. For example, with 10^8 trajectories we get the value of $-6.0515 \cdot 10^{-3}$ and statistical error of $1.087 \cdot 10^{-4}$ (less than 2%).

Also we could estimate difference with FFTCap results by norm: let A—FFTCap result matrix, B—WoH result matrix for 10^7 trajectories from each conductor, then $\frac{\|A - B\|_F}{\|A\|_F} \approx 0.009$, where $\|A\|_F$ is Frobenius norm of matrix A.

Table 10. Mutual capacitance of 9×9 conductive pins. Time and memory usage.

	FFTCap	WoH, 10^7
Time, min	10	34
Memory, Mb	239	12

4.6. Mutual Capacitance of Rectangular Parallelepipeds in Dielectric Shells

In Tables 11 and 12, the results of mutual capacitance estimation using FC2 (correspondent sphere discretization was made using the spheregen tool from [21] with 10 and 15 panels per side) and Walk on Hemispheres for three conductive rectangular parallelepipeds in parallelepipedic dielectrics (Figure 12) are presented.

Conductor size:	$10 \times 10 \times 1$;
Dielectric size:	$12 \times 12 \times 3$;
C_1 origin:	$(1, 1, 1)$;
$Diel_1$ origin:	$(0, 0, 0)$;
ε_1:	2;
C_2 origin:	$(1, 1, 4)$;
$Diel_2$ origin:	$(0, 0, 3)$;
ε_2:	4;
C_3 origin:	$(1, 1, 7)$;
$Diel_3$ origin:	$(0, 0, 6)$;
ε_3:	3;
Shell offset:	0.25;
β:	0.5;
"External" sphere radius:	19.714;
δ:	10^{-8}.

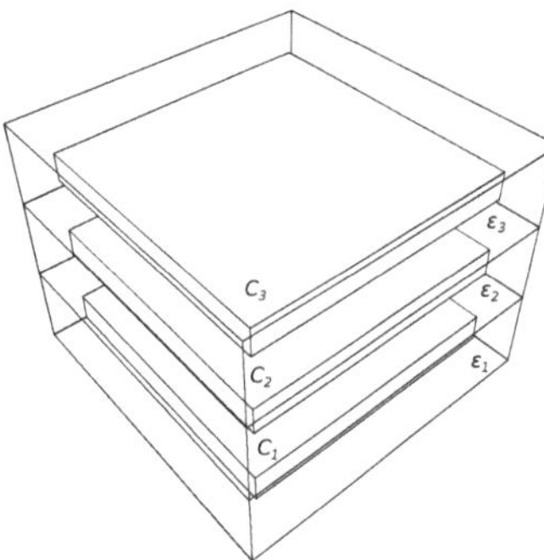

Figure 12. Rectangular parallelepipeds in dielectric shells.

Table 11. Mutual capacitance of rectangular parallelepipeds in dielectric shells, $C_{i,j}/4\pi\varepsilon_0$.

i, j	FC2, $n = 10$	FC2, $n = 15$	WoH, 10^6	Δ	WoH, 10^8	Δ
1, 1	17.491	17.468	17.650	$8.663 \cdot 10^{-1}$	17.452	$8.661 \cdot 10^{-2}$
2, 1	−14.252	−14.247	−14.144	$2.377 \cdot 10^{-1}$	−14.218	$2.374 \cdot 10^{-2}$
2, 2	34.100	34.092	34.381	1.736	34.030	$1.737 \cdot 10^{-1}$
3, 1	−0.848	−0.855	−0.875	$5.204 \cdot 10^{-2}$	−0.861	$5.164 \cdot 10^{-3}$
3, 2	−18.284	−18.264	−18.094	$2.898 \cdot 10^{-1}$	−18.223	$2.899 \cdot 10^{-2}$
3, 3	21.787	21.766	21.827	1.314	21.676	$1.314 \cdot 10^{-1}$

As before, we can compare the RWHC results with the FC2 estimation. In Table 11, the results are not matched between FC2 and RWHC with 10^8 trajectories when $i = 2$, $j = 1$. Because WoH results are matched for 10^6 and 10^8 trajectories and FC2 results with 15 panels per side are closer to our estimation than results with 10 panels per side, we could assume that this discrepancy is related to an FC2 estimation error, as in example Section 4.3.

Table 12. Mutual capacitance of rectangular parallelepipeds in dielectric shells. Time and memory usage.

	FC2, $n = 10$	FC2, $n = 15$	WoH, 10^8
Time, s	4	8	3840
Memory, Mb	270	591	11

4.7. Mutual Capacitance of "Woven Bus"

In Tables 13 and 14, the results of mutual capacitance estimation using FFTCap [18] (correspondent discretization was made using wovengen tool from [21] with 10 panels per side) and Walk on Hemispheres for 9×9 woven bus [7] (Figure 13) are presented.

Width:	1;
Segment length:	8;
Distance between wovens:	1;
Shell offset:	0.1;
β:	0.5;
δ:	10^{-8}.

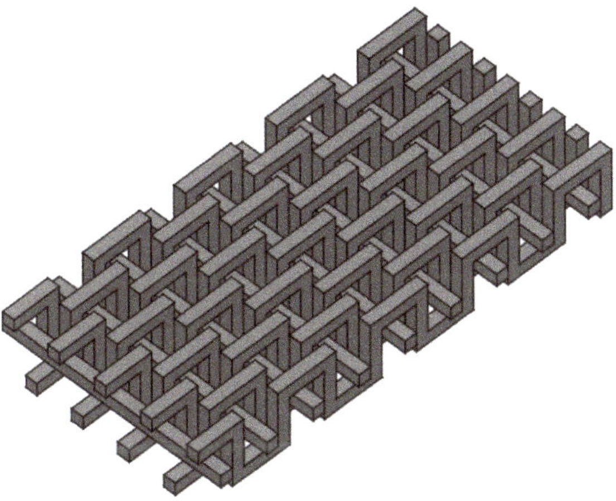

Figure 13. 9×9 woven bus.

Table 13. Mutual capacitance of "woven bus", $C_{i,j}/4\pi\varepsilon_0$.

	FFTCap	**WoH, 10^6**	**Δ**	**WoH, 10^8**	**Δ**
1,1	8.7662	8.7338	$3.565 \cdot 10^{-1}$	8.7480	$3.566 \cdot 10^{-2}$
1,2	$-5.1204 \cdot 10^{-2}$	$-5.3310 \cdot 10^{-2}$	$8.066 \cdot 10^{-3}$	$-5.1310 \cdot 10^{-2}$	$8.013 \cdot 10^{-4}$
1,18	-1.0981	-1.1145	$5.534 \cdot 10^{-2}$	-1.0996	$5.476 \cdot 10^{-3}$

The results in Table 13 match. The difference with FFTCap results also could be evaluated by norm: let A—FFTCap result matrix, B—WoH result matrix for 10^8 trajectories from each conductor, then $\frac{\|A - B\|_F}{\|A\|_F} \approx 0.001$.

Table 14. Mutual capacitance of "woven bus". Time and memory usage.

	FFTCap	**RWHC, 10^8**
Time, min	16	176
Memory, Mb	1863	12

5. Conclusions

We developed some new numerical algorithms for extracting capacitances. These algorithms do not use the approximation of the Laplace operator by its difference counterpart. Their computational error is determined by the sum of the statistical error and the value of the estimator bias. The statistical error is determined in the course of calculations. The systematic error of the estimator is equal to the error when we approximate the potential at points lying near the boundary of the conductor by values at the boundary. This error is controlled by the parameter δ.

The Random Walk on Spheres algorithm is universal in the case of a homogeneous dielectric. It works for conductors with any geometry.

The Random Walk on Hemispheres is applied when dielectric interfaces are polyhedral. In cases when surfaces of the conductors are also polyhedral, the algorithm gives unbiased statistical estimators of the capacitances. The accuracy of this algorithm is equal to the statistical error of the estimators, which is easily determined in the course of calculations.

The Modified Random Walk on Hemispheres algorithm works for convex dielectric interfaces.

Computational experiments show that the algorithms are effective. For systems where capacitances are calculated analytically [3], it is shown that the accuracy of the Monte Carlo approximation is within the statistical error (see Tables 1–3). In more complex examples, to prove that the Monte Carlo estimation results are correct, we have matched them with the results of the calculation of the capacitances using the non-Monte Carlo methods implemented in the FastCap2 and FFTCap programs [18] (see Tables 5, 7, 9, 11 and 13). The algorithm also works correctly in cases when the ratio of the permittivities is 100 or more (see Table 3).

Monte Carlo simulation times for different cases were presented with numerical results, but these are not final and could be improved upon, even with the same configuration of PC, by using another implementation of the pseudo random number generator, for example, or using a function for calculating distance that is optimized for a particular task. For example, by using another implementation of pseudo random number generator, or using function for calculating distance, that is optimized for particular task.

Author Contributions: Investigation, A.K., A.S. All authors contributed equally to this work. All authors have read and agreed to the published version of the manuscript.

Funding: The research was funded by Russian Science Foundation grant 19-11-00020.

Institutional Review Board Statement: Not applicable.

Informed Consent Statement: Not applicable.

Data Availability Statement: Data sharing is not applicable to this article.

Conflicts of Interest: The authors declare no conflict of interest.

References

1. Kamon, M.; McCormick, S.; Shepard, K. Interconnect parasitic extraction in the digital IC design methodology. In Proceedings of the 1999 IEEE/ACM International Conference on Computer-Aided Design. Digest of Technical Papers (Cat. No.99CH37051), San Jose, CA, USA, 7–11 November 1999; pp. 223–230. [CrossRef]
2. Bezrukov, A.; Rusakov, A.; Tkachev, D.; Khapaev, M. Methods of the parasitic extraction of interconnect in the integral circuits. In *Problems of Perspective Microelectronic Systems Development*; Stempkovsky, A., Ed.; Institute for Design Problems in Microelectronics of Russian Academy of Sciences (IPPM RAS): Moscow, Russia, 2005; pp. 45–50. (In Russian)
3. Smythe, W. *Static and Dynamic Electricity*, 2nd ed.; McGraw-Hill: New York, NY, USA; Toronto, ON, Canada; London, UK, 1950.
4. Yu, W.; Song, M.; Yang, M. Advancements and Challenges on Parasitic Extraction for Advanced Process Technologies. In Proceedings of the 2021 26th Asia and South Pacific Design Automation Conference (ASP-DAC), Tokyo, Japan, 18–21 January 2021; pp. 841–846.
5. Nabors, K.; Kim, S.; White, J. Fast capacitance extraction of general three-dimensional structures. *IEEE Trans. Microw. Theory Tech.* **1992**, *40*, 1496–1506. [CrossRef]
6. Nabors, K.; White, J. Multipole-accelerated capacitance extraction algorithms for 3-D structures with multiple dielectrics. *IEEE Trans. Circuits Syst. Fundam. Theory Appl.* **1992**, *39*, 946–954. [CrossRef]
7. Tausch, J.; White, J. Capacitance extraction of 3-D conductor systems in dielectric media with high-permittivity ratios. *IEEE Trans. Microw. Theory Tech.* **1999**, *47*, 18–26. [CrossRef]
8. Iverson, R.B.; Le Coz, Y.L. A floating random-walk algorithm for extracting electrical capacitance. *Math. Comput. Simul.* **2001**, *55*, 59–66. [CrossRef]

9. Simonov, N.A.; Mascagni, M. Random walk algorithms for estimating electrostatic properties of large molecules. In *The International Conference on Computational Mathematics*; Mikhailov, G.A., Il'in, V.P., Laevsky, Y.M., Eds.; ICM&MG Publisher: Novosibirsk, Russia, 2004; pp. 352–358.
10. Kuznetsov, A.N.; Sipin, A.S. Universal Monte-Carlo algorithm for extracting electrical capacitancies. *Mat. Model.* **2009**, *21*, 41–52. (In Russian)
11. Kuznetsov, A.N. Calculation of mutual capacitances for system of conductors in dielectric media using "walk on hemispheres". *Mat. Model.* **2015**, *27*, 86–95. (In Russian)
12. Yu, W.; Zhuang, H.; Zhang, C.; Hu, G.; Liu, Z. RWCap: A Floating Random Walk Solver for 3-D Capacitance Extraction of Very-Large-Scale Integration Interconnects. *IEEE Trans.-Comput.-Aided Des. Integr. Circuits Syst.* **2013**, *32*, 353–366. [CrossRef]
13. Xu, Z.; Zhang, C.; Yu, W. Floating Random Walk-Based Capacitance Extraction for General Non-Manhattan Conductor Structures. *IEEE Trans.-Comput.-Aided Des. Integr. Circuits Syst.* **2017**, *36*, 120–133. [CrossRef]
14. Yang, M.; Yu, W. Floating Random Walk Capacitance Solver Tackling Conformal Dielectric With On-the-Fly Sampling on Eight-Octant Transition Cubes. *IEEE Trans.-Comput.-Aided Des. Integr. Circuits Syst.* **2020**, *39*, 4935–4943. [CrossRef]
15. Mascagni, M.; Simonov, N.A. Random walk on the boundary methods for computing reaction rate and capacitance. In *The International Conference on Computational Mathematics*; Mikhailov, G., Il'in, V., Laevsky, Y., Eds.; ICM&MG Publisher: Novosibirsk, Russia, 2002; pp. 238–242.
16. Bernal, F.; Acebrón, J.A.; Anjam, I. A Stochastic Algorithm Based on Fast Marching for Automatic Capacitance Extraction in Non-Manhattan Geometries. *SIAM J. Imaging Sci.* **2014**, *7*, 2657–2674. [CrossRef]
17. Phillips, J.R.; White, J.K. A precorrected-FFT method for electrostatic analysis of complicated 3-D structures. *IEEE Trans.-Comput.-Aided Des. Integr. Circuits Syst.* **1997**, *16*, 1059–1072. [CrossRef]
18. Computer Codes Produced and Supported by the RLE Computational Prototyping Group. Available online: https://www.rle.mit.edu/cpg/research_codes.htm (accessed on 3 January 2011).
19. Ermakov, S.M.; Nekrutkin, V.V.; Sipin, A.S. *Random Processes for Classical Equations of Mathematical Physics*; Kluwer Academic Publishers: Dordrecht, The Netherlands; Boston, MA, USA; London, UK, 1989. [CrossRef]
20. Ermakov, S.M.; Sipin, A.S. The "walk in hemispheres" process and its applications to solving boundary value problems. *Vestn. St. Petersburg Univ. Math.* **2009**, *42*, 155–163. [CrossRef]
21. Fast Field Solvers FastCap2. Available online: https://www.fastfieldsolvers.com/fastcap2.htm (accessed on 26 July 2021).

Article

Resource Retrial Queue with Two Orbits and Negative Customers

Ekaterina Lisovskaya, Ekaterina Fedorova, Radmir Salimzyanov and Svetlana Moiseeva *

Institute of Applied Mathematics and Computer Science, National Research Tomsk State University, 634050 Tomsk, Russia; ekaterina_lisovs@mail.ru (E.L.); moiskate@mail.ru (E.F.); radmir.salimzyanov@stud.tsu.ru (R.S.)
* Correspondence: smoiseeva@mail.ru; Tel.: +7-913-815-3262

Abstract: In this paper, a multi-server retrial queue with two orbits is considered. There are two arrival processes of positive customers (with two types of customers) and one process of negative customers. Every positive customer requires some amount of resource whose total capacity is limited in the system. The service time does not depend on the customer's resource requirement and is exponentially distributed with parameters depending on the customer's type. If there is not enough amount of resource for the arriving customer, the customer goes to one of the two orbits, according to his type. The duration of the customer delay in the orbit is exponentially distributed. A negative customer removes all the customers that are served during his arrival and leaves the system. The objects of the study are the number of customers in each orbit and the number of customers of each type being served in the stationary regime. The method of asymptotic analysis under the long delay of the customers in the orbits is applied for the study. Numerical analysis of the obtained results is performed to show the influence of the system parameters on its performance measures.

Keywords: retrial queue; negative customers; resource heterogeneous queue; asymptotic analysis

Citation: Lisovskaya, E.; Fedorova, E.; Salimzyanov, R.; Moiseeva, S. Resource Retrial Queue with Two Orbits and Negative Customers. *Mathematics* **2022**, *10*, 321. https://doi.org/10.3390/math10030321

Academic Editors: Alexander Zeifman, Victor Korolev and Alexander Sipin

Received: 25 November 2021
Accepted: 18 January 2022
Published: 20 January 2022

1. Introduction

The theory of queuing systems with repeated calls (Retrial Queue) is an important section of the modern teletraffic theory, the relevance of which is due to wide practical applications, such as the performance evaluation and design of broadcast, radio, and cellular networks, as well as local networks with the random multiple access protocols. In the monographs [1–3], the detailed survey of recent queuing models applications in telecommunication, modern computer networks, and information systems are presented.

The retry phenomenon is the integral feature of data transmission systems, and this phenomenon ignored in theoretical research can lead to significant errors in engineering decisions. Many multimedia and service applications on subscriber devices can automatically generate such requests, without any relative restrictions. Such unaccounted traffic consumes the channel resource in excess of the planned one. On the network sections, overflows begin to appear, that leads to the service rejection; thus, it generates more repeated calls again.

A large number of publications has been devoted to the study of retrial queues. The most extensive reviews of significant results, up to 2008, are presented in the monographs [4,5].

Queuing models with negative customers [6–8], or G-queues, are useful models for the analysis of multiprocessor computer systems, neural networks, communication systems, and manufacturing [9,10]. In its simplest version, a negative customer (negative arrival) has the effect of a positive (ordinary) customer(s) being deleted according to some strategy. For example:

- a negative arrival eliminates all the customers in the system or in its part, e.g., under the service or in the buffer (catastrophes);

- a negative arrival removes a customer from the system, e.g., from the server, from the head of the queue, or its end;
- a negative arrival breaks the service device, etc.

Negative arrivals are interpreted as viruses, orders or inhibitor signals, etc. The detailed overview of G-queues is presented in Reference [11]. Retrial queuing systems with negative arrivals have been considered, for example, in References [12–15]. The effects of negative customers considered in these papers are close to the effect of breakdowns. Retrial queues with breakdowns have been studied in References [16–19].

Other features of modern data transmission systems are random amounts of transmitted data and requests for additional resources. Often, in telecommunication systems, calls come from different sources and have different service time characteristics and different priorities, or they need more than one service device, etc. These features make the system analysis more complex. Identifying these aspects and analyzing their influence on the systems allow to optimize networks for loss reduction. Such mathematical models are called queuing systems with a random volume of the customers or resource queuing systems [20–28]. Resource queuing systems are applied in modern wireless communication networks, cloud computing systems, technical devices, or next-generation data transmission networks. It is known that, in classical queuing theory, the evaluation of almost all performance characteristics leads us to the analysis of a stochastic process of the number of customers in the system. However, it is insufficient if we would like to determine a buffer space capacity of a communication network's node which guarantees small losses of transmitted data [20,21,23]. Incoming customers can request some resources (for example, the amount of memory). The requests may be random or deterministic. In queuing models, the total amount of resource is usually limited by a constant value $R > 0$, which is called the buffer space capacity of the system. The buffer space is occupied by a customer at the arrival epoch and is entirely released at the service finish epoch. If value R is finite, it leads to additional losses of customers. The complexity of the study of resource systems is due to an universal approach does not exist. We use asymptotic methods [15,27,29], which give asymptotic expressions of the studied system characteristics that are acceptable for practical usage. A retrial queuing system with limited processor sharing (close to resource systems) is considered in Reference [30].

In the paper, the model under study is a non-classical retrial queuing system with non-homogeneous customers. The main feature of the research is the consideration of possible failures in the system. In the paper, we apply the theory of G-queues [6] for modeling breakdowns in real networks. A breakdown is represented by negative customers arriving at the queuing system and removing all served customers if any is present. So, we research the queuing system with all mentioned above features: repeated calls, negative customers, and resource. Such a model can be applied, for example, for 5G New Radio systems. The key feature and the main problem of the 5G New Radio network is that people themselves, cars, buildings, etc., are signal blockers, which is the cause of service interruptions. In this reasoning, the scientific community has the task of analyzing the performance of these systems and improving it in the future. Currently, there are known studies of "basic" mathematical models of such networks. However, due to introducing this technology into the daily lives of subscribers, we need to propose and analyze of the most appropriate mathematical models.

The considered mathematical model is described in Section 2. In Section 3, the method of asymptotic analysis is proposed and applied for the study. The numerical analysis is presented in Section 4. It includes a comparison of asymptotic and simulated distributions, numerical examples for various values of the model parameters, and calculation of some performance characteristics, such as the probability of the first time rejecting (or, in other words, it is a joining probability) and buffer space utilization. The problems and discussions about the applicability of the obtained approximations are presented in the conclusion.

2. Mathematical Model

Let us consider the multi-server retrial queue (Figure 1) with two positive and one negative arrival processes. We call customers arrived in the k-th positive arrival process customers of the k-th type ($k \in \{1,2\}$). All the processes are Poisson with parameters λ_1, λ_2 (for the first and the second type of positive customers), and α (for the negative ones). Positive customers need the service, and the service laws are exponential with parameters μ_1 and μ_2 for the first and the second type of customers, respectively. In addition, each positive customer requires a deterministic amount of resource (x_1 or x_2, respectively); therefore, a customer occupies some amount of resources during his service. The service time does not depend on customer's resource requirement. The service unit has a limited capacity of resources, which equals to R. The number of servers N is limited but large enough. If the customer cannot be served at the arrival moment (there is not enough amount of the resource), it goes to the corresponding orbit (some virtual place). The duration of the customers delay in the orbits is distributed exponentially with parameters σ_1 and σ_2, respectively. The negative customer deletes all customers being served in the service unit at his moment of arrival.

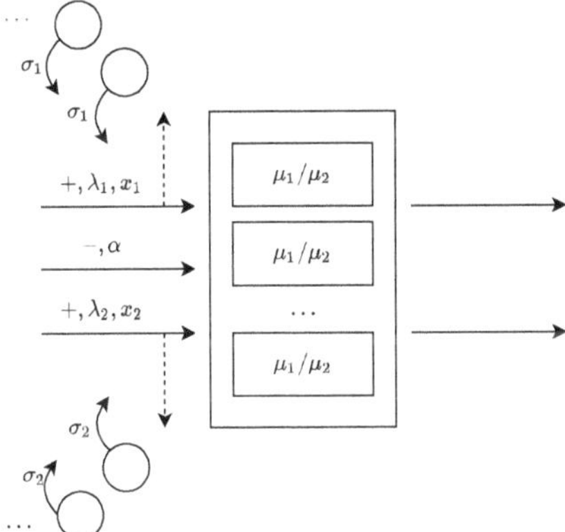

Figure 1. Resource retrial queue with two orbits and negative customers.

The goal of the paper is to study four-dimensional stochastic process

$$X(t) = (N_1(t), N_2(t), I_1(t), I_2(t)),$$

where $N_k(t)$ is the number of k-type customers in the service unit at the moment t, and $I_k(t)$ is the number of customers in the k-th orbit at the moment t. Then, the state space has the form:

$$\mathbb{X} = \{(n_1, n_2, i_1, i_2) : x_1 n_1 + x_2 n_2 \leq R, \quad i_k \geq 0, \quad k = 1,2\},$$

where n_k is a value of the process $N_k(t)$, and i_k is a value of the process $I_k(t)$, where $k = 1,2$.

Traditionally, continuous-time Markov chains can be represented as a transition graph. In Figure 2, we depict such a graph for the considered Markov chain $X(t)$. The ovals on the graph represent the states, and the arrows show the possible transitions and their intensities. In addition, next to each state, we show the condition under which it exists, knowing that the central state (any state) on the graph is $(n_1, n_2, i_1, i_2) \in \mathbb{X}$.

Let us consider in more detail the possible events that cause a change in the state of the considered Markov chain:

- the negative customer arrival with intensity α, so the number of customers from both types being serviced becomes equal to 0 (transitions $(k_1, k_2, i_1, i_2) \to (n_1, n_2, i_1, i_2)$ and $(n_1, n_2, i_1, i_2) \to (0, 0, i_1, i_2)$);
- the first type arrival with intensity λ_1:

  - if there is a sufficient resource amount for the customer, it gets serviced; therefore, the number of customers of the first type being serviced increases by 1 (transitions $(n_1, n_2, i_1, i_2) \to (n_1 + 1, n_2, i_1, i_2)$ and $(n_1 - 1, n_2, i_1, i_2) \to (n_1, n_2, i_1, i_2)$);
  - otherwise, the customer goes to the orbit and waits for the retrial; therefore, the number of customers in the first orbit increases by 1 (transitions $(n_1, n_2, i_1, i_2) \to (n_1, n_2, i_1 + 1, i_2)$ and $(n_1, n_2, i_1 - 1, i_2) \to (n_1, n_2, i_1, i_2)$);
- the end of first type customer service with an intensity $n_1\mu_1$; therefore, the number of customers of the first type being serviced is reduced by 1 (transitions $(n_1, n_2, i_1, i_2) \to (n_1 - 1, n_2, i_1, i_2)$ and $(n_1 + 1, n_2, i_1, i_2) \to (n_1, n_2, i_1, i_2)$);
- the successful retrial to service of the first type customer with an intensity $i_1\sigma_1$; therefore, the number of customers in the first orbit is reduced by 1, and the number of customers of the first type being serviced is reduced by 1 (transitions $(n_1, n_2, i_1, i_2) \to (n_1 + 1, n_2, i_1 - 1, i_2)$ and $(n_1 - 1, n_2, i_1 + 1, i_2) \to (n_1, n_2, i_1, i_2)$);
- similarly, we have transitions for the second type customers.

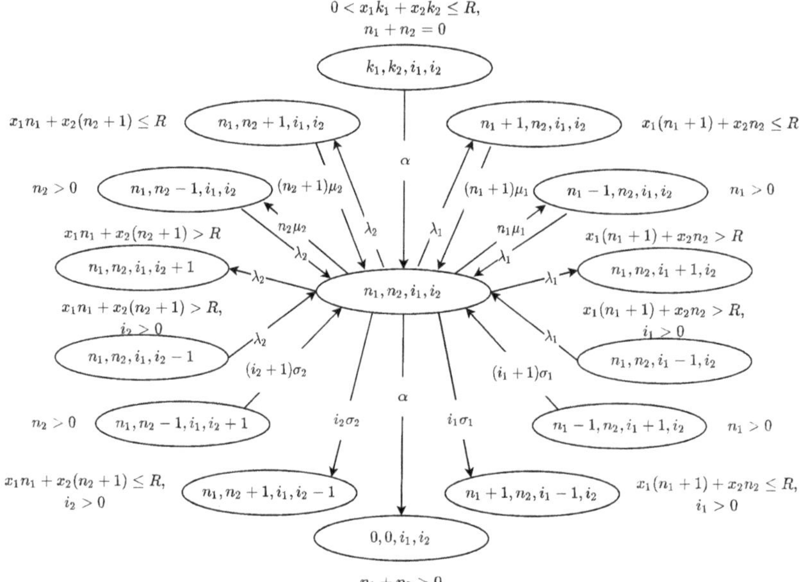

Figure 2. Graph of the input/output transitions of central state.

So, we will find the stationary probabilities

$$p(n_1, n_2, i_1, i_2) = Pr\{N_1(t) = n_1, N_2(t) = n_2, I_1(t) = i_1, I_2(t) = i_2\}.$$

Let us write the following system of equations for them:

$$\begin{aligned}
p(n_1,n_2,i_1,i_2)[\lambda_1+\lambda_2+n_1\mu_1+n_2\mu_2+i_1\sigma_1 I((n_1+1)x_1+n_2x_2\le R,i_1>0)+\\
i_2\sigma_2 I(n_1x_1+(n_2+1)x_2\le R,i_2>0)+\alpha I(n_1+n_2\ne 0)]=\\
\lambda_1 p(n_1-1,n_2,i_1,i_2)I(n_1>0)+\lambda_2 p(n_1,n_2-1,i_1,i_2)I(n_2>0)+\\
\lambda_1 p(n_1,n_2,i_1-1,i_2)I((n_1+1)x_1+n_2x_2>R,i_1>0)+\\
\lambda_2 p(n_1,n_2,i_1,i_2-1)I(n_1x_1+(n_2+1)x_2>R,i_2>0)+\\
(n_1+1)\mu_1 p(n_1+1,n_2,i_1,i_2)I((n_1+1)x_1+n_2x_2\le R)+\\
(n_2+1)\mu_2 p(n_1,n_2+1,i_1,i_2)I(n_1x_1+(n_2+1)x_2\le R)+\\
(i_1+1)\sigma_1 p(n_1-1,n_2,i_1+1,i_2)I(n_1>0)+(i_2+1)\sigma_2 p(n_1,n_2-1,i_1,i_2+1)I(n_2>0)\\
+\alpha\sum_{(k_1,k_2,i_1,i_2)\in\mathbb{X}} p(k_1,k_2,i_1,i_2)I(k_1+k_2\ne 0)I(n_1+n_2=0),
\end{aligned} \quad (1)$$

where

$$I(A)=\begin{cases}1, \text{if } A \text{ is true},\\ 0, \text{if } A \text{ is false}.\end{cases}$$

From System (1), we write several equations for different states.

(a) For $n_1x_1+n_2x_2=0$:

$$\begin{aligned}
p(0,0,i_1,i_2)[\lambda_1+\lambda_2+i_1\sigma_1+i_2\sigma_2]=\\
p(1,0,i_1,i_2)\mu_1+p(0,1,i_1,i_2)\mu_2+\alpha\sum_{(k_1,k_2,i_1,i_2)\in\mathbb{X}} p(k_1,k_2,i_1,i_2)I(k_1+k_2\ne 0).
\end{aligned}$$

(b) For $[(n_1+1)x_1+n_2x_2\le R]\cap[n_1x_1+(n_2+1)x_2\le R]$:

$$\begin{aligned}
p(n_1,n_2,i_1,i_2)[\lambda_1+\lambda_2+n_1\mu_1+n_2\mu_2+i_1\sigma_1+i_2\sigma_2+\alpha]=\\
p(n_1-1,n_2,i_1,i_2)\lambda_1 I(n_1>0)+p(n_1,n_2-1,i_1,i_2)\lambda_2 I(n_2>0)+\\
p(n_1+1,n_2,i_1,i_2)(n_1+1)\mu_1+p(n_1,n_2+1,i_1,i_2)(n_2+1)\mu_2+\\
p(n_1-1,n_2,i_1+1,i_2)(i_1+1)\sigma_1 I(n_1>0)+\\
p(n_1,n_2-1,i_1,i_2+1)(i_2+1)\sigma_2 I(n_2>0).
\end{aligned}$$

(c) For $[(n_1+1)x_1+n_2x_2>R]\cap[n_1x_1+(n_2+1)x_2>R]$:

$$\begin{aligned}
p(n_1,n_2,i_1,i_2)[\lambda_1+\lambda_2+n_1\mu_1+n_2\mu_2+\alpha]=\\
p(n_1-1,n_2,i_1,i_2)\lambda_1 I(n_1>0)+p(n_1,n_2-1,i_1,i_2)\lambda_2 I(n_2>0)+\\
p(n_1,n_2,i_1-1,i_2)\lambda_1 I(i_1>0)+p(n_1,n_2,i_1,i_2-1)\lambda_2 I(i_2>0)+\\
p(n_1-1,n_2,i_1+1,i_2)(i_1+1)\sigma_1 I(n_1>0)+\\
p(n_1,n_2-1,i_1,i_2+1)(i_2+1)\sigma_2 I(n_2>0).
\end{aligned}$$

(d) For $[(n_1+1)x_1+n_2x_2\le R]\cap[n_1x_1+(n_2+1)x_2>R]$:

$$\begin{aligned}
p(n_1,n_2,i_1,i_2)[\lambda_1+\lambda_2+n_1\mu_1+n_2\mu_2+i_1\sigma_1+\alpha]=\\
p(n_1,n_2-1,i_1,i_2)\lambda_2 I(n_2>0)+p(n_1,n_2,i_1,i_2-1)\lambda_2 I(i_2>0)+\\
p(n_1+1,n_2,i_1,i_2)(n_1+1)\mu_1+p(n_1-1,n_2,i_1+1,i_2)(i_1+1)\sigma_1 I(n_1>0)+\\
p(n_1,n_2-1,i_1,i_2+1)(i_2+1)\sigma_2 I(n_2>0).
\end{aligned}$$

(e) For $[(n_1+1)x_1+n_2x_2 > R] \cap [n_1x_1+(n_2+1)x_2 \leq R]$:

$$p(n_1,n_2,i_1,i_2)[\lambda_1+\lambda_2+n_1\mu_1+n_2\mu_2+i_1\sigma_2+\alpha] = p(n_1-1,n_2,i_1,i_2)\lambda_1 I(n_1>0)+ p(n_1,n_2-1,i_1,i_2)\lambda_2 I(n_2>0)+p(n_1,n_2,i_1-1,i_2)\lambda_1 I(i_1>0)+ p(n_1,n_2+1,i_1,i_2)(n_2+1)\mu_2+p(n_1-1,n_2,i_1+1,i_2)(i_1+1)\sigma_1 I(n_1>0)+ p(n_1,n_2-1,i_1,i_2+1)(i_2+1)\sigma_2 I(n_2>0).$$

Because of $0 \leq i_k < \infty$, Equations (a)–(e) have infinite dimension. So, for solving of difference equations System (a)–(e), we use the method of characteristic transforms. This method allows to find solutions of complex equations in queuing theory in more simple way. In the paper, we introduce the following partial characteristic functions:

$$H(n_1,n_2,u_1,u_2) = \sum_{i_1=0}^{\infty} e^{ju_1i_1} \sum_{i_2=0}^{\infty} e^{ju_2i_2} p(n_1,n_2,i_1,i_2), \tag{2}$$

where $j=\sqrt{-1}$.

Note that $h(u_1,u_2) = \sum_{n_1=0}^{N}\sum_{n_2=0}^{N} H(n_1,n_2,u_1,u_2) = \mathrm{E}\{e^{ju_1i_1+ju_2i_2}\}$ is a characteristic function of the two-dimensional process $(I_1(t), I_2(t))$ of the number of customers in the orbits.

Using Notation (2), Equations (a)–(e) are rewritten as follows:

(a) For $n_1x_1+n_2x_2=0$:

$$[\lambda_1+\lambda_2]H(0,0,u_1,u_2) - j\sigma_1\frac{\partial H(0,0,u_1,u_2)}{\partial u_1} - j\sigma_2\frac{\partial H(0,0,u_1,u_2)}{\partial u_2} = \mu_1 H(1,0,u_1,u_2)+\mu_2 H(0,1,u_1,u_2)+\alpha\sum_{(k_1,k_2,i_1,i_2)\in\mathbb{X}} H(k_1,k_2,u_1,u_2)I(k_1+k_2\neq 0).$$

(b) For $[(n_1+1)x_1+n_2x_2 \leq R] \cap [n_1x_1+(n_2+1)x_2 \leq R]$:

$$[\lambda_1+\lambda_2+n_1\mu_1+n_2\mu_2+\alpha]H(n_1,n_2,u_1,u_2) - j\sigma_1\frac{\partial H(n_1,n_2,u_1,u_2)}{\partial u_1} - j\sigma_2\frac{\partial H(n_1,n_2,u_1,u_2)}{\partial u_2} = \lambda_1 H(n_1-1,n_2,u_1,u_2)+\lambda_2 H(n_1,n_2-1,u_1,u_2)+ (n_1+1)\mu_1 H(n_1+1,n_2,u_1,u_2)+(n_2+1)\mu_2 H(n_1,n_2+1,u_1,u_2)- j\sigma_1 e^{-ju_1}\frac{\partial H(n_1-1,n_2,u_1,u_2)}{\partial u_1} - j\sigma_2 e^{-ju_2}\frac{\partial H(n_1,n_2-1,u_1,u_2)}{\partial u_2}.$$

(c) For $[(n_1+1)x_1+n_2x_2 > R] \cap [n_1x_1+(n_2+1)x_2 > R]$:

$$[\lambda_1+\lambda_2+n_1\mu_1+n_2\mu_2+\alpha]H(n_1,n_2,u_1,u_2) = \lambda_1 H(n_1-1,n_2,u_1,u_2)+\lambda_2 H(n_1,n_2-1,u_1,u_2)I(n_2>0)+ \lambda_1 e^{ju_1}H(n_1,n_2,u_1,u_2)+\lambda_2 e^{ju_2}H(n_1,n_2,u_1,u_2)- j\sigma_1 e^{-ju_1}\frac{\partial H(n_1-1,n_2,u_1,u_2)}{\partial u_1} - j\sigma_2 e^{-ju_2}\frac{\partial H(n_1,n_2-1,u_1,u_2)}{\partial u_2}.$$

(d) For $[(n_1+1)x_1+n_2x_2 \leq R] \cap [n_1x_1+(n_2+1)x_2 > R]$:

$$[\lambda_1+\lambda_2+n_1\mu_1+n_2\mu_2+\alpha]H(n_1,n_2,u_1,u_2) - j\sigma_1\frac{\partial H(n_1,n_2,u_1,u_2)}{\partial u_1} = \lambda_1 H(n_1-1,n_2,u_1,u_2)+\lambda_2 H(n_1,n_2-1,u_1,u_2)+ \lambda_2 e^{ju_2}H(n_1,n_2,u_1,u_2)+(n_1+1)\mu_1 H(n_1+1,n_2,u_1,u_2)- j\sigma_1 e^{-ju_1}\frac{\partial H(n_1-1,n_2,u_1,u_2)}{\partial u_1} - j\sigma_2 e^{-ju_2}\frac{\partial H(n_1,n_2-1,u_1,u_2)}{\partial u_2}.$$

(e) For $[(n_1+1)x_1+n_2x_2 > R] \cap [n_1x_1+(n_2+1)x_2 \leq R]$:

$$[\lambda_1+\lambda_2+n_1\mu_1+n_2\mu_2+\alpha]H(n_1,n_2,u_1,u_2) - j\sigma_2\frac{\partial H(n_1,n_2,u_1,u_2)}{\partial u_2} = \lambda_1 H(n_1-1,n_2,u_1,u_2)+\lambda_2 H(n_1,n_2-1,u_1,u_2)+ \lambda_1 e^{ju_1}H(n_1,n_2,u_1,u_2)+(n_2+1)\mu_2 H(n_1,n_2+1,u_1,u_2)- j\sigma_1 e^{-ju_1}\frac{\partial H(n_1-1,n_2,u_1,u_2)}{\partial u_1} - j\sigma_2 e^{-ju_2}\frac{\partial H(n_1,n_2-1,u_1,u_2)}{\partial u_2}.$$

Let us denote the matrix $\mathbf{H}(u_1,u_2) = \{H(n_1,n_2,u_1,u_2)\}_{n_1,n_2}$. So, the following equation can be written:

$$(\mathbf{A}+\lambda_1 e^{ju_1}\mathbf{B_1}+\lambda_2 e^{ju_2}\mathbf{B_2})\mathbf{H}(u_1,u_2)+ j\sigma_1(\mathbf{C_1}-e^{-ju_1}\mathbf{D_1})\frac{\partial \mathbf{H}(u_1,u_2)}{\partial u_1}+j\sigma_2(\mathbf{C_2}-e^{-ju_2}\mathbf{D_2})\frac{\partial \mathbf{H}(u_1,u_2)}{\partial u_2} = 0, \tag{3}$$

where $\mathbf{A}, \mathbf{B_1}, \mathbf{B_2}, \mathbf{C_1}, \mathbf{C_2}, \mathbf{D_1}, \mathbf{D_2}$ are the followings operators:

$$\mathbf{A}\mathbf{H}(u_1,u_2) = \begin{cases} -[\lambda_1+\lambda_2]H(n_1,n_2,u_1,u_2)+\mu_1 H(n_1+1,n_2,u_1,u_2)+ \\ \mu_2 H(n_1,n_2+1,u_1,u_2)+\alpha \sum\limits_{(k_1,k_2,i_1,i_2)\in\mathbb{X}} H(k_1,k_2,u_1,u_2)I(k_1+k_2\neq 0), \text{ (a)} \\ \\ -[\lambda_1+\lambda_2+n_1\mu_1+n_2\mu_2+\alpha]H(n_1,n_2,u_1,u_2)+ \\ \lambda_1 H(n_1-1,n_2,u_1,u_2)+\lambda_2 H(n_1,n_2-1,u_1,u_2)+ \\ (n_1+1)\mu_1 H(n_1+1,n_2,u_1,u_2)+(n_2+1)\mu_2 H(n_1,n_2+1,u_1,u_2), \text{ (b)} \\ \\ -[\lambda_1+\lambda_2+n_1\mu_1+n_2\mu_2+\alpha]H(n_1,n_2,u_1,u_2)+ \\ \lambda_1 H(n_1-1,n_2,u_1,u_2)+\lambda_2 H(n_1,n_2-1,u_1,u_2), \text{ (c)} \\ \\ -[\lambda_1+\lambda_2+n_1\mu_1+n_2\mu_2+\alpha]H(n_1,n_2,u_1,u_2)+ \\ \lambda_1 H(n_1-1,n_2,u_1,u_2)+\lambda_2 H(n_1,n_2-1,u_1,u_2)+ \\ (n_1+1)\mu_1 H(n_1+1,n_2,u_1,u_2), \text{ (d)} \\ \\ -[\lambda_1+\lambda_2+n_1\mu_1+n_2\mu_2+\alpha]H(n_1,n_2,u_1,u_2)+ \\ \lambda_1 H(n_1-1,n_2,u_1,u_2)+\lambda_2 H(n_1,n_2-1,u_1,u_2)+ \\ (n_2+1)\mu_2 H(n_1,n_2+1,u_1,u_2). \text{ (e)} \end{cases}$$

$$\mathbf{B_1}\mathbf{H}(u_1,u_2) = \begin{cases} 0, & \text{(a), (b), (d)} \\ H(n_1,n_2,u_1,u_2). & \text{(c), (e)} \end{cases}$$

$$\mathbf{B_2}\mathbf{H}(u_1,u_2) = \begin{cases} 0, & \text{(a), (b), (e)} \\ H(n_1,n_2,u_1,u_2). & \text{(c), (d)} \end{cases}$$

$$\mathbf{C_1H}(u_1,u_2)=\begin{cases}H(n_1,n_2,u_1,u_2), & \text{(a), (b), (d)}\\ 0. & \text{(c), (e)}\end{cases}$$

$$\mathbf{C_2H}(u_1,u_2)=\begin{cases}H(n_1,n_2,u_1,u_2), & \text{(a), (b), (e)}\\ 0. & \text{(c), (d)}\end{cases}$$

$$\mathbf{D_1H}(u_1,u_2)=\begin{cases}0, & \text{(a)}\\ H(n_1-1,n_2,u_1,u_2). & \text{(b),(c),(d),(e)}\end{cases}$$

$$\mathbf{D_2H}(u_1,u_2)=\begin{cases}0, & \text{(a)}\\ H(n_1,n_2-1,u_1,u_2). & \text{(b),(c),(d),(e)}\end{cases}$$

Operator **E** is the sum of equations for all values of n_1, n_2. Obviously,

$$\mathbf{E}(\mathbf{A}+\lambda_1\mathbf{B_1}+\lambda_2\mathbf{B_2})=0,$$

$$\mathbf{E}(\mathbf{C_1}-\mathbf{D_1})=0,\quad \mathbf{E}(\mathbf{C_2}-\mathbf{D_2})=0.$$

So, we have the following equation:

$$\mathbf{E}(\lambda_1(e^{ju_1}-1)\mathbf{B_1}+\lambda_2(e^{ju_2}-1)\mathbf{B_2})\mathbf{H}(u_1,u_2)+ \\ j\sigma_1(1-e^{-ju_1})\mathbf{ED_1}\frac{\partial\mathbf{H}(u_1,u_2)}{\partial u_1}+j\sigma_2(1-e^{-ju_2})\mathbf{ED_2}\frac{\partial\mathbf{H}(u_1,u_2)}{\partial u_2}=0. \tag{4}$$

3. Asymptotic Analysis Method

Since it is not possible to find an explicit form of the solution of equations system (3) and (4), we propose the asymptotic analysis method [15,27,29]. In the paper, we will use the asymptotic condition of the long delay (when $\sigma_1 \to 0$ and $\sigma_2 \to 0$). The practical mean of the long delay condition is that the service time is much less than the time of repeated call.

The algorithm of the proposed method includes the following steps.

1. Deriving of the asymptotic mean of the considered process:
   (a) introducing of an infinitesimal parameter ε and an asymptotic function notation (5);
   (b) rewriting of Equations (3) and (4) for the asymptotic notations;
   (c) deriving of a limit solution of the asymptotic equations for $\varepsilon \to 0$;
   (d) using the inverse substitutions, we obtain the form of the first-order asymptotic characteristic function (10), which gives the value of the asymptotic mean of the considered process.
2. Deriving of the asymptotic variance of the considered process:
   (a) using the result of the first-order asymptotic analysis 1.(d), we rewrite the characteristic function as (11);
   (b) rewriting of Equations (3) and (4) for this notations;
   (c) introducing of an infinitesimal parameter ε^2 and new asymptotic function notation (13);
   (d) rewriting of equations obtained in 2.(b) for the asymptotic notations;
   (e) approximating asymptotic functions by its 2th-degree Maclaurin series with respect to ε as (14);
   (f) deriving of a limit solution of the asymptotic equations for $\varepsilon \to 0$;
   (g) using the inverse substitutions, we obtain the form of the second-order asymptotic characteristic function, which gives the value of the asymptotic variance of the considered process.
3. Combining the results of 1.(d) and 2.(g), we obtain the final form of asymptotic characteristic function (21).

3.1. First-Order Asymptotics

First of all, we denote $\sigma_k = \gamma_k\sigma$, where $\sigma \to 0$ and $\gamma_k = const$. In the first-order asymptotic analysis method, we use the following notations:

$$\sigma = \varepsilon, \quad u_k = \varepsilon w_k, \quad \mathbf{H}(u_1, u_2) = \mathbf{F}(w_1, w_2, \varepsilon), \tag{5}$$

where ε is infinitesimal, and $\mathbf{F}(w_1, w_2, \varepsilon)$ is an asymptotic function.

Substituting Notations (5) into Equations (3) and (4), we obtain the following asymptotic equations:

$$\begin{cases} (\mathbf{A} + \lambda_1 e^{j\varepsilon w_1}\mathbf{B_1} + \lambda_2 e^{j\varepsilon w_2}\mathbf{B_2})\mathbf{F}(w_1, w_2, \varepsilon) + \\ j\gamma_1(\mathbf{C_1} - e^{-j\varepsilon w_1}\mathbf{D_1})\dfrac{\partial \mathbf{F}(w_1, w_2, \varepsilon)}{\partial w_1} + j\gamma_2(\mathbf{C_2} - e^{-j\varepsilon w_2}\mathbf{D_2})\dfrac{\partial \mathbf{F}(w_1, w_2, \varepsilon)}{\partial w_2} = 0, \\ \mathbf{E}(\lambda_1(e^{j\varepsilon w_1} - 1)\mathbf{B_1} + \lambda_2(e^{j\varepsilon w_2} - 1)\mathbf{B_2})\mathbf{F}(w_1, w_2, \varepsilon) + \\ j\gamma_1\mathbf{E}\mathbf{D_1}(1 - e^{-j\varepsilon w_1})\dfrac{\partial \mathbf{F}(w_1, w_2, \varepsilon)}{\partial w_1} + j\gamma_2\mathbf{E}\mathbf{D_2}(1 - e^{-j\varepsilon w_2})\dfrac{\partial \mathbf{F}(w_1, w_2, \varepsilon)}{\partial w_2} = 0. \end{cases} \tag{6}$$

Let us make limits $\lim\limits_{\varepsilon \to 0} \mathbf{F}(w_1, w_2, \varepsilon) = \mathbf{F}(w_1, w_2)$. Then, System (6) has the form:

$$\begin{cases} (\mathbf{A} + \lambda_1\mathbf{B_1} + \lambda_2\mathbf{B_2})\mathbf{F}(w_1, w_2) + j\gamma_1(\mathbf{C_1} - \mathbf{D_1})\dfrac{\partial \mathbf{F}(w_1, w_2)}{\partial w_1} + \\ j\gamma_2(\mathbf{C_2} - \mathbf{D_2})\dfrac{\partial \mathbf{F}(w_1, w_2)}{\partial w_2} = 0, \\ \mathbf{E}(\lambda_1 w_1\mathbf{B_1} + \lambda_2 w_2\mathbf{B_2})\mathbf{F}(w_1, w_2) + j\gamma_1 w_1\mathbf{E}\mathbf{D_1}\dfrac{\partial \mathbf{F}(w_1, w_2)}{\partial w_1} + \\ j\gamma_2 w_2\mathbf{E}\mathbf{D_2}\dfrac{\partial \mathbf{F}(w_1, w_2)}{\partial w_2} = 0. \end{cases} \tag{7}$$

Obviously, the solution of Equation (7) has the form

$$\mathbf{F}(w_1, w_2) = \mathbf{R}\exp\{jw_1\kappa_1 + jw_2\kappa_2\}, \tag{8}$$

where $\mathbf{R} = [R(n_1, n_2)]$ is the matrix of the stationary probabilities of the states of process $(N_1(t), N_2(t))$; κ_1, κ_2 are normalized means of the number of customers in the orbit, which are calculated from the following equations (obtained by substituting (8) in (7)):

$$\begin{cases} [(\mathbf{A} + \lambda_1\mathbf{B_1} + \lambda_2\mathbf{B_2}) - \kappa_1\gamma_1(\mathbf{C_1} - \mathbf{D_1}) - \kappa_2\gamma_2(\mathbf{C_2} - \mathbf{D_2})]\mathbf{R} = 0, \\ \mathbf{E}[\lambda_1\mathbf{B_1} - \kappa_1\gamma_1\mathbf{D_1}]\mathbf{R} = 0, \\ \mathbf{E}[\lambda_2\mathbf{B_2} - \kappa_2\gamma_2\mathbf{D_2}]\mathbf{R} = 0, \\ \mathbf{E}\mathbf{R} = 1. \end{cases} \tag{9}$$

After returning to Substitutions (5), we have obtained the first-order approximation of the characteristic function

$$\mathbf{H}(u_1, u_2) = \mathbf{F}(w_1, w_2, \varepsilon) \approx \mathbf{F}(w_1, w_2) = \mathbf{R}\exp\Big\{jw_1\frac{\kappa_1}{\sigma} + jw_2\frac{\kappa_2}{\sigma}\Big\}, \tag{10}$$

where $\dfrac{\kappa_1}{\sigma}$ and $\dfrac{\kappa_2}{\sigma}$ are normalized means of processes $I_1(t)$ and $I_2(t)$.

3.2. Second-Order Asymptotics

The first step of the second-order asymptotic analysis is rewriting the characteristic functions using the result of the first-order asymptotics (10) as follows:

$$\mathbf{H}(u_1, u_2) = \mathbf{H}^{(2)}(u_1, u_2) \cdot \exp\Big\{j\frac{u_1}{\sigma_1}\gamma_1\kappa_1 + j\frac{u_2}{\sigma_2}\gamma_2\kappa_2\Big\}, \tag{11}$$

where $\mathbf{H}^{(2)}(u_1,u_2)$ is matrix of characteristic functions of two-dimensional stochastic centered process $\left\{I_1(t)-\frac{\kappa_1}{\sigma}, I_2(t)-\frac{\kappa_2}{\sigma}\right\}$.

Substituting (11) into Equations (3) and (4), we have

$$\begin{cases} (\mathbf{A}+\lambda_1 e^{ju_1}\mathbf{B_1}+\lambda_2 e^{ju_2}\mathbf{B_2})\mathbf{H}^{(2)}(u_1,u_2)+ \\ j\sigma_1(\mathbf{C_1}-e^{-ju_1}\mathbf{D_1})\frac{\partial \mathbf{H}^{(2)}(u_1,u_2)}{\partial u_1}-\gamma_1\kappa_1(\mathbf{C_1}-e^{-ju_1}\mathbf{D_1})\mathbf{H}^{(2)}(u_1,u_2)+ \\ j\sigma_2(\mathbf{C_2}-e^{-ju_2}\mathbf{D_2})\frac{\partial \mathbf{H}^{(2)}(u_1,u_2)}{\partial u_2}-\gamma_2\kappa_2(\mathbf{C_2}-e^{-ju_2}\mathbf{D_2})\mathbf{H}^{(2)}(u_1,u_2)=0, \\ \mathbf{E}(\lambda_1(e^{ju_1}-1)\mathbf{B_1}+\lambda_2(e^{ju_2}-1)\mathbf{B_2})\mathbf{H}^{(2)}(u_1,u_2)+ \\ j\sigma_1(1-e^{-ju_1})\mathbf{ED_1}\frac{\partial \mathbf{H}^{(2)}(u_1,u_2)}{\partial u_1}-\gamma_1\kappa_1(1-e^{-ju_1})\mathbf{ED_1}\mathbf{H}^{(2)}(u_1,u_2)+ \\ j\sigma_2(1-e^{-ju_2})\mathbf{ED_2}\frac{\partial \mathbf{H}^{(2)}(u_1,u_2)}{\partial u_2}-\gamma_2\kappa_2(1-e^{-ju_2})\mathbf{ED_2}\mathbf{H}^{(2)}(u_1,u_2)=0. \end{cases} \tag{12}$$

The next step of the analysis is introducing the following notations (as in Section 3.1):

$$\sigma_k=\sigma\gamma_k, \quad \sigma=\varepsilon^2, \quad u_k=\varepsilon\cdot w_k, \quad \mathbf{H}^{(2)}(u_1,u_2)=\mathbf{F}^{(2)}(w_1,w_2,\varepsilon). \tag{13}$$

Substituting (13) into Equation (12), we obtain the following system of asymptotic equations:

$$\begin{cases} (\mathbf{A}+\lambda_1 e^{j\varepsilon w_1}\mathbf{B_1}+\lambda_2 e^{j\varepsilon w_2}\mathbf{B_2})\mathbf{F}^{(2)}(w_1,w_2,\varepsilon)+ \\ j\varepsilon\gamma_1(\mathbf{C_1}-e^{-j\varepsilon w_1}\mathbf{D_1})\frac{\partial \mathbf{F}^{(2)}(w_1,w_2,\varepsilon)}{\partial w_1}-\gamma_1\kappa_1(\mathbf{C_1}-e^{-j\varepsilon w_1}\mathbf{D_1})\mathbf{F}^{(2)}(w_1,w_2,\varepsilon)+ \\ j\varepsilon\gamma_2(\mathbf{C_2}-e^{-j\varepsilon w_2}\mathbf{D_2})\frac{\partial \mathbf{F}^{(2)}(w_1,w_2,\varepsilon)}{\partial w_2}-\gamma_2\kappa_2(\mathbf{C_2}-e^{-j\varepsilon w_2}\mathbf{D_2})\mathbf{F}^{(2)}(w_1,w_2,\varepsilon)=0, \\ \mathbf{E}(\lambda_1(e^{j\varepsilon w_1}-1)\mathbf{B_1}+\lambda_2(e^{j\varepsilon w_2}-1)\mathbf{B_2})\mathbf{F}^{(2)}(w_1,w_2,\varepsilon)+ \\ j\varepsilon\gamma_1(1-e^{-j\varepsilon w_1})\mathbf{ED_1}\frac{\partial \mathbf{F}^{(2)}(w_1,w_2,\varepsilon)}{\partial w_1}-\gamma_1\kappa_1(1-e^{-j\varepsilon w_1})\mathbf{ED_1}\mathbf{F}^{(2)}(w_1,w_2,\varepsilon)+ \\ j\varepsilon\gamma_2(1-e^{-j\varepsilon w_2})\mathbf{ED_2}\frac{\partial \mathbf{F}^{(2)}(w_1,w_2,\varepsilon)}{\partial w_2}-\gamma_2\kappa_2(1-e^{-j\varepsilon w_2})\mathbf{ED_2}\mathbf{F}^{(2)}(w_1,w_2,\varepsilon)=0. \end{cases} \tag{14}$$

The solution of System (14) will be found in the following multiplicative form:

$$\mathbf{F}^{(2)}(w_1,w_2,\varepsilon)=\Phi(w_1,w_2)\cdot(\mathbf{R}+j\varepsilon w_1\mathbf{f_1}+j\varepsilon w_2\mathbf{f_2})+O(\varepsilon^2), \tag{15}$$

where $\Phi(w_1,w_2)$ is an unknown scalar function, and $\mathbf{f_1}$, $\mathbf{f_2}$ are unknown operators, which will be found further.

Substituting solution (15) into System (14) and using Maclaurin series, we have

$$\begin{cases} (\mathbf{A}+\lambda_1(1+j\varepsilon w_1)\mathbf{B_1}+\lambda_2(1+j\varepsilon w_2)\mathbf{B_2})\Phi(w_1,w_2)(\mathbf{R}+j\varepsilon w_1\mathbf{f_1}+j\varepsilon w_2\mathbf{f_2})+ \\ j\varepsilon\gamma_1(\mathbf{C_1}-(1-j\varepsilon w_1)\mathbf{D_1})\left[\frac{\partial\Phi(w_1,w_2)}{\partial w_1}(\mathbf{R}+j\varepsilon w_1\mathbf{f_1}+j\varepsilon w_2\mathbf{f_2})+\Phi(w_1,w_2)j\varepsilon\mathbf{f_1}\right]- \\ \gamma_1\kappa_1(\mathbf{C_1}-((1-j\varepsilon w_1))\mathbf{D_1})\Phi(w_1,w_2)(\mathbf{R}+j\varepsilon w_1\mathbf{f_1}+j\varepsilon w_2\mathbf{f_2})+ \\ j\varepsilon\gamma_2(\mathbf{C_2}-(1-j\varepsilon w_2)\mathbf{D_2})\left[\frac{\partial\Phi(w_1,w_2)}{\partial w_2}(\mathbf{R}+j\varepsilon w_1\mathbf{f_1}+j\varepsilon w_2\mathbf{f_2})+\Phi(w_1,w_2)j\varepsilon\mathbf{f_2}\right]- \\ \gamma_2\kappa_2(\mathbf{C_2}-((1-j\varepsilon w_2))\mathbf{D_2})\Phi(w_1,w_2)(\mathbf{R}+j\varepsilon w_1\mathbf{f_1}+j\varepsilon w_2\mathbf{f_2})=0, \\ \mathbf{E}\left(\lambda_1\left(j\varepsilon w_1+\frac{(j\varepsilon w_1)^2}{2}\right)\mathbf{B_1}+\lambda_2\left(j\varepsilon w_2+\frac{(j\varepsilon w_2)^2}{2}\right)\mathbf{B_2}\right)\Phi(w_1,w_2)\times \\ (\mathbf{R}+j\varepsilon w_1\mathbf{f_1}+j\varepsilon w_2\mathbf{f_2})+j\varepsilon\gamma_1\left(j\varepsilon w_1-\frac{(j\varepsilon w_1)^2}{2}\right)\mathbf{ED_1}\times \\ \left[\frac{\partial\Phi(w_1,w_2)}{\partial w_1}(\mathbf{R}+j\varepsilon w_1\mathbf{f_1}+j\varepsilon w_2\mathbf{f_2})+\Phi(w_1,w_2)j\varepsilon\mathbf{f_1}\right]- \\ \gamma_1\kappa_1\left(j\varepsilon w_1-\frac{(j\varepsilon w_1)^2}{2}\right)\mathbf{ED_1}\Phi(w_1,w_2)\cdot(\mathbf{R}+j\varepsilon w_1\mathbf{f_1}+j\varepsilon w_2\mathbf{f_2})+ \\ j\varepsilon\gamma_2\left(j\varepsilon w_2-\frac{(j\varepsilon w_2)^2}{2}\right)\mathbf{ED_2}\left[\frac{\partial\Phi(w_1,w_2)}{\partial w_2}(\mathbf{R}+j\varepsilon w_1\mathbf{f_1}+j\varepsilon w_2\mathbf{f_2})+\Phi(w_1,w_2)j\varepsilon\mathbf{f_2}\right]- \\ \gamma_2\kappa_2\left(j\varepsilon w_2-\frac{(j\varepsilon w_2)^2}{2}\right)\mathbf{ED_2}\Phi(w_1,w_2)\cdot(\mathbf{R}+j\varepsilon w_1\mathbf{f_1}+j\varepsilon w_2\mathbf{f_2})=0. \end{cases}$$

Make some transformations and obtain $\varepsilon \to 0$.

$$\begin{cases} ((\lambda_1 w_1\mathbf{B_1}+\lambda_2 w_2\mathbf{B_2})\mathbf{R}+(\mathbf{A}+\lambda_1\mathbf{B_1}+\lambda_2\mathbf{B_2})(w_1\mathbf{f_1}+w_2\mathbf{f_2}))+ \\ \gamma_1\frac{\partial\Phi(w_1,w_2)/\partial w_1}{\Phi(w_1,w_2)}(\mathbf{C_1}-\mathbf{D_1})\mathbf{R}-\gamma_1\kappa_1(w_1\mathbf{D_1}\mathbf{R}+(\mathbf{C_1}-\mathbf{D_1})(w_1\mathbf{f_1}+w_2\mathbf{f_2}))+ \\ \gamma_2\frac{\partial\Phi(w_1,w_2)/\partial w_2}{\Phi(w_1,w_2)}(\mathbf{C_2}-\mathbf{D_2})\mathbf{R}-\gamma_2\kappa_2(w_2\mathbf{D_2}\mathbf{R}+(\mathbf{C_2}-\mathbf{D_2})(w_1\mathbf{f_1}+w_2\mathbf{f_2}))=0, \\ \mathbf{E}\left(\frac{(jw_1)^2}{2}\lambda_1\mathbf{B_1}\mathbf{R}+\frac{(jw_2)^2}{2}\lambda_2\mathbf{B_2}\mathbf{R}+w_1^2\lambda_1\mathbf{B_1}\mathbf{f_1}+\right. \\ \quad\left. w_1w_2\lambda_2\mathbf{B_2}\mathbf{f_1}+w_1w_2\lambda_1\mathbf{B_1}\mathbf{f_1}+w_2^2\lambda_2\mathbf{B_2}\mathbf{f_2}\right)+ \\ \gamma_1 w_1\frac{\partial\Phi(w_1,w_2)/\partial w_1}{\Phi(w_1,w_2)}\mathbf{ED_1}\mathbf{R}-\gamma_1\kappa_1\mathbf{ED_1}\left(-\frac{(jw_1)^2}{2}\mathbf{R}+w_1^2\mathbf{f_1}+w_1w_2\mathbf{f_2}\right)+ \\ \gamma_2 w_2\frac{\partial\Phi(w_1,w_2)/\partial w_2}{\Phi(w_1,w_2)}\mathbf{ED_2}\mathbf{R}-\gamma_2\kappa_2\mathbf{ED_2}\left(-\frac{(jw_2)^2}{2}\mathbf{R}+w_1w_2\mathbf{f_1}+w_2^2\mathbf{f_2}+\right)=0. \end{cases} \tag{16}$$

The solution of System (16) has the form:

$$\Phi(w_1,w_2)=\exp\left(\frac{(jw_1)^2}{2}K_{11}+jw_1jw_2K_{12}+\frac{(jw_2)^2}{2}K_{22}\right), \tag{17}$$

where K_{11} and K_{22} are normalized variances of the stochastic process $(I_1(t), I_2(t))$, and K_{12} is their normalized covariance. For the K_{11}, K_{12}, and K_{22} finding, we substitute (17) into System (16):

$$\begin{cases} ((\lambda_1 w_1\mathbf{B_1}+\lambda_2 w_2\mathbf{B_2})\mathbf{R}+(\mathbf{A}+\lambda_1\mathbf{B_1}+\lambda_2\mathbf{B_2})(w_1\mathbf{f_1}+w_2\mathbf{f_2}))- \\ \gamma_1(w_1K_{11}+w_2K_{12})(\mathbf{C_1}-\mathbf{D_1})\mathbf{R}-\gamma_1\kappa_1(w_1\mathbf{D_1}\mathbf{R}+(\mathbf{C_1}-\mathbf{D_1})(w_1\mathbf{f_1}+w_2\mathbf{f_2}))+ \\ \gamma_2(w_1K_{12}+w_2K_{22})(\mathbf{C_2}-\mathbf{D_2})\mathbf{R}-\gamma_2\kappa_2(w_2\mathbf{D_2}\mathbf{R}+(\mathbf{C_2}-\mathbf{D_2})(w_1\mathbf{f_1}+w_2\mathbf{f_2}))=0, \\ \mathbf{E}\left(\frac{w_1^2}{2}\lambda_1\mathbf{B_1}\mathbf{R}+\frac{w_2^2}{2}\lambda_2\mathbf{B_2}\mathbf{R}+w_1^2\lambda_1\mathbf{B_1}\mathbf{f_1}+w_1w_2\lambda_2\mathbf{B_2}\mathbf{f_1}+w_1w_2\lambda_1\mathbf{B_1}\mathbf{f_1}+w_2^2\lambda_2\mathbf{B_2}\mathbf{f_2}\right)+ \\ \gamma_1 w_1(w_1K_{11}+w_2K_{12})\mathbf{ED_1}\mathbf{R}-\gamma_1\kappa_1\mathbf{ED_1}(-\frac{w_1^2}{2}\mathbf{R}+w_1^2\mathbf{f_1}+w_1w_2\mathbf{f_2})+ \\ \gamma_2 w_2(w_1K_{12}+w_2K_{22})\mathbf{ED_2}\mathbf{R}-\gamma_2\kappa_2\mathbf{ED_1}(-\frac{w_2^2}{2}\mathbf{R}+w_1w_2\mathbf{f_1}+w_2^2\mathbf{f_2})=0. \end{cases} \tag{18}$$

For (18) solving, let us write the factors for different powers of w_1, w_2:

$$\begin{cases} (\lambda_1\mathbf{B_1} - \gamma_1\kappa_1\mathbf{D_1} - \gamma_1 K_{11}(\mathbf{C_1} - \mathbf{D_1}) - \gamma_2 K_{12}(\mathbf{C_2} - \mathbf{D_2}))\mathbf{R} + \\ (\mathbf{A} + \lambda_1\mathbf{B_1} + \lambda_2\mathbf{B_2} - \gamma_1\kappa_1(\mathbf{C_1} - \mathbf{D_1}) - \gamma_2\kappa_2(\mathbf{C_2} - \mathbf{D_2}))\mathbf{f_1} = 0, \\ (\lambda_2\mathbf{B_2} - \gamma_2\kappa_2\mathbf{D_2} - \gamma_1 K_{11}(\mathbf{C_2} - \mathbf{D_2}) - \gamma_2 K_{12}(\mathbf{C_1} - \mathbf{D_1}))\mathbf{R} + \\ (\mathbf{A} + \lambda_1\mathbf{B_1} + \lambda_2\mathbf{B_2} - \gamma_1\kappa_1(\mathbf{C_1} - \mathbf{D_1}) - \gamma_2\kappa_2(\mathbf{C_2} - \mathbf{D_2}))\mathbf{f_2} = 0, \\ \mathbf{E}[(\lambda_1\mathbf{B_1} + \gamma_1(\kappa_1 - 2K_{11})\mathbf{D_1})\mathbf{R} + 2(\lambda_1\mathbf{B_1} - \gamma_1\kappa_1\mathbf{D_1})\mathbf{f_1}] = 0, \\ \mathbf{E}[(\lambda_2\mathbf{B_2} + \gamma_2(\kappa_2 - 2K_{22})\mathbf{D_2})\mathbf{R} + 2(\lambda_2\mathbf{B_2} - \gamma_2\kappa_2\mathbf{D_2})\mathbf{f_2}] = 0, \\ \mathbf{E}[-K_{12}(\gamma_1\mathbf{D_1} + \gamma_2\mathbf{D_2})\mathbf{R} + (\lambda_2\mathbf{B_2} - \gamma_2\kappa_2\mathbf{D_2})\mathbf{f_1} + (\lambda_1\mathbf{B_1} - \gamma_1\kappa_1\mathbf{D_1})\mathbf{f_2}] = 0. \end{cases} \quad (19)$$

The system of the first and second equations in (19) is the heterogeneous system of linear algebraic equations with respect to $\mathbf{f_1}$ and $\mathbf{f_2}$. The determinant of the matrix of the system is equal to zero, while the rank of the extended matrix is equal to the rank of the matrix of coefficients, i.e., the system has many solutions. Comparing the first and second equations of systems (19) and the first equation of (9) (in the first-order asymptotics), we can write:

$$\mathbf{f_1} = C\mathbf{R} + \mathbf{g_1}, \quad \mathbf{f_2} = C\mathbf{R} = \mathbf{g_2},$$

where

$$\mathbf{E}\mathbf{g_1} = 0, \quad \mathbf{E}\mathbf{g_2} = 0.$$

So, we have the following system for $\mathbf{g_1}$ and $\mathbf{g_2}$ finding:

$$\begin{cases} (\lambda_1\mathbf{B_1} - \gamma_1\kappa_1\mathbf{D_1} - \gamma_1 K_{11}(\mathbf{C_1} - \mathbf{D_1}) - \gamma_2 K_{12}(\mathbf{C_2} - \mathbf{D_2}))\mathbf{R} + \\ (\mathbf{A} + \lambda_1\mathbf{B_1} + \lambda_2\mathbf{B_2} - \gamma_1\kappa_1(\mathbf{C_1} - \mathbf{D_1}) - \gamma_2\kappa_2(\mathbf{C_2} - \mathbf{D_2}))\mathbf{g_1} = 0, \\ (\lambda_2\mathbf{B_2} - \gamma_2\kappa_2\mathbf{D_2} - \gamma_1 K_{12}(\mathbf{C_1} - \mathbf{D_1}) - \gamma_2 K_{22}(\mathbf{C_2} - \mathbf{D_2}))\mathbf{R} + \\ (\mathbf{A} + \lambda_1\mathbf{B_1} + \lambda_2\mathbf{B_2} - \gamma_1\kappa_1(\mathbf{C_1} - \mathbf{D_1}) - \gamma_2\kappa_2(\mathbf{C_2} - \mathbf{D_2}))\mathbf{g_2} = 0, \\ \mathbf{E}[(\lambda_1\mathbf{B_1} + \gamma_1(\kappa_1 - 2K_{11})\mathbf{D_1})\mathbf{R} + 2(\lambda_1\mathbf{B_1} - \gamma_1\kappa_1\mathbf{D_1})\mathbf{g_1}] = 0, \\ \mathbf{E}[(\lambda_2\mathbf{B_2} + \gamma_2(\kappa_2 - 2K_{22})\mathbf{D_2})\mathbf{R} + 2(\lambda_2\mathbf{B_2} - \gamma_2\kappa_2\mathbf{D_2})\mathbf{g_2}] = 0, \\ \mathbf{E}[-K_{12}(\gamma_1\mathbf{D_1} + \gamma_2\mathbf{D_2})\mathbf{R} + (\lambda_2\mathbf{B_2} - \gamma_2\kappa_2\mathbf{D_2})\mathbf{g_1} + (\lambda_1\mathbf{B_1} - \gamma_1\kappa_1\mathbf{D_1})\mathbf{g_2}] = 0. \end{cases} \quad (20)$$

After returning to Substitutions (13), we obtain that the second-order approximation of the characteristic function has the following form:

$$\mathbf{H}(u_1, u_2) = \mathbf{H}^{(2)}(u_1, u_2) \cdot \exp\left\{ju_1\frac{\kappa_1}{\sigma} + ju_2\frac{\kappa_2}{\sigma}\right\} \approx \mathbf{F}^{(2)}\left(\frac{u_1}{\sigma}, \frac{u_2}{\sigma}\right) \cdot \exp\left\{ju_1\frac{\kappa_1}{\sigma} + ju_2\frac{\kappa_2}{\sigma}\right\}.$$

So, the asymptotic characteristic function of the considered multi-dimensional process has the following matrix form:

$$\mathbf{H}(u_1, u_2) = \mathbf{R} \cdot \exp\left\{ju_1\frac{\kappa_1}{\sigma} + ju_2\frac{\kappa_2}{\sigma} + \frac{(ju_1)^2}{2}\frac{K_{11}}{\sigma} + ju_1ju_2\frac{K_{12}}{\sigma} + \frac{(ju_2)^2}{2}\frac{K_{22}}{\sigma}\right\}. \quad (21)$$

Therefore, two-dimensional process $(I_1(t), I_2(t))$ of number of customers in the orbits is asymptotically Gaussian with means' vector $\left\{\frac{\kappa_1}{\sigma}, \frac{\kappa_1}{\sigma}\right\}$ and covariance matrix

$$\mathbf{cov} = \begin{pmatrix} \frac{K_{11}}{\sigma} & \frac{K_{12}}{\sigma} \\ \frac{K_{12}}{\sigma} & \frac{K_{22}}{\sigma} \end{pmatrix}.$$

4. Numerical Examples

In this section, we present the comparison of asymptotic and simulated distributions for some values of the model parameters. In addition, some performance characteristics of the model are evaluated.

For the numerical examples presentation, we assume the following values of the retrial queue parameters:

$$\lambda_1 = 1, \mu_1 = 1, x_1 = 1,$$

$$\lambda_2 = 0.3, \mu_2 = 1, x_2 = 3,$$

$$\alpha = 0.1, R = 5, \sigma_1 = 2 \cdot \sigma, \sigma_2 = 1 \cdot \sigma, \sigma = 0.01.$$

We have software applications for the considering model simulation and computing the asymptotic results.

We use the discrete-event simulation. The state of the system is changed by events, such as:

- the arrival of the first customer,
- the arrival of the second customer,
- the arrival of the negative customer,
- the service end of the first customer,
- the service end of the second customer,
- the customer attempt to access the service unit from the first orbit, and
- the customer attempt to access the service unit from the second orbit.

The condition for stopping the simulation is the service end at least 10^7 customers of each arrival process.

Note that the structural parameters of the system, such as $R > x_1, x_2$, can take any values, and, with growth R regarding x_1, x_2, the number of system states increases, which entails an increase in the time spent on analytical calculating the probability distribution. The proposed parameter values allow for clearly demonstrating the dimension of the system and the influence of the system load parameters, such as $\alpha, \lambda_k, \mu_k, k = 1, 2$, on the main performance characteristics.

First, we compare asymptotic and simulated distributions for various values of σ (the infinitesimal variable of the asymptotic analysis). In Figure 3, the two-dimensional asymptotic probability distribution of the number of customers in orbits is presented for $\sigma = 0.01$.

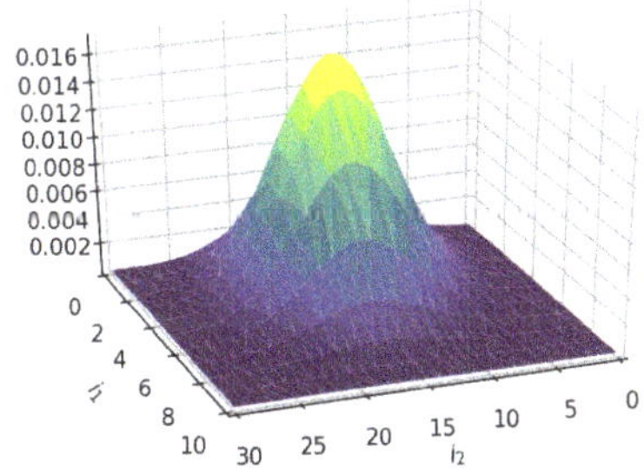

Figure 3. The two-dimensional asymptotic probability distribution of the number of customers in the orbits.

In Figures 4–6, you may find the comparison of the asymptotic and simulated one dimensional distributions for various values of σ. Curves 1 and 2 are the probability distributions of the number of customers in the first and the second orbits, respectively.

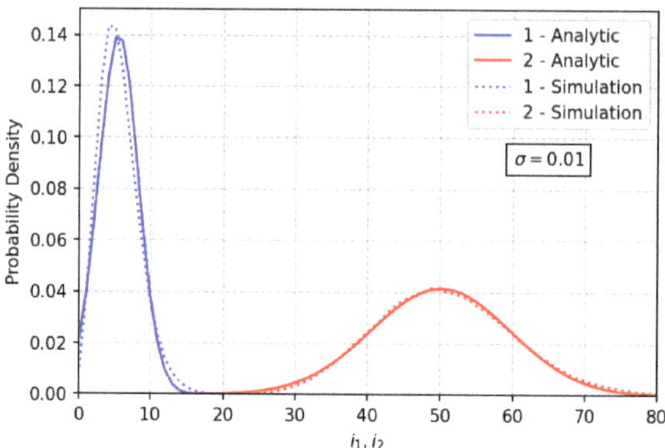

Figure 4. Comparison of the asymptotic and simulated distributions for $\sigma = 0.01$.

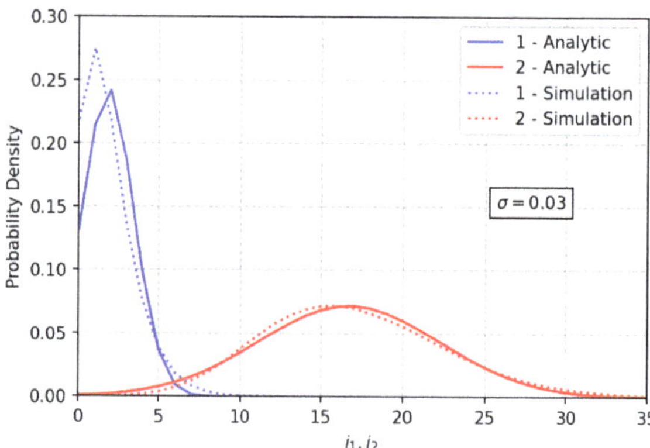

Figure 5. Comparison of the asymptotic and simulated distributions for $\sigma = 0.03$.

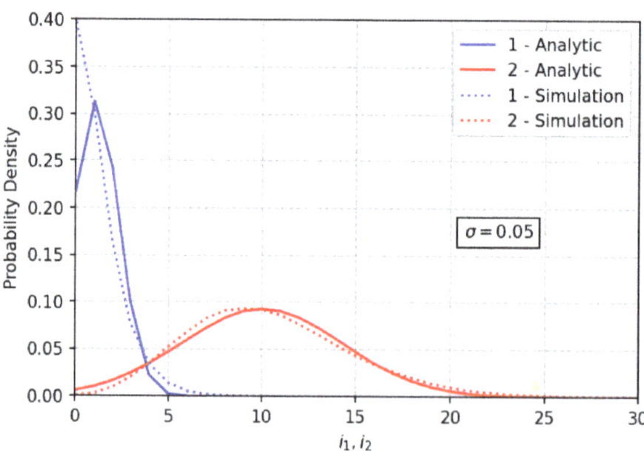

Figure 6. Comparison of the asymptotic and simulated distributions for $\sigma = 0.05$.

The mean and the variance of considered processes are calculated (Figures 7 and 8).

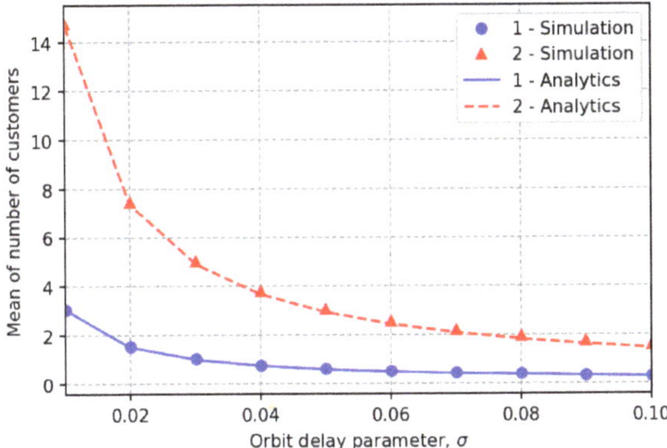

Figure 7. Asymptotic and simulate means.

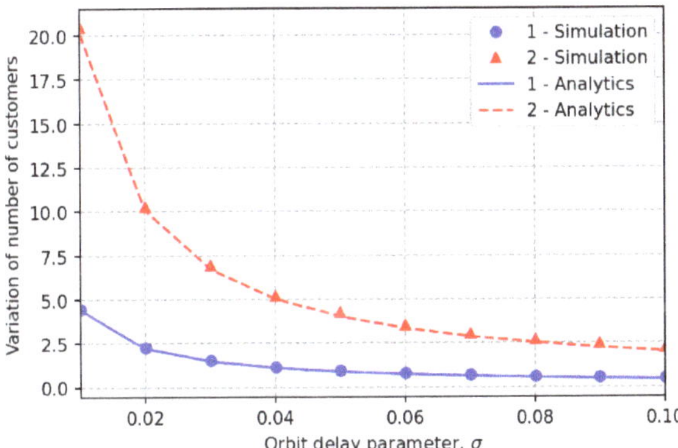

Figure 8. Asymptotic and simulate variances.

In addition, the relative errors of the asymptotic means and variances (in comparison with the simulated results) are computed (Tables 1 and 2).

Table 1. Table of relative errors of means.

σ	**0.01**	**0.02**	**0.03**	**0.04**	**0.05**	**0.06**	**0.07**	**0.08**	**0.09**	**0.10**
δ_1	0.007	0.024	0.028	0.036	0.052	0.056	0.070	0.076	0.084	0.092
δ_2	0.006	0.016	0.024	0.033	0.038	0.051	0.050	0.063	0.074	0.078

Table 2. Table of relative errors of variances.

σ	**0.01**	**0.02**	**0.03**	**0.04**	**0.05**	**0.06**	**0.07**	**0.08**	**0.09**	**0.10**
δ_1	0.013	0.031	0.036	0.042	0.063	0.069	0.084	0.093	0.103	0.107
δ_2	0.018	0.020	0.036	0.031	0.054	0.057	0.057	0.078	0.084	0.096

By analyzing Tables 1 and 2, we conclude that the asymptotic formula can be used for $\sigma < 0.05$.

Another measure of distributions comparison is Kolmogorov distance

$$\delta = \max_{i \geq 0} \left| \sum_{l=0}^{i} [\tilde{p}(l) - p(l)] \right|,$$

where $p(l)$ is the probability distribution of the processes $I_1(t)$ or $I_2(t)$ calculated using the asymptotic formula, and $\tilde{p}(l)$ is the corresponding empiric distribution based on the simulation. Values of the Kolmogorov distances for our example are presented in Table 3.

Table 3. The Kolmogorov distances between asymptotic and simulated distributions.

σ	0.01	0.02	0.03	0.04	0.05	0.06	0.07	0.08	0.09	0.10
δ_1	0.099	0.165	0.190	0.229	0.249	0.253	0.251	0.241	0.231	0.220
δ_2	0.014	0.030	0.045	0.062	0.076	0.090	0.098	0.110	0.120	0.128

Another direction of the numerical analysis is the computations of the model performance characteristics. The main practical feature is the joining probability P_k (the probability that the arrival customer goes to an orbit):

$$P_k = \sum_{(n_1,n_2)\in\mathbb{B}_k} R(n_1,n_2),$$

where

$$\mathbb{B}_1 = \{(n_1,n_2) : b_1(n_1+1) + b_2 n_2 > R\}, \mathbb{B}_2 = \{(n_1,n_2) : b_1 n_1 + b_2(n_2+1) > R\}.$$

In the following tables, the dependence P_k on the system parameters values is presented.

From Figures 9 and 10, we see that parameter α does not have much impact on joining probability P, and we need to take into account only the system load (λ/μ) for choosing the amount of resource R. Tables 4–7 can be used in practical goals.

Table 4. Values of joining probability P_1.

R/α	0.1	0.05	0.01
4	0.273	0.295	0.315
5	0.098	0.107	0.115
6	0.069	0.074	0.079
7	0.038	0.041	0.044
8	0.015	0.016	0.018

Table 5. Values of joining probability P_2.

R/α	0.1	0.05	0.01
4	0.643	0.676	0.704
5	0.500	0.525	0.547
6	0.284	0.303	0.320
7	0.151	0.162	0.172
8	0.097	0.104	0.111

Table 6. Values of joining probability P_1.

R/λ_2	0.1	0.05	0.01
4	0.059	0.159	0.273
5	0.020	0.057	0.098
6	0.007	0.031	0.069
7	0.002	0.016	0.038
8	0.0009	0.006	0.015

Table 7. Values of joining probability P_2.

R/λ_2	0.1	0.05	0.01
4	0.319	0.484	0.643
5	0.155	0.328	0.500
6	0.069	0.176	0.284
7	0.026	0.083	0.151
8	0.010	0.045	0.097

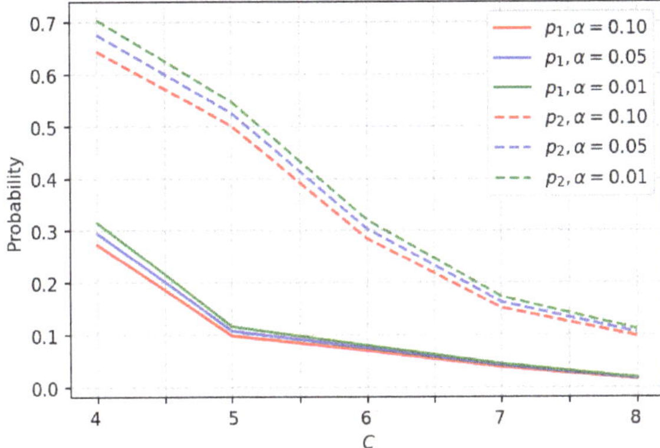

Figure 9. Dependence of P on α.

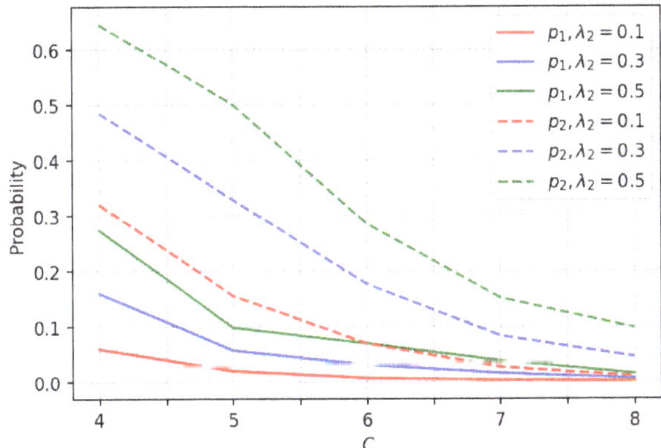

Figure 10. Dependence of P on λ_2.

5. Conclusions

In this paper, the resource multi-server retrial queue with different rates of the service laws for different types of arrival customers and with breakdowns of servers modeling by the negative arrival process is analyzed. The capacity of the system resources is limited. The stationary distribution of the number of customers in the orbits and in the service unit is obtained under the asymptotic condition of the long delay in orbits. The comparison of the asymptotic results with simulation ones shows the high accuracy of our approach. Finally, we provide numerical examples to show the impact of the model parameters on some performance characteristics. Such analysis for the queuing system with a limited total amount of resources can be used in the process of communication network node design.

In the future, we plan to study multi-server resource retrial queuing systems with non-Poisson arrival processes and random values of the required resource. In addition, study of queuing systems with the service time depending on the required resource seems interesting.

Author Contributions: Conceptualization, E.L., E.F. and S.M.; methodology, E.L., R.S., E.F. and S.M.; software, E.L., R.S.; writing, original draft preparation, E.F.; writing, review and editing, E.L., R.S., E.F. and S.M.; supervision S.M. All authors have read and agreed to the published version of the manuscript.

Funding: This research received no external funding.

Institutional Review Board Statement: Not applicable.

Informed Consent Statement: Not applicable.

Data Availability Statement: Not applicable.

Conflicts of Interest: The authors declare no conflict of interest.

References

1. Dudin, A.N.; Klimenok, V.I.; Vishnevsky, V.M. *The Theory of Queuing Systems with Correlated Flows*; Springer: Berlin/Heidelberg, Germany, 2020; 410p.
2. Lakatos, L.; Szeidl, L.; Miklos, T. *Introduction to Queueing Systems with Telecommunication Applications*; Springer: Boston, MA, USA, 2013; 388p.
3. Bocharov, P.; D'Apice, C.; Pechinkin, A.; Salerno, S. *Queueing Theory*; VSP: Utrecht, The Netherlands; Boston, MA, USA, 2004; 446p.
4. Artalejo, J.R.; Gomez-Corral, A. *Retrial Queueing Systems: A Computational Approach*; Springer: Berlin/Heidelberg, Germany, 2008; 318p.
5. Falin, G.I.; Templeton, J.G.C. *Retrial Queues*; Chapman and Hall: London, UK, 1997; 328p.
6. Gelenbe, E. Product form networks with negative and positive customers. *J. Appl. Prob.* **1991**, *28*, 656–663. [CrossRef]
7. Gelenbe, E. Harrison, P.; Pitel, E. The M/G/1 queue with negative customers. *Adv. Appl. Probab.* **1996**, *28*, 540–566.
8. Bocharov, P.P.; Vishnevskii, V.M. G-Networks: Development of the theory of multiplicative networks. *Automat. Rem. Control* **2003**, *64*, 46–74.
9. Chao, X.; Xiu, L.; Masakiyo, M.; Michael, P. *Queueing Networks: Customers, Signals and Product Form Solutions*; Wiley: Hoboken, NJ, USA, 1999; 464p.
10. Matalytsky, M.; Zajac, P. Application of HM-network with positive and negative claims for finding of memory volumes in information systems. *Appl. Math. Comput. Mech.* **2019**, *18*, 41–51. [CrossRef]
11. Do, T.V. Bibliography on G-networks, negative customers and applications. *Math. Comput. Model.* **2011**, *53*, 205–212. [CrossRef]
12. Anisimov, V.; Artalejo, J.R. Analysis of Markov multiserver retrial queues with negative arrivals. *Queueing Syst.* **2001**, *39*, 157–182. [CrossRef]
13. Shin, Y.W. Multi-server retrial queue with negative customers and disasters. *Queueing Syst.* **2007**, *55*, 223–237. [CrossRef]
14. Kirupa, K.; Chandrika, K. Unreliable Batch Arrival Retrial Queue with Negative Arrivals, Multi-Types Of Heterogeneous Service, Feedback and Randomized J Vacations. *Int. J. Comput. Appl.* **2015**, *127*, 38–42.
15. Fedorova, E.A.; Nazarov, A.A.; Farkhadov, M.P. Asymptotic Analysis of the $MMPP/M/1$ Retrial Queue with Negative Calls under the Heavy Load Condition. *Izv. Saratov Univ. Math. Mech. Inform.* **2020**, *20*, 534–547. [CrossRef]
16. Dudin, A.; Dudina, O.; Dudin, S.; Samouylov, K. Analysis of Single-Server Multi-Class Queue with Unreliable Service, Batch Correlated Arrivals, Customers Impatience, and Dynamical Change of Priorities. *Mathematics* **2021**, *9*, 1257. [CrossRef]
17. Atencia, I.; Galán-García, J.L. Sojourn Times in a Queueing System with Breakdowns and General Retrial Times. *Mathematics* **2021**, *9*, 2882. [CrossRef]
18. Melikov, A.; Aliyeva, S.; Sztrik, J. Retrial Queues with Unreliable Servers and Delayed Feedback. *Mathematics* **2021**, *9*, 2415. [CrossRef]
19. Nazarov, A.; Phung-Duc, T.; Paul, S.; Lizyura, O.; Shulgina, K. Central Limit Theorem for an M/M/1/1 Retrial Queue with Unreliable Server and Two-Way Communication. In *Information Technologies and Mathematical Modelling. Queueing Theory and Applications. ITMM 2020. Communications in Computer and Information Science*; Springer: Berlin/Heidelberg, Germany, 2021; Volume 1391, pp. 120–130.
20. Tikhonenko, O.; Ziółkowski, M. Queueing systems with non–homogeneous customers and infinite sectorized memory space. *Commun. Comput. Inf. Sci.* **2019**, *1039*, 316–329.
21. Tikhonenko, O.; Ziółkowski, M. Queueing Systems with Random Volume Customers and their Performance Characteristics. *JIOS* **2021**, *45*, 21–38. [CrossRef]
22. Naumov, V.; Samouylov, K. Resource System with Losses in a Random Environment. *Mathematics* **2021**, *9*, 2685. [CrossRef]
23. Naumov, V.; Samuilov, K. Analysis of networks of the resource queuing systems. *Autom. Remote.* **2018**, *79*, 822–829. [CrossRef]
24. Efrosinin, D. Stepanova, N. Sztrik, J. Algorithmic Analysis of Finite-Source Multi-Server Heterogeneous Queueing Systems. *Mathematics* **2021**, *9*, 2624. [CrossRef]

25. Lisovskaya, E.; Pankratova, E.; Moiseeva, S.; Pagano, M. Analysis of a Resource-Based Queue with the Parallel Service and Renewal Arrivals, Lecture Notes in Computer Science. *Distrib. Comput. Commun. Netw.* **2020**, *12563*, 335–349.
26. Bushkova, T.; Moiseeva, S.; Moiseev, A.; Sztrik, J.; Lisovskaya, E.; Pankratova, E. Using Infinite-server Resource Queue with Splitting of Requests for Modeling Two-channel Data Transmission. *Methodol. Comput. Appl. Probab.* **2021**, 1–20. [CrossRef]
27. Bushkova, T.; Galileyskaya, A.; Lisovskaya, E.; Pankratova, E.; Moiseeva, S. Multi-service resource queue with the multy-component Poisson arrivals. *Glob. Stoch. Anal.* **2021**, *8*, 97–109.
28. Moskaleva, F.; Lisovskaya, E.; Gaidamaka, Y. Resource Queueing System for Analysis of Network Slicing Performance with QoS-Based Isolation. In *Information Technologies and Mathematical Modelling. Queueing Theory and Applications. ITMM 2020. Communications in Computer and Information Science*; Springer: Berlin/Heidelberg, Germany, 2021; Volume 1391, pp. 198–211.
29. Nazarov, A.; Moiseev, A.; Phung-Duc, T.; Paul, S. Diffusion Limit of Multi-Server Retrial Queue with Setup Time. *Mathematics* **2020**, *8*, 2232. [CrossRef]
30. Dudin, S.; Dudin, A.; Dudina, O.; Samouylov, K. Analysis of a Retrial Queue with Limited Processor Sharing Operating in the Random Environment. In *Wired/Wireless Internet Communications. WWIC 2017. Lecture Notes in Computer Science*; Springer: Berlin/Heidelberg, Germany, 2017; Volume 10372, pp. 38–49.

Article

Bounds on the Rate of Convergence for $M_t^X/M_t^X/1$ Queueing Models

Alexander Zeifman [1,2,3,4,*], **Yacov Satin** [1] and **Alexander Sipin** [1]

1 Department of Applied Mathematics, Vologda State University, Lenina 15, 160000 Vologda, Russia; yacovi@mail.ru (Y.S.); cac1909@mail.ru (A.S.)
2 Institute of Informatics Problems, Federal Research Center "Computer Science and Control" of the Russian Academy of Sciences, Vavilova 44-2, 119333 Moscow, Russia
3 Moscow Center for Fundamental and Applied Mathematics, Moscow State University, Leninskie Gory 1, 119991 Moscow, Russia
4 Vologda Research Center of the Russian Academy of Sciences, 160014 Vologda, Russia
* Correspondence: a_zeifman@mail.ru

Abstract: We apply the method of differential inequalities for the computation of upper bounds for the rate of convergence to the limiting regime for one specific class of (in)homogeneous continuous-time Markov chains. Such an approach seems very general; the corresponding description and bounds were considered earlier for finite Markov chains with analytical in time intensity functions. Now we generalize this method to locally integrable intensity functions. Special attention is paid to the situation of a countable Markov chain. To obtain these estimates, we investigate the corresponding forward system of Kolmogorov differential equations as a differential equation in the space of sequences l_1.

Keywords: inhomogeneous continuous-time Markov chain; weak ergodicity; rate of convergence; sharp bounds; differential inequalities; forward Kolmogorov system

Citation: Zeifman, A.; Satin, Y.; Sipin, A. Bounds on the Rate of Convergence for $M_t^X/M_t^X/1$ Queueing Models. *Mathematics* **2021**, 9, 1752. https://doi.org/10.3390/math9151752

Academic Editor: Tuan Phung-Duc

Received: 6 July 2021
Accepted: 23 July 2021
Published: 25 July 2021

1. Introduction

In this paper we consider the problem of finding the upper bounds for the rate of convergence for some (in)homogeneous continuous-time Markov chains.

To obtain these estimates, we investigate the corresponding forward system of Kolmogorov differential equations.

Consideration is given to classic inhomogeneous birth–death processes and to special inhomogeneous chains with transitions intensities, which do not depend on the current state. Namely, let $\{X(t),\ t \geq 0\}$ be an inhomogeneous continuous-time Markov chain with the state space $\mathcal{X} = \{0, 1, 2, \dots,\}$. Denote by $p_{ij}(s,t) = P\{X(t) = j | X(s) = i\}, i, j \geq 0, 0 \leq s \leq t$, the transition probabilities of $X(t)$ and by $p_i(t) = P\{X(t) = i\}$—the probability that $X(t)$ is in state i at time t. Let $\mathbf{p}(t) = (p_0(t), p_1(t), \dots,)^T$ be probability distribution vector at instant t. Throughout the paper it is assumed that in a small time interval h the possible transitions and their associated probabilities are

$$p_{ij}(t,t+h) = \begin{cases} q_{ij}(t)h + \alpha_{ij}(t,h), & \text{if } j \neq i, \\ 1 - \sum\limits_{k \in \mathcal{X}, k \neq i} q_{ik}(t)h + \alpha_i(t,h), & \text{if } j = i, \end{cases}$$

where $\sup_{i \geq 0} \sum_{j \geq 0} |\alpha_{ij}(t,h)| = o(h)$, for any $t \geq 0$. We also suppose that the transition intensities $q_{ij}(t) \geq 0$ are arbitrary non-random functions of t, locally integrable on $[0, \infty)$ and, moreover, that there exists a positive number L such that

$$\sup_{i \in \mathcal{X}} \left(\sum_{k \in \mathcal{X}, k \neq i} q_{ik}(t) \right) \leq L < \infty, \tag{1}$$

for almost all $t \geq 0$. Then the probabilistic dynamics of the process $X(t)$ is given by the forward Kolmogorov system

$$\frac{d}{dt}\mathbf{p}(t) = A(t)\mathbf{p}(t), \tag{2}$$

where $A(t)$ is the transposed intensity matrix i.e., $a_{ij}(t) = q_{ji}(t), i, j \in \mathcal{X}$.

We can consider (2) as the differential equation with bounded operator function in the space of sequences l_1 (see details, for instance in [1]) and apply all results of [2].

Throughout this paper by $\|\cdot\|$ (or by $\|\cdot\|_1$ if ambiguity is possible) we denote the l_1-norm, i.e., $\|\mathbf{p}(t)\| = \sum_{i\in\mathcal{X}} |p_i(t)|$ and $\|A(t)\| = \sup_{j\in\mathcal{X}} \sum_{i\in\mathcal{X}} |a_{ij}(t)|$. Let Ω be a set of all stochastic vectors, i.e., l_1 vectors with non-negative coordinates and unit norm. Then $\|A(t)\| \leq 2L$ for almost all $t \geq 0$, and $\mathbf{p}(s) \in \Omega$ implies $\mathbf{p}(t) \in \Omega$ for any $0 \leq s \leq t$.

Recall that a Markov chain $X(t)$ is called *weakly ergodic*, if $\|\mathbf{p}^*(t) - \mathbf{p}^{**}(t)\| \to 0$ as $t \to \infty$ for any initial conditions $\mathbf{p}^*(0)$ and $\mathbf{p}^{**}(0)$, where $\mathbf{p}^*(t)$ and $\mathbf{p}^{**}(t)$ are the corresponding solutions of (2).

We consider, as in [3], the four classes of of Markov chains $X(t)$ with the following transition intensities:

(i) $q_{ij}(t) = 0$ for any $t \geq 0$ if $|i-j| > 1$ and both $q_{i,i+1}(t) = \lambda_i(t)$ and $q_{i,i-1}(t) = \mu_i(t)$ may depend on i;

(ii) $q_{i,i-k}(t) = 0$ for $k > 1$, $q_{i,i-1}(t) = \mu_i(t)$ may depend on i; and $q_{i,i+k}(t)$, $k \geq 1$, depend only on k and does not depend on i;

(iii) $q_{i,i+k}(t) = 0$ for $k > 1$, $q_{i,i+1}(t) = \lambda_i(t)$ may depend on i; and $q_{i,i-k}(t)$, $k \geq 1$, depend only on k and does not depend on i;

(iv) both $q_{i,i-k}(t)$ and $q_{i,i+k}(t)$, $k \geq 1$, depend only on k and do not depend on i.

Each such process can be considered as the queue-length process for the corresponding queueing system $M_t^X/M_t^X/1$.

Then type (i) transitions describe Markovian queues with possibly state-dependent arrival and service intensities (for example, the classic $M_n(t)/M_n(t)/1$ queue); type (ii) transitions allow consideration of Markovian queues with state-independent batch arrivals and state-dependent service intensity; type (iii) transitions lead to Markovian queues with possible state-dependent arrival intensity and state-independent batch service; type (iv) transitions describe Markovian queues with state-independent batch arrivals and batch service. We can refer to them as a $M_t^X/M_t^X/1$ queueing model following the original paper [4], see also [3,5,6].

The paper is organized as follows. Section 2 introduces a description of the problem. Section 3 considers the explicit form of the reduced intensity matrices. In Section 4, we obtain upper bounds for the rate of convergence. Section 5 concludes the paper.

2. Preliminaries

The problem of estimating the rate of convergence, like the very fact of convergence, is very important for studying the long-run (limiting) behavior of continuous time Markov chains with time varying intensities, see detailed discussion, examples and references in [7]. The simplest and most convenient for studying the rate of convergence to the limiting regime is the method of the logarithmic norm, see, for example [1,3,8].

However, there are situations in which this approach does not give good results.

Next, we show the possibility of using a different approach in such cases, namely the method of differential inequalities.

Another (but similar) approach is to use piecewise-line Lyapunov functions, see, for example, [9–12].

Consider here the two simplest examples of bounding the rate of convergence for differential equations.

Let firstly

$$\frac{d\mathbf{x}}{dt} = P\mathbf{x}$$

be a system of differential equations with $\mathbf{x} = (x_1, x_2)^T$, and $P = \begin{pmatrix} -5 & 8 \\ 2 & -5 \end{pmatrix}$. Put $d_1 = 1, d_2 = 2$, and $\mathbf{z} = D\mathbf{x} = (d_1x_1, d_2x_2)^T$. Then $\frac{d\mathbf{z}}{dt} = DPD^{-1}\mathbf{z}$, both column sums for $P^* = DPD^{-1}$ equal to -1. Hence the logarithmic norm $\gamma(P^*) = \sup_i \left(p^*_{ii} + \sum_{j \neq i} p^*_{ji} \right)$ equals -1, and we obtain a sharp upper bound on the rate of convergence $\|\mathbf{z}(t)\| \leq e^{-t}\|\mathbf{z}(0)\|$. Such a situation is typical if the matrix of the considered system is essentially non-negative (i.e., all off-diagonal elements are non-negative for any $t \geq 0$). Note that the corresponding eigenvalues of P are $-1, -9$.

On the other hand, let $P = \begin{pmatrix} -3 & 8 \\ -2 & -3 \end{pmatrix}$. Then corresponding eigenvalues of P are $-3 \pm 4i$. On the other hand, the "weighting" logarithmic norm P is not less than 1. In principle, here it is also possible to reduce the matrix to the exact value of the logarithmic norm (-3), see [2], but the corresponding transformation will be complex and difficult to implement. The best result (Ce^{-3t}) here can be obtained using the Lyapunov function (which does not work well in a countable situation), but the use of differential inequalities gives us an estimate like $Ce^{-(3+\epsilon)t}$ for any positive ϵ, see the corresponding description below, in Section 4. This approach deals with the sums of the columns for various combinations of the signs of the coordinates of the solutions of the system; it is described further in Section 4. It was first proposed in our recent papers; see [3] for the case of finite Markov chain with analytical (in t) intensities.

In this paper, it is shown that this method can be applied in a more general situation of locally integrable intensities, and, which is most difficult, for a countable chain that does not lend itself to direct reasoning and requires rather fine approximation estimates.

3. Explicit Forms of the Reduced Intensity Matrices

Due to the normalization condition $p_0(t) = 1 - \sum_{i \geq 1} p_i(t)$, we can rewrite the system (2) as follows:

$$\frac{d}{dt}\mathbf{z}(t) = B(t)\mathbf{z}(t) + \mathbf{f}(t), \tag{3}$$

where

$$\mathbf{f}(t) = (a_{10}(t), a_{20}(t), \dots)^T, \; \mathbf{z}(t) = (p_1(t), p_2(t), \dots)^T,$$

$$B(t) = \begin{pmatrix} a_{11}-a_{10} & a_{12}-a_{10} & \cdots & a_{1r}-a_{10} & \cdots \\ a_{21}-a_{20} & a_{22}-a_{20} & \cdots & a_{2r}-a_{20} & \cdots \\ \cdots & \cdots & \cdots & \cdots & \cdots \\ a_{r1}-a_{r0} & a_{r2}-a_{r0} & \cdots & a_{rr}-a_{r0} & \cdots \\ \vdots & \vdots & \vdots & \vdots & \ddots \end{pmatrix}. \tag{4}$$

Let $\mathbf{y}(t) = \mathbf{z}^*(t) - \mathbf{z}^{**}(t)$ be the difference of two solutions of system (3), and $\mathbf{y}(t) = (y_1(t), y_2(t), \dots,)^T$. Then, in contrast to the coordinates of the vector $\mathbf{p}(t)$, the coordinates of the vector $\mathbf{y}(t)$ have arbitrary signs.

Consider now the 'homogeneous' system

$$\frac{d}{dt}\mathbf{y}(t) = B(t)\mathbf{y}(t), \tag{5}$$

corresponding to (3). As was firstly noticed in [13], it is more convenient to study the rate of convergence using the transformed version $B^*(t)$ of $B(t)$ given by $B^*(t) = TB(t)T^{-1}$, where T is the upper triangular matrix of the form

$$T = \begin{pmatrix} 1 & 1 & 1 & \cdots & 1 & \cdots \\ 0 & 1 & 1 & \cdots & 1 & \cdots \\ 0 & 0 & 1 & \cdots & 1 & \cdots \\ \vdots & \vdots & \vdots & \ddots & \cdots & \\ 0 & 0 & 0 & \cdots & 1 & \cdots \\ \cdots & \cdots & \cdots & \cdots & \cdots & \cdots \end{pmatrix}. \tag{6}$$

Let $\mathbf{u}(t) = T\mathbf{y}(t)$. Then the system (5) can be rewritten in the form

$$\frac{d}{dt}\mathbf{u}(t) = B^*(t)\mathbf{u}(t), \tag{7}$$

where $\mathbf{u}(t) = (u_1(t), u_2(t), \dots)^T$ is the vector with the coordinates of arbitrary signs. If one of the two matrices $B^*(t)$ or $B(t)$ is known, the other is also (uniquely) defined.

The approach based on the differential inequalities (see [3]) seems to be the most general. On the other hand, if $B^*(t)$ is essentially non-negative (i.e., all off-diagonal elements are non-negative for any $t \geq 0$), then the method based on the logarithmic norm gives the same results, but in a much more visual form, see [3].

Let us write out the form of the matrix $B^*(t)$ for each class of chains; in more detail, the corresponding transformations can be seen in [3].

For $X(t)$ belonging to class (i) (inhomogeneous birth–death process) one has

$$B^*(t) = TB(t)T^{-1} =$$

$$\begin{pmatrix} -(\lambda_0+\mu_1) & \mu_1 & 0 & \cdots & 0 & \cdots & \cdots \\ \lambda_1 & -(\lambda_1+\mu_2) & \mu_2 & \cdots & 0 & \cdots & \cdots \\ \ddots & \ddots & \ddots & \ddots & \ddots & \cdots & \\ 0 & \cdots & \cdots & \lambda_{r-1} & -(\lambda_{r-1}+\mu_r) & \mu_r & \cdots \\ \cdots & \cdots & \cdots & \cdots & \cdots & \cdots & \cdots \end{pmatrix}. \tag{8}$$

For $X(t)$ belonging to class (ii) (which corresponds to the queueing system with batch arrivals and single services), one has

$$B^*(t) = TB(t)T^{-1} = \begin{pmatrix} a_{11} & \mu_1 & 0 & \cdots & 0 \\ a_1 & a_{22} & \mu_2 & \cdots & 0 \\ a_2 & a_1 & a_{33} & \mu_3 & \cdots \\ \ddots & \ddots & \ddots & \ddots & \ddots \\ \ddots & \ddots & \ddots & \ddots & \ddots \end{pmatrix}. \tag{9}$$

For $X(t)$ belonging to class (iii) (which corresponds to the queueing system with single arrivals and group services), one has $B^*(t) = TB(t)T^{-1} =$

$$\begin{pmatrix} -(\lambda_0+b_1) & b_1-b_2 & b_2-b_3 & \cdots & \cdots \\ \lambda_1 & -(\lambda_1+\sum\limits_{i\le 2} b_i) & b_1-b_3 & \cdots & \cdots \\ \ddots & \ddots & \ddots & \ddots & \ddots \\ 0 & \cdots & \cdots & \lambda_{r-1} & -(\lambda_{r-1}+\sum\limits_{i\le r} b_i)\cdots \\ \ddots & \ddots & \ddots & \ddots & \ddots \end{pmatrix}. \tag{10}$$

Finally, for $X(t)$ belonging to class (iv) (which corresponds to the queueing system with state-independent batch arrivals and group services), one has

$$B^* = TB(t)T^{-1} = \begin{pmatrix} a_{11} & b_1 - b_2 & b_2 - b_3 & \cdots & \cdots & \\ a_1 & a_{22} & b_1 - b_3 & \cdots & \cdots & \\ & & & & & \\ \ddots & \ddots & \ddots & \ddots & \ddots & \\ a_{r-1} & \cdots & \cdots & a_1 & a_{rr} & \cdots \\ \cdots & \cdots & \cdots & \cdots & \cdots & \cdots \end{pmatrix}, \tag{11}$$

where

$$T^{-1} = \begin{pmatrix} 1 & -1 & 0 & \cdots & 0 & \cdots \\ 0 & 1 & -1 & \cdots & 0 & \cdots \\ 0 & 0 & 1 & \cdots & 0 & \cdots \\ \vdots & \vdots & \vdots & \ddots & \cdots & \\ 0 & 0 & 0 & \cdots & 1 & \cdots \\ \cdots & \cdots & \cdots & \cdots & \cdots & \cdots \end{pmatrix}.$$

Remark 1. *Generally speaking, for models of the first and second classes, the matrix $B^*(t)$ is always essentially non-negative; at the same time, for models of the third and fourth classes, this requires some additional assumptions. Under essential non-negativity of $B^*(t)$ all bounds on the rate of convergence can be obtained via logarithmic norm, see [3]. However, in the general case, this approach may not work, and the method of differential inequalities described in our previous papers, see [3,14], would be more effective.*

Thus, in this paper we will consider chains of the third and fourth classes with a countable state space. For simplicity of calculations, we will additionally assume that the size of the simultaneously arriving and/or servicing group of customers does not exceed some fixed number, say R, i.e., that all $q_{ij}(t) = 0$ for $|i-j| > R$ and any $t \geq 0$.

Let $\{d_i,\ i \geq 1\}$ be a sequence of non-zero numbers such that $\inf_k |d_k| = d > 0$. Denote by $D = diag(d_1, d_2, \dots)$ the corresponding diagonal matrix, with the off-diagonal elements equal to zero. Let $\mathbf{w}(t) = D\mathbf{u}(t)$ in (7), then we obtain the following equation

$$\frac{d}{dt}\mathbf{w}(t) = B^{**}(t)\mathbf{w}(t), \tag{12}$$

where

$$B^{**}(t) = DB^*(t)D^{-1} = \left(b_{ij}^{**}(t)\right)_{i,j\geq 1}. \tag{13}$$

If we write out $B^*(t) = \left(b_{ij}^*(t)\right)_{i,j\geq 1}$, then

$$b_{ij}^{**}(t) = \frac{d_i}{d_j} b_{ij}^*(t), \quad |i-j| \leq R, \tag{14}$$

and our assumption implies $b_{ij}^{**}(t) = b_{ij}^*(t) = 0$ for any $t \geq 0$ if $|i-j| > R$.

4. Upper Bounds on the Rate of Convergence

Let us first consider a general finite system of linear differential equations, which we will write in the form

$$\frac{d}{dt}\mathbf{x}(t) = B^*(t)\mathbf{x}(t),\ t \geq 0, \tag{15}$$

where $\mathbf{x}(t) = (x_1(t), \dots, x_S(t))^T$, and let D now be the corresponding *finite* diagonal matrix.

The simplest situation with analytical (in t) coefficients $b_{ij}^*(t)$ has been studied in [3,14,15]. The method of estimating under such assumption is based on the fact that, in this case, on any finite interval, each coordinate has a finite number of sign changes, which means that the semiaxis can be divided into intervals, on each of which the signs of the coordinates are constant. Consider such an (t_1, t_2). Choose the signs of d_k-s so that all $d_k x_k(t) > 0$. Hence $\|\mathbf{w}(t)\| = \|\mathbf{x}(t)\|_D = \sum_{k=1}^{S} d_k x_k(t) \geq d\|\mathbf{x}(t)\|_1$ can be considered as the corresponding norm.

Let $\sum_{i=1}^{S} b_{ij}^{**}(t) \leq -\alpha_D(t)$, for any j, then

$$\frac{d}{dt}\|\mathbf{w}(t)\| = \frac{d(\sum_k w_k)}{dt} = \sum_{i,j} b_{ij}^{**}(t) w_j(t) \leq -\alpha_D(t)\|\mathbf{w}(t)\|. \tag{16}$$

Then

$$\|\mathbf{w}(t)\| = \|D\mathbf{x}(t)\|_1 \leq e^{-\int_s^t \alpha_D(\tau)d\tau}\|D\mathbf{x}(s)\|_1, \quad t_1 < s < t < t_2, \tag{17}$$

for the corresponding matrix D and corresponding function $\alpha_D(t)$. Hence, we have

$$\|\mathbf{x}(t)\|_1 \leq \frac{\max |d_k|}{\min |d_m|} e^{-\int_s^t \alpha_D(\tau)d\tau}\|\mathbf{x}(s)\|_1, \tag{18}$$

for any $t_1 < s < t < t_2$, and by continuity, for all $t_1 \leq s < t \leq t_2$.

Let now s, t be arbitrary, $0 \leq s \leq t < \infty$. Then for any interval with fixed signs of coordinates we have bound (18) with the corresponding D and $\alpha_D(t)$. Let now $\alpha^*(t) = \min \alpha_D(t)$, and $d^*(S) = d^* = \max \frac{|d_k|}{|d_m|}$, where the minimum and maximum are taken over all possible combinations of coordinate signs of the solution $\mathbf{x}(t)$, for *any* $0 \leq s \leq t$. Then we obtain the following general estimate

$$\|\mathbf{x}(t)\|_1 \leq d^*(S) e^{-\int_s^t \alpha^*(\tau)d\tau}\|\mathbf{x}(s)\|_1, \tag{19}$$

Let there exist positive numbers M, β such that

$$e^{-\int_s^t \alpha^*(\tau)\,d\tau} \leq M e^{-\beta(t-s)}, \quad 0 \leq s \leq t. \tag{20}$$

Consider now an arbitrary interval $[0, t^*]$; if our original coefficients are locally integrable, they can be approximated arbitrarily accurately by a continuous functions. In turn, a continuous function can be approximated arbitrarily accurately by an analytic function. As a result, instead of the integrable $B^*(t)$, we obtain an analytic $\bar{B}^*(t)$, such that

$$\int_0^{t^*} \|B^*(\tau) - \bar{B}^*(\tau)\|d\tau \leq \varepsilon. \tag{21}$$

Denote now by $W(t,s)$ and $\bar{W}(t,s)$ the Cauchy operators for (15) and the respective system with matrix $\bar{B}^*(t)$. Then, if (20) holds, in accordance with Lemma 3.2.3 [2] (see [2], pp. 110–111) we obtain

$$\begin{aligned} \|W(t,s) - \bar{W}(t,s)\| &\leq M d^* e^{-\beta(t-s)}\left(e^{Md^* \int_s^t \|B^*(\tau) - \bar{B}^*(\tau)\|d\tau} - 1\right) \\ &\leq M d^* e^{-\beta(t-s)}\left(e^{Md^*\varepsilon} - 1\right). \end{aligned} \tag{22}$$

Hence we have the following statement.

Lemma 1. *Let all $b_{ij}^*(t)$ be locally integrable on $[0, \infty)$. Let inequality (20) hold. Then*

$$\|\mathbf{x}(t)\|_1 \leq d^*(S) M e^{-\beta(t-s)}\|\mathbf{x}(s)\|_1, \tag{23}$$

for any solution of (15) and any $0 \leq s \leq t$.

Let us now return to a countable system (7) and consider the corresponding truncated system

$$\frac{d}{dt}\mathbf{u}(n,t) = B^*(n,t)\mathbf{u}(n,t), \tag{24}$$

where $B^*(n,t) = \left(b^*_{ij}(t)\right)^n_{i,j=1}$.

Below we will identify the finite vector with entries $(a_1,\ldots,a_n)$ and the infinite vector with the same first n coordinates and the others equal to zero.

Rewrite system (24) as

$$\frac{d}{dt}\mathbf{u}(n,t) = B^*(t)\mathbf{u}(n,t) + (B^*(n,t) - B^*(t))\mathbf{u}(n,t). \tag{25}$$

Denote by $V(t,s)$ and $V(n,t,s)$ the Cauchy operators for (7) and (24), respectively.

Suppose that $n > S$, and that, in addition

$$\mathbf{u}(0) = \mathbf{u}(n,0) = \mathbf{u}(S,0), \quad \|\mathbf{u}(0)\|_1 \leq 1. \tag{26}$$

Then one has from (7)

$$\mathbf{u}(t) = V(t)\mathbf{u}(0) = V(t)\mathbf{u}(n,0). \tag{27}$$

On the other hand, from (25) we have

$$\mathbf{u}(n,t) = V(t)\mathbf{u}(n,0) + \int_0^t V(t,\tau)(B^*(n,\tau) - B^*(\tau))\mathbf{u}(n,\tau)\,d\tau. \tag{28}$$

Hence in any norm we obtain the bound

$$\|\mathbf{u}(t) - \mathbf{u}(n,t)\| \leq \int_0^t \|V(t,\tau)\|\|(B^*(n,\tau) - B^*(\tau))\mathbf{u}(n,\tau)\|\,d\tau. \tag{29}$$

Denote $\sup \frac{|d_k|}{|d_m|} = \hat{d} < \infty$, where supremum is taken over all possible combinations of coordinate signs of the solution $\mathbf{u}(t)$ of (7), under assumption $|k-m| = 1$.

Put now $D^* = diag(d^*(1), d^*(2), \ldots)$.

Note that according to (14) the matrix $B^{**}(t)$ has nonzero entries only on the main diagonal and at most R diagonals above and below it. Then

$$\|B^*(t)\|_{1D^*} = \|B^{**}(t)\|_1 \leq K = 2L\hat{d}^R, \tag{30}$$

for almost all $t \geq 0$. Then

$$\|V(t,s)\|_{1D^*} \leq e^{K(t-s)} \leq e^{Kt^*}. \tag{31}$$

On the other hand, all elements of the first $n-R$ columns of the matrix $(B^*(n,\tau) - B^*(\tau))$ are zeros for any $\tau \geq 0$. Hence, all the first $n-R$ coordinates of the corresponding vector $(B^*(n,\tau) - B^*(\tau))\mathbf{u}(n,\tau)$ are also zeros too, and

$$\|(B^*(n,\tau) - B^*(\tau))\mathbf{u}(n,\tau)\|_{1D^*} \leq K \sum_{k=n-R}^{n} \hat{d}^k |u_k(t)|. \tag{32}$$

Put $D^{**} = diag(\hat{d}^2, \hat{d}^4, \ldots)$ and $\mathbf{w}^*(t) = D^{**}\mathbf{u}(t)$.

Then, instead of (30) and (31) we have

$$\|B^*(t)\|_{1D^{**}} = \|D^{**}B^*{D^{**}}^{-1}(t)\|_1 \leq K^* = 2L\hat{d}^{2R}, \tag{33}$$

and

$$\|V(t,s)\|_{1D^{**}} \leq e^{K^*(t-s)} \leq e^{K^*t^*}, \tag{34}$$

respectively.

Then

$$\|\mathbf{u}(n,t)\|_{1D^{**}} = \sum_{k=1}^{n} d^{2k}|u_k(n,t)| \le e^{K^*t^*} \sum_{k=1}^{S} d^{2k}|u_k(n,0)| \le e^{K^*t^*} d^{2S} \sum_{k=1}^{S} |u_k(n,0)|. \quad (35)$$

Then (35) and (26) imply the bound

$$d^{2n-2R} \sum_{k=n-R}^{n} |u_k(n,t)| \le \sum_{k=1}^{n} d^{2k}|u_k(n,t)| \le e^{K^*t^*} d^{2S}. \quad (36)$$

Then

$$\sum_{k=n-R}^{n} d^{k}|u_k(n,t)| \le d^{n} \sum_{k=n-R}^{n} |u_k(n,t)| \le e^{K^*t^*} d^{2S+2R-n}. \quad (37)$$

Finally, for the right-hand side of (29) we have the bound

$$\int_0^t \|V(t,\tau)\|_{1D^*} \|(B^*(n,\tau) - B^*(\tau))\mathbf{u}(n,\tau)\|_{1D^*}\, d\tau \le e^{Kt^*} Kt^* e^{K^*t^*} d^{2S+2R-n}, \quad (38)$$

which tends to be zero at $n \to \infty$.

Hence we have the following statement.

Lemma 2. *Let assumptions of Lemma 1 be fulfilled for any S. Then, under assumption (26), and for any fixed $\varepsilon > 0$, $t^* > 0$, we obtain $\|\mathbf{u}(t) - \mathbf{u}(n,t)\|_{1D^*} < \varepsilon$ for sufficiently large n, for any $t \in [0, t^*]$.*

As a result, Lemmas 1 and 2 guarantee an estimate of the form

$$\|\mathbf{u}(t)\|_1 \le Me^{-\beta t}\|\mathbf{u}(0)\|_{1D^*}. \quad (39)$$

Consider now two arbitrary solutions $\mathbf{p}^*(t)$ and $\mathbf{p}^{**}(t)$ of the forward Kolmogorov system (2) with the corresponding initial conditions $\mathbf{p}^*(0)$ and $\mathbf{p}^{**}(0)$. Denote by $\mathbf{p}_0^*(t)$ and $\mathbf{p}_0^{**}(t)$ the respective vector functions with coordinates $1, 2, \dots$ (i.e., without zero coordinates).

One can write $\mathbf{u}(t) = T(\mathbf{p}_0^*(t) - \mathbf{p}_0^{**}(t))$. Then (see for instance [8]), the following inequality holds: $\|\mathbf{p}^*(t) - \mathbf{p}^{**}(t)\|_1 \le \frac{2}{d}\|\mathbf{u}(t)\|_1$.

Finally we obtain the following statement.

Theorem 1. *Let the assumptions of Lemma 1 hold for any natural S. Then $X(t)$ is weakly ergodic and the following bound on the rate of convergence holds:*

$$\|\mathbf{p}^*(t) - \mathbf{p}^{**}(t)\|_1 \le \frac{2M}{d} e^{-\beta t}\|\mathbf{p}_0^*(0) - \mathbf{p}_0^{**}(0)\|_{1D^*}. \quad (40)$$

Remark 2. *A specific model (which belongs to both classes* (iii) *and* (iv)*) was investigated in [16] by the method described here.*

Namely, in this paper, the queueing model with possible transitions and respective intensities of single arrival $\lambda(t)$ and service of group of two customers $\mu(t)$ was considered. Hence

$$B^*(t) =$$

$$= \begin{pmatrix} -\lambda(t) & -\mu(t) & \mu(t) & 0 & 0 & 0 & \cdots \\ \lambda(t) & -(\lambda(t)+\mu(t)) & 0 & \mu(t) & 0 & 0 & \cdots \\ 0 & \lambda(t) & -(\lambda(t)+\mu(t)) & 0 & \mu(t) & 0 & \cdots \\ 0 & 0 & \lambda(t) & -(\lambda(t)+\mu(t)) & 0 & \mu(t) & \cdots \\ 0 & 0 & 0 & \lambda(t) & -(\lambda(t)+\mu(t)) & 0 & \cdots \\ \ddots & \ddots & \ddots & \ddots & \ddots & \ddots & \ddots \\ \cdots & \cdots & \cdots & \cdots & \cdots & \cdots & \cdots \end{pmatrix}.$$

Let $\delta > 1$ be a positive number. Put
$d_1 = 1, d_2 = 1/\delta, d_k = \delta^{k-2}, k \geq 3$ if all coordinates of solutions are positive;
$|d_k| = \delta^{k-1}, k \geq 1$ otherwise.
Then one has,

$$\alpha^*(t) \geq \min\Big[\lambda(t)\Big(1-\delta^{-1}\Big), \mu(t)(1+\delta)- \\ \lambda(t)\Big(\delta^2-1\Big), \mu(t)\Big(1-\delta^{-1}\Big) - \lambda(t)(\delta-1)\Big]. \tag{41}$$

Moreover, $d = \delta^{-1}, \hat{d} = \delta, d_k^* = \delta^{k-1}$, for $k \geq 1$.

In particular, if the process $X(t)$ is homogeneous i.e., $\lambda(t) = \lambda$ and $\mu(t) = \mu$ are positive numbers, then $\int_0^\infty \alpha^*(t)dt = +\infty$ is equivalent to $\alpha^* > 0$ and this is equivalent to $0 < \lambda < \mu$. Put $\delta = \sqrt{\frac{\mu}{\lambda}}$. Hence,

$$\alpha^* = \min\left[\left(\sqrt{\mu}-\sqrt{\lambda}\right)^2, \lambda\left(1-\sqrt{\frac{\lambda}{\mu}}\right)\right]. \tag{42}$$

In the paper [16] the specific example with periodic intensities was considered. Namely, let $\lambda(t) = 2 + \sin 2\pi t$ and $\mu(t) = 4 - \cos 2\pi t$. Put $\delta = \frac{11}{10}$. Then, $\int_0^1 \alpha^*(t)\,dt \geq \frac{1}{22} > 0$, $X(t)$ is exponentially weakly ergodic and has the 1-periodic limiting mean (a Markov chain has the limiting mean $m(t)$, if $\lim_{t\to\infty}(m(t) - E(t,k)) = 0$ for any k, $E(t,k)$ is the mathematical expectation of $X(t)$ under initial condition $X(0) = k$). Now, applying the known truncation technique (see the detailed discussion and bounds in [8]), one can compute all probability characteristics of the queue-length process $X(t)$. Some of the corresponding graphs are shown in Figures 1–4; see detailed discussion in [16].

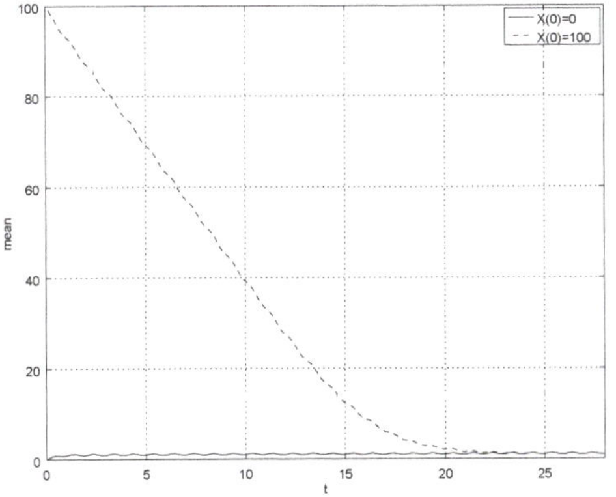

Figure 1. The mean $E(t,0)$ and $E(t,100)$ for $t \in [0,28]$, this figure shows the rate of convergence.

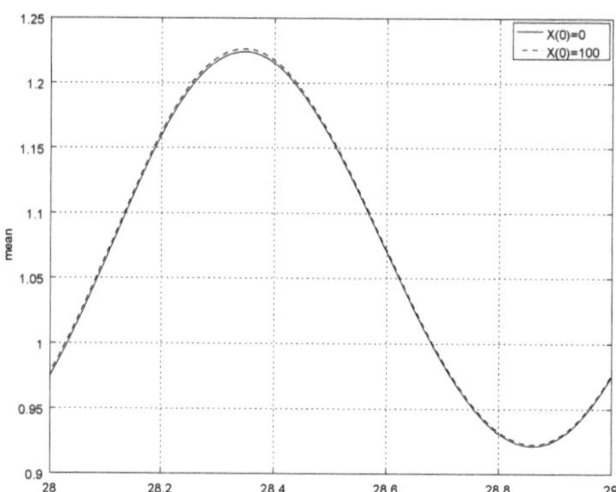

Figure 2. The mean $E(t,0)$ and $E(t,100)$ for $t \in [28,29]$, this figure shows approximation of the limiting mean.

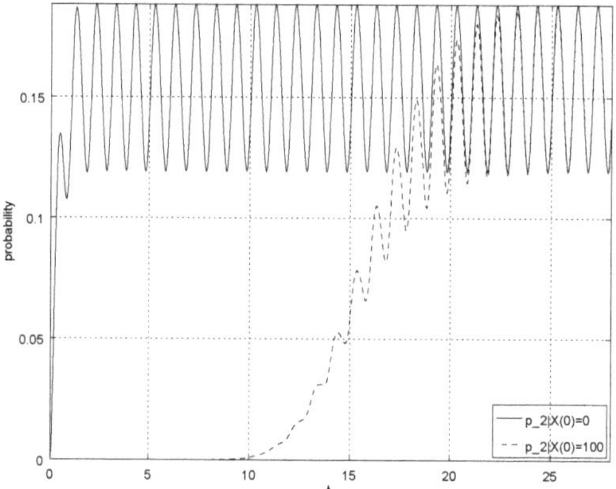

Figure 3. Probability $p_2(t)$ for $t \in [0,28]$ and initial conditions $X(0) = 0$ and $X(0) = 100$; this figure shows the rate of convergence.

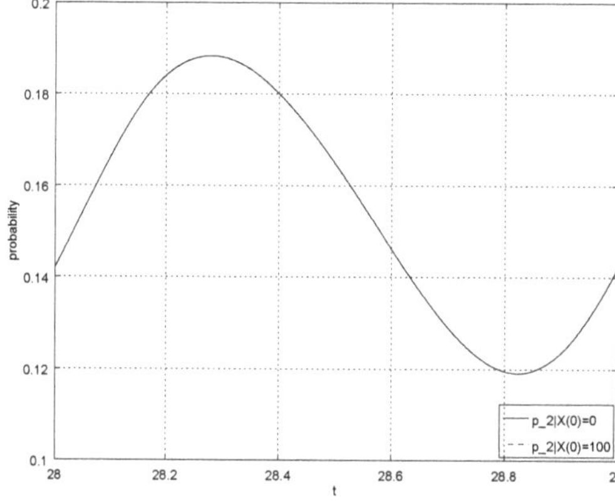

Figure 4. Probability $p_2(t)$ for $t \in [28,29]$ and initial conditions $X(0) = 0$ and $X(0) = 100$; this figure shows approximation of the limiting probability $p_2(t)$.

5. Conclusions

In this paper, we have substantiated one of the most general methods for studying the rate of convergence to limit characteristics for weakly ergodic Markov chains with continuous time. Namely, the applicability of the method of differential inequalities for countable inhomogeneous processes in the case of a nonsmooth dependence of intensities as functions of time is shown. Thus, studying models with continuous time from the theory of queues, biology, physics and other sciences, and obtaining guaranteed estimates of the rate of convergence, we can both make sure that the influence of the initial conditions of the system disappears with increasing time, and build the main characteristics of the system to control them.

Author Contributions: Investigation, A.Z., Y.S. and A.S. All authors contributed equally to this work. All authors have read and agreed to the published version of the manuscript.

Funding: This research was supported by Russian Science Foundation under grant 19-11-00020.

Institutional Review Board Statement: Not applicable.

Informed Consent Statement: Not applicable.

Data Availability Statement: Data sharing is not applicable to this article.

Conflicts of Interest: The author declares no conflict of interest.

References

1. Zeifman, A. On the Study of Forward Kolmogorov System and the Corresponding Problems for Inhomogeneous Continuous-Time Markov Chains. *Springer Proc. Math. Stat.* **2020**, *333*, 21–39.
2. Daleckii, J.L.; Krein, M.G. *Stability of Solutions of Differential Equations in Banach Space*; American Mathematical Soc.: Providence, RI, USA, 2002; Volume 43.
3. Zeifman, A.; Satin, Y.; Kryukova, A.; Razumchik, R.; Kiseleva, K.; Shilova, G. On the Three Methods for Bounding the Rate of Convergence for some Continuous-time Markov Chains. *Int. J. Appl. Math. Comput. Sci.* **2020**, *30*, 251–266.
4. Nelson, R.; Towsley, D.; Tantawi, A. Performance Analysis of Parallel Processing Systems. *IEEE Trans. Softw. Eng.* **1988**, *14*, 532–540. [CrossRef]
5. Li, J.; Zhang, L. $M^X/M/c$ Queue with catastrophes and state-dependent control at idle time. *Front. Math. China* **2017**, *12*, 1427–1439. [CrossRef]
6. Satin, Y.; Zeifman, A.; Korotysheva, A. On the Rate of Convergence and Truncations for a Class of Markovian Queueing Systems. *Theory Probab. Appl.* **2013**, *57*, 529–539. [CrossRef]
7. Zeifman, A.; Satin, Y.; Kovalev, I.; Razumchik, R.; Korolev, V. Facilitating Numerical Solutions of Inhomogeneous Continuous Time Markov Chains Using Ergodicity Bounds Obtained with Logarithmic Norm Method. *Mathematics* **2021**, *9*, 42. [CrossRef]
8. Zeifman, A.; Satin, Y.; Korolev, V.; Shorgin, S. On truncations for weakly ergodic inhomogeneous birth and death processes. *Int. J. Appl. Math. Comput. Sci.* **2014**, *24*, 503–518. [CrossRef]
9. Bertsimas, D.; Gamarnik, D.; Tsitsiklis, J.N. Performance of multiclass Markovian queueing networks via piecewise linear Lyapunov functions. *Ann. Appl. Probab.* **2001**, *11*, 1384–1428. [CrossRef]
10. Blanchini, F.; Giordano, G. Piecewise-linear Lyapunov functions for structural stability of biochemical networks. *Automatica* **2014**, *50*, 2482–2493. [CrossRef]
11. Bobyleva, O.N. Piecewise-linear Lyapunov functions for linear stationary systems. *Autom. Remote Control* **2002**, *63*, 540–549. [CrossRef]
12. Orlov, Y. *Nonsmooth Lyapunov Analysis in Finite and Infinite Dimensions*; Springer International Publishing: Cham, Switzerland, 2020.
13. Zeifman, A. Some properties of the loss system in the case of varying intensities. *Autom. Remote Control* **1989**, *50*, 107–113.
14. Kryukova, A.; Oshushkova, V.; Zeifman, A.; Satin, Y. Application of Method of Differential Inequalities to Bounding the Rate of Convergence for a Class of Markov Chains. *Springer Proc. Math. Stat.* **2020**, *333*, 95–103.
15. Zeifman, A.; Kiseleva, K.; Satin, Y.; Kryukova, A.; Korolev, V. On a Method of Bounding the Rate of Convergence for Finite Markovian Queues. In Proceedings of the 10th International Congress on Ultra Modern Telecommunications and Control Systems and Workshops (ICUMT), Moscow, Russia, 5–9 November 2018; pp. 1–5.
16. Satin, Y.; Zeifman, A.; Kryukova, A. On the Rate of Convergence and Limiting Characteristics for a Nonstationary Queueing Model. *Mathematics* **2019**, *7*, 678. [CrossRef]

MDPI
St. Alban-Anlage 66
4052 Basel
Switzerland
Tel. +41 61 683 77 34
Fax +41 61 302 89 18
www.mdpi.com

Mathematics Editorial Office
E-mail: mathematics@mdpi.com
www.mdpi.com/journal/mathematics

www.ingramcontent.com/pod-product-compliance
Lightning Source LLC
LaVergne TN
LVHW070101170726
843515LV00007B/1587

* 9 7 8 3 0 3 6 5 3 8 1 5 0 *